Intelligent Systems, Control and Automation: Science and Engineering

Volume 75

Intelligent Systems, Control and Automation: Science and Engineering, the new name for the series formerly known as Microprocessor-Based and Intelligent Systems Engineering Bookseries, has published around thirty books on developments in this interesting field that borders on so many disciplines and has so many practical applications: human-like biomechanics, humanoid robotics, intelligent control of robotic systems, robot interaction with different kinds of environments, automation, computer and control engineering, but also neural networks, fuzzy logic, genetic algorithms, nonlinear dynamics and control, and of course theoretical mechanics. This wide range of topics, approaches, perspectives and applications is reflected in a large readership of researchers and practioners in various fields, as well as graduate students who want to learn more on a given subject.

As a reflection of the changing world of scientific progress, a change in series title has been effectuated. The topic has evolved over the years and the following title is felt to be a more accurate fit for the current content of the books that are being published.

The current Series Editor is Professor S. G. Tzafestas, National Technical University of Athens, Greece. He is assisted by an Editorial Advisory Board who help to select the most interesting and cutting edge manuscripts for the series.

Springer and Professor Tzafestas welcome book ideas from authors. Potential authors who wish to submit a book proposal should contact Nathalie Jacobs, Publishing Editor, Springer (Dordrecht), e-mail: Nathalie.Jacobs@springer.com

Indexed by SCOPUS and Springerlink.

More information about this series at http://www.springer.com/series/6259

Photographer: Markku Pirttimaa

Jacques Periaux • Felipe Gonzalez
Dong Seop Chris Lee

Evolutionary Optimization and Game Strategies for Advanced Multi-Disciplinary Design

Applications to Aeronautics and UAV Design

Springer

Jacques Periaux
International Center
for Numerical Methods
Barcelona
Spain

Felipe Gonzalez
Science and Engineering Faculty
Queensland University of Technology
Brisbane
Queensland
Australia

Dong Seop Chris Lee
DMC R&D Center
Samsung Electronics
Suwon
Kyonggi-do
Korea, Republic of (South Korea)

ISSN 2213-8986 ISSN 2213-8994 (electronic)
Intelligent Systems, Control and Automation: Science and Engineering
ISBN 978-94-017-7698-1 ISBN 978-94-017-9520-3 (eBook)
DOI 10.1007/978-94-017-9520-3

Springer Dordrecht Heidelberg New York London

Softcover reprint of the hardcover 1st edition 2015

Printed on acid-free paper

Springer is part of Springer Science+Business Media (www.springer.com)

"This book is dedicated to my son KiYoung Alexander Lee and my daughter JungBin Elisha Lee. Also to my parents SangJin Lee and YoungBok Kim whom I cannot thank enough for all the support and love you have given me."

To my beloved late wife Chantal, and our uniquely talented and accomplished children Sébastien, Daphne, Virgile and Bergame,
Jacques Periaux

This book is dedicated to my wife and parents for their kindness and devotion. Also to Ashleigh, Chloe, Leo and Louie.
Felipe Gonzalez

"This book is dedicated to my wife Sun-Hee Lee, our son KiYoung Alexander Lee and our daughter JungBin Elisha Lee who are my inspiration in everything I do and every choice I make, and also to my parents SangJin Lee and Young Bok Kim who always supported me in every endeavour."
Dong Seop Chris LEE

Foreword

Ignorance more frequently begets confidence than does knowledge: it is those who know little, and not those who know much, who so positively assert that this or that problem will never be solved by science.

–Charles Darwin

The German physicist Max Plank hypothesized that important scientific innovations become accepted because their opponents gradually die out, not because they gradually win over opponents. Evolutionary algorithms certainly had no shortage of opponents in the optimization community when they were first introduced in the early 1990s. We now see that Evolutionary algorithms are a maturing technology well beyond the bleeding edge and in the realm of the practitioner.Evolutionary algorithms for multi-disciplinary design optimization have been successfully applied in a variety of areas that include: industrial design by parameterization, network design by construction, scheduling, routing, database mining, control systems, time series prediction, artificial intelligence, molecular design and artificial life systems. The current wide- spread use of Evolutionary Algorithms is not a consequence of a high rate of mortality in the optimization community as Dr. Planck would suggest but a result of the dramatic increase in computational capacity and Information Technology implementation. This capacity has transformed what formerly was an interesting theoretical construct into a useful (engineering design) tool.

The issue at hand is not theoretical and conceptual development of Evolutionary Algorithms but developing the expertise for their implementation.The present volume by Felipe Gonzalez, Dong Seop Chris Lee, and Jacques Periaux is dedicated to the development of this expertise. It describes the role that evolution based tools play with game theoretic strategy in the solution or set of solutions as well as multi-disciplinary optimization in a variety of applications. Although the applications focus on aerospace, the methods translate to other fields including surface and marine transport, UAV systems, biomedical engineering, financial engineering, communications, and artificial intelligence. This volume is directed towards entry level scientists, engineers, and technologists in industry and governmental agencies as well in the academy. Optimization is a contact, not a spectator, sport. Several aeronautical applications with increasing complexity are presented. These are accompanied by a series of "hands- on" test problems are which are proposed as exercises to the reader.

November, 2013 William Fitzgibbon
College of Technology, University of Houston
Houston, USA

Foreword

The present volume has three authors, the only one I know personally being Jacques Périaux. For me, Jacques Périaux is many persons in one: a former student, a personal friend, the co-author of many articles, and has been for many years the inspiration of many of my own investigations.

Jacques Périaux is a most remarkable individual and scientist: during a long and remarkable career in Aeronautics and Aerospace industry in France at Dassault Aviation, Jacques Périaux has been actively involved in designing airplanes, and in closing the gap between Academia and Industry, firmly believing that engineers can learn from academics, and conversely. A constant attitude of Jacques Périaux has been to look at new methods developed in universities and research centers, in France and abroad, for their potential applicability to problems from Industry, quite often suggesting new problems to look at, and ideas to investigate. To give evidence of J. Périaux's vision related to this book, let us mention the fact that, almost twenty years ago, he anticipated the power of evolution algorithms for the solution of those multi-objective and multi-physics design problems occurring in Aeronautics and Aerospace.

Jacques' initiative in that direction met with some scepticism from a number of well-known members of the Computational Mechanics community, but time has proved that his vision was right. The good news is that Jacques Periaux has decided, with the help of two young collaborators, Felipe Gonzalez and Dong Seop Chris Lee, to publish a book (this one) about evolution algorithms, such as the genetic ones, multi-objective optimization, game theory, and more, about the solution of design problems from Aeronautics. The reader will find herein a wealth of methods, examples, convergence studies, and recipes. One of the main values of this book is that practitioners from areas other than Aeronautics will have no difficulty in adapting and implementing the various methods discussed here to the solutions of their specific problems. This book is not easy to read, deep thought being inherent to the topics it addresses, but its reading is facilitated by its excellent structure and the large number of very interesting real-life examples it considers. Those motivated by sufficient drive will learn a lot about a unique combination of methodologies applied in a growing number of areas of Science and Engineering.

June, 2013 Roland Glowinski
French National Academy of Technology
Jyväskylä
Finland

Preface

Computer programs that "evolve" in ways that resemble natural selection can solve complex problems even their creators do not fully understand...

John Holland, Scientific American, July 1992

John Holland of the University of Michigan is commonly regarded as the father of Genetic Algorithms. The story begins in 1993.

Genetic Algorithms belong to a larger class of algorithms called Evolutionary Algorithms that generate solutions to optimization problems using techniques inspired by natural evolution. John Holland of the University of Michigan is commonly regarded as the father of Genetic Algorithms. Holland perceived a connection between biology and computation. Machines as well as animals could be trained to adapt to their environment over successive generations with the principles of evolution being the engines of the adaption. The genetic algorithm begins with a population having randomly generated characteristics. Population members are evaluated by some method and then merged to produce a child with traits that are a combination of its parent's characteristics.

Our story, which begins in 1993, concerns the translation of Holland's ideas from theoretical computational science to industrial design, in particular aeronautical design. Pierre Bohn, Director of "Division des Etudes Avancées" of Dassault Aviation anticipated that Genetic Algorithms could be used to improve the performance of aero-engines and suggested that I investigate new evolution based optimization techniques called Genetic Algorithms. In the 90's the Dassault Scientific Center became a hotbed of activity for development of Advanced Evolutionary Software, many of which are used in this book. In this stimulating environment a group of visitors, colleagues, post doctoral fellows and PhD students actively contributed to the effort. Included in the group are: Bertrand Mantel, CHEN Hong Quan, Mourad Sefrioui, Eric Whitney, TANG Zhili, ZHOU Chun Hua and Jun Feng WANG. Special recognition is due to the late Bertrand Mantel, for his enthusiastic numerical experimentation in the area of robust Genetic Algorithms.

Murphy's first law says that nothing is as easy as it first looks. The fact that Genetic Algorithms are simple to implement and frequently succeed at generating solutions of high fitness when applied to practical problems first led me to believe that further investigation would lead to a rapid completion of several existing optimization methods proposed to those working in the area of Aerodynamic Design. Despite the remarkable potential of Genetic Algorithms their application to design in industrial environments, for several reasons, proved more complicated than I had anticipated:

- Many aeronautical applications involve nonlinear Partial Differential Equations (PDEs) for the mathematical modeling of fluids, structures or waves. This makes the evaluation of a candidate solution in a detailed design expensive.
- The needs of aeronautical industries were moving very rapidly to automated multi-disciplinary design optimization (MDO) in the field of (green) digital aircraft design and manufacturing;
- In order to account for uncertainties in flight conditions or geometrical parameters for maintaining reliable performance, robust stochastic optimization of real life problems was becoming a first priority that designers had to take into consideration.

In the time period between 1996 and 2006 Dassault Aviation launched several research projects in collaboration with the Laboratoire Informatique Paris 6 (LIP6) and the Aerospace Engineering Department of the University of Sydney (USYD) from 1996 to 2005. The outcomes of these projects showed significant progress in both efficiency and quality and led to a new class of advanced evolutionary algorithms for multi-objective optimization problems in aeronautics. Several new methodologies were introduced including successively hierarchical topology, asynchronous parallelism, game strategies and more recently hybridized game coalitions.

Over the course of the last ten years, Felipe Gonzalez, Dong Seop Chris Lee and I have given many seminars, workshop presentations and conference lectures on Genetic Algorithms within the context of aerodynamical design. On the basis of this activity we decided to create the present volume, whose raison d'être is to provide physicists, engineers and computer scientists with advanced evolutionary methods and tools useful to the numerical solution of multi-objective optimization problems in Aeronautical applications. In particular we wished to introduce entry level, neophyte, scientists and engineers to Genetic Algorithms, thus providing them with not only computing guidelines but examples or test problems of increasing complexity and difficulty.

We hope to make the scientific and industrial community aware of the potential of both Evolutionary Algorithms and Game Theory. Evolutionary Algorithms form the basis for robust decision making software that searches solutions in hilly spaces. Game Theory offers mathematically well-defined strategies for solving multi-objective optimization problems. In an effort to optimize efficiency within the context of industrial applications, we provide a detailed description of methodologies that combine game theoretical strategy with Evolutionary Algorithms. We conclude with an Appendix which offers the reader several practical, "hands-on" test problems together with descriptions of software and guidelines for capturing, presenting and discussing solutions. It is our expectation that these test problems will provide readers a pathway for solving their own optimization problems.

The applications of innovative advanced evolutionary tools are not limited strictly to aeronautics. Applications may be found in complex multi-physics optimization problems drawn from a wide variety of areas such as biomedical engineering, surface transport, energy, and green technologies, among others.

After an introductory chapter which outlines the requisite background we sequentially address the following topics:

- Requisite Background
- Evolutionary Methods
- Multi-objective Evolutionary Algorithms and Game Theory
- Advanced Evolutionary Algorithms
- Multi-disciplinary Design/Robust Design
- Introduction to Algorithms for Numerical Design and Optimization
- Single Objective Model Test Problems
- Multi-Objective Model Test Problems
- Multi -Disciplinary Design Optimization
- Robust Multi-Disciplinary Design Optimization
- Morphing Design Optimization

We provide a summarizing conclusion and then end with the aforementioned appendix that contains the hands on optimization test problems with guidelines on how to apply the methods and tools that have been described in the text.

Most of the numerical experiments have been conducted in collaboration with my co-authors Dr. Felipe Gonzalez and Dr. Dong Seop Chris Lee, both of whom I met during their course of PhD study at the University of Sydney. Computations performed by two of my PhD students J. Leskinen and Ms. Wang Hong, at the University of Jyvaskyla have been useful in the presentation of some test problems of the Appendix which can be found in the Finnish Database http://www.jucri.jyu.fi. Several numerical experiments have been performed by accessing and using software from many well-known scientists. We express our thanks to Frederic Hecht and Olivier Pironneau (FreeFem++), A. Jameson (Flow22), Mark Drela (XFOIL and MSES), Bijan Mohammadi (NSC2KE), and the Navy (POFACET).

John Lennon and Paul McCarthy penned the phrase “I get by with a little help from my friends” in 1967. Continuing in this vein I would like to acknowledge and extend my warmest thanks to my friends and colleagues. First and foremost, J.L. Lions has continuously shown me at the University Paris 6 and later at Dassault Aviation along my career the critical role of Applied Mathematics in complex problems in Aeronautics Industry; Roland Glowinski provided a foreword to this book and over the course of forty years has provided mentorship, continuous encouragement, and inspiration. William Fitzgibbon has also provided a foreword and personal encouragement and guidance for proper word usage and syntax. Olivier Pironneau has been always present when needed in difficult moments of my life with high spirit and at important decisions of my scientific and industrial career.

This book would not have existed without the support and encouragement at Dassault Aviation received from CEOs of the company and many colleagues and friends of the Theoretical Aerodynamics Department . I am particularly indebted to my colleagues and friends, Pierre Perrier who taught me with talent and vision aerodynamics applied to the design of civil, military and space aircraft and also to Patrick Pinganeau and Bruno Stoufflet, Division of the Scientific Strategy Directors for their encouragement in the creation and animation of a Pole Scientifique at Dassault Aviation co-chaired with Olivier Pironneau, UPMC .

Different chapters of this book have been written at the following institutions: the University of Pierre et Marie Curie (UPMC), INRIA Rocquencourt and INRIA Sophia Antipolis, the University of Sydney, the University of Queensland, the University of Milano, the University of Jyvaskyla, the International Centre of Numerical Methods in Engineering (CIMNE), the University of Las Palmas de Gran Canaria (CEANI and IUSIANI), the Von Karman Institute (VKI), the National Technical University of Athens (NTUA), The Kyoto Institute of Technology (KIT), Chiba University, the University of Houston and the Nanjing University of Aeronautics & Astronautics (NUAA). The authors express special thanks to V. Capasso, B. Chetverushkin, G. Ganascia, H. Deconink, J.-A. Desideri, W. Fitzgibbon, B. Galvan, K. Giannakoglou, R. Glowinski, GUO Xiaoping, JIN Quanyuan, H. Kawarada, Y. Kuznetsov, Y. Maday, K. Morinishi, E. Onate, K. Papailiou, K. Srinivas, R. Walker, G. Winter, N. Satofuka, T. Verstraete, and ZHA Rong, ZHAO Ning for their hospitality and the facilities provided during many stays or educational courses. A special thanks to P. Neittaanmaki for his long and enjoyable hospitality at the Univ. of Jyvaskyla and for his continuous encouragement to complete the material of the book.

We would also to thank Eric J. Whitney, A. Abbas, J. Alonso, G. Auchmuty, W. Annicchiarico, R. Bank, T. Burczynski, G. Bugeda, M. Cerrolaza, C.A. Coello Coello, P. Cuesta, Damp, K.Deb, A. Dervieux, M. Galan+, N. Gauger, D. Goldberg, D. Greiner, M. Hafez, F. Hecht, C. Hirsch, H. Kawarada, H. Suito, W. Haase, M. Holst, R. Hoppe, A. Jameson, I. Kroo, Y. Kuznetsov, R. Makinen, G. Meurant, Z. Michalevich, K. Miettinen, T. Nguyen, S. Obayashi, P. Perrier, V. Pediroda, O. Pironneau, C. Poloni, Ning Qin, D. Quagliarella, M. Schoenauer, K. Srinivas, J. Toivanen whose collaboration through fruitful discussions were essential for many of the results presented in the book.

My motivation in Computational Multi- physics Design Optimization in Aeronautics has been stimulated for almost two decades by European Commission projects of DG Research Aeronautics coordinated by CIMNE. The recent series of EU-China projects with Aerochina, Aerochina2 and more recently GRAIN and GRAIN2 are examples of collaborative teams focused on new methods and tools for advanced design in Aeronautics. I am particularly grateful to Dr. D. Knoerzer, a colleague and friend, for his continuous interest shown in Computational Fluid Dynamics and Multi-disciplinary Design Optimization and also to Dr. L. Breslin, Head of Unit Aeronautics for his involvement in EU-China cooperation and his interest in numerical methods, particularly at the ECCOMAS 2004 Congress in Jyvaskyla, Finland.

We are grateful to Ms Kati Valpe, Ms Marja Leina Rantalainen and Ms Anu Penttila, from the University of Jyvaskyla, for providing us with the beautiful picture of the front page of the book— Swans in a “high lift aerodynamic configuration” during landing on a Finnish lake—and also for checking many figures and helping in many ways the completion of this book.

Finally, I would like personally to express my gratitude to Ms. Nathalie Jacobs, Springer Publishing Editor and her staff and also to Prof. E. Onate, CIMNE Director and Editor of the Series Lectures Notes in Numerical Methods in Engineering and Sciences for their understanding, patience and encouragement during the difficult 2007–2008 period of personal tragedy and trauma.

Jacques Periaux and his co authors Felipe Gonzalez and Donseop Chris Lee
April 2014

Jyvaskyla, Finland

Contents

Chapter 1
Introduction

The content of this book describes algorithms and applications of robust evolutionary methods for multi-objective and multi-disciplinary design problems in aeronautics and UAV design. The chapters cover methods that couple robust evolutionary optimisers, parallel computing, asynchronous evaluation strategies, game theory and hierarchical topology of fidelity solvers that reduce the computational expense of multi-objective and multi-disciplinary design optimisation processes. The methods and framework described are applicable to single and multi-objective, inverse or direct, complex engineering problems that can be highly non-linear, multi-modal involved approximations, non-differentiable, with convex, non-convex or discontinuous Pareto optimal fronts.

The goals of the authors include addressing *real-world problems* in engineering involving three types of complexities:

- Instead of a single optimal solution, a *trade-off* between conflicting objectives arises due to manufacturing, technical and human constraints. Without preference information none of these trade-off solutions can be said to be better than the others.
- The *search space* for an optimal solution is often *complex* and can involve non-linearities, multi-modal and traditional search methods which may often encounter difficulties or eventually fail.
- *Real-world problems* usually involve *multiple disciplines*.

A designer (or a team of designers) in an engineering company or a research institute is (are) usually presented with problems that involve several complexities, objectives and multi-physics; therefore a systematic approach, referred to as Multidisciplinary Design Optimisation (MDO), that accounts for these complexities is required. The traditional approach in aeronautical design and MDO has been the use of deterministic optimisers. These methods are efficient in finding *local but not global optimal* solutions if objective functions and constraints are differentiable. However, robust alternative numerical tools are required if a broader application of the optimiser is desired, or if the problem is multi-modal, involves approximations,

J. Periaux et al., *Evolutionary Optimization and Game Strategies for Advanced Multi-Disciplinary Design,* Intelligent Systems, Control and Automation: Science and Engineering 75, DOI 10.1007/978-94-017-9520-3_1

is non-differentiable or involves multiple objectives and strongly non-linear physics; as is usually the case in the design of complex systems in aeronautics.

A relatively recent technique for optimisation is the use of Evolutionary Algorithms (EAs). EAs are based on Darwinian principles of evolution—*the so-called survival of the fittest* ", where populations of individuals evolve in a search space and adapt to the environment through the use of three different mechanisms such as *mutation, crossover and selection.* EAs require no derivatives or gradients of the objective function and have the capability of finding global optimum solutions amongst many local optima in hilly landscapes. They are easily executed using parallel computing techniques and can be adapted to arbitrary analysis codes without major modification. Another major advantage of EAs is that they can tackle multi-objective problems directly. Together these characteristics provide EAs substantial advantages over conventional deterministic approaches.

The aim of this book is to illustrate, identify and develop robust methods and algorithms and their coupling with *game strategies* applicable to a wide variety of problems in aeronautics. The methods are implemented in a generic framework by integrating several components so that the designer or team of designers can focus on the problem itself. These methods and framework are developed to automatically evolve and refine a candidate design such as an airfoil, a nozzle, a wing or an aircraft shape by changing the design variables, which characterise what might be called an individual. At the end of the optimisation process, the optimal set of individuals reflects suitable designs from which the designer can select and progress into further *detailed* steps of the design procedure. The challenge is to make the concept genuinely useful in a generic sense and adapted to different design problems. The key to this is the way methods and framework integrate with different components of the optimizer and the choice to encode—*the genotype*—design variables, to treat constraints and select good fitness functions (for example, takeoff-weight, fuel-weight, and aerodynamic performance).

The content of this book aims:

- To illustrate and develop several methods and set up a framework that can be used to solve single, multi-objective, multi-disciplinary *conceptua*l and *preliminary* and *detailed* design optimisation problems related to manned and unmanned aircraft systems (UAS). The family of problems that can be solved with these methods can be highly non-linear, multi-modal, involve approximations, be non-differentiable or involve multiple objectives and physics, with convex, non-convex or discontinuous Pareto optimal fronts;
- To illustrate how to develop a robust generic framework which assists the designer in the task of multi-objective and multi-disciplinary design optimisation in aeronautics. This framework comprises a GUI using object-oriented principles, a series of robust multi-objective and multi-disciplinary design optimisation algorithms coupled with games strategies and the coupling of several modules for parallel computing, post-processing, and aeronautical systems design;
- To illustrate the benefits and practicality of the methods for representative types of problems in aeronautical systems design. These optimisation problems fall

into the categories of single and aerodynamic multi-element based configuration, multi-disciplinary and multi-objective wing design, and multidisciplinary and multi-objective civil aircraft and unmanned air vehicles (UAVs) design.

The book presents a literature survey, covering three main research areas, namely optimisation methods, multi-disciplinary design optimisation and evolutionary algorithms combined with game strategies. Then, the main requirements and implementation of evolutionary methods and game strategies are described and mathematical test cases are considered. Finally, real-world applications to single, multi-objective aerofoil, wing and aircraft design optimisation problems are presented, analysed and discussed.

The content of this book provides the reader advantages and flexibility to implement evolutionary methods in terms of efficiency and quality dealing for solving multi-objective and multi-disciplinary design optimisation problems linked to several aircraft systems. In particular, practical applications to airfoil, wing and aircraft design using a combination of low-to-medium-to-high fidelity solvers are implemented and evaluated. Green challenges in the next decade driven by continued growth air transport—*safe, secure, fast, affordable and environmentally friendly*—need to extend these evolutionary approaches to more than ever *complex systems* with higher numbers of design variables, mature Design of Experiment (DOE) studies and advanced morphing design techniques using active devices and smart materials.

1.1 Background

In today's world of advancing technology, engineers are faced with the problem of designing increasingly complicated multi = disciplinary systems. This is a difficult task, as these systems not only involve different kinds of physics, but also a large number of variables and a series of objectives and constraints. At the same time, engineers need to optimise and address several requirements which include the reduction of the time cycle spent on design and reducing the cost of research and development, while improving the performance, reliability, quality and safety of the product or process under consideration.

These complex interactions have generated a growing interest in the area of multi-objective and Multi-disciplinary Design Optimisation (MDO). In multi-objective optimisation the designer is interested not only in a single global optimal solution but in a set of solutions that represent a trade-off between the different objectives. MDO refers to an approach that formalises the design process accounting for the interaction amongst the different physics involved, while optimising for a number of objectives and constraints.

When applied to aeronautics, the necessity of optimisation and MDO is evident, when knowing that even a very small improvement in weight or a reduction in aerodynamic drag have a tremendous impact on the overall performance of the design.

The application of MDO to the design and optimisation in aeronautics for instance includes a number of disciplines (aerodynamics, structures, propulsion, aeroacoustics, ...) are present and interact. The task of the designer or team of designers is to develop a solution that conforms to all disciplines while guaranteeing the requirements and constraints. When optimisation is intended, the different objectives (for example, aerodynamic performance, purchase price, take-off weight) need to be considered *(on a vector basis)* in order to find an optimal solution or set of non-dominated solutions.

A common approach for optimisation is the use of aggregating functions in which different weights are assigned to each objective. The problem with this approach is that the weight for each objective needs to be known in advance. Another approach is to compute or produce a set of solutions in what will be referred to in this book as a Pareto optimal front or surface. This Pareto optimal front represents the optimal set of non-dominated solutions and the trade-off between the objectives and disciplines involved. The process of MDO involves the use of several analysis tools such as Computational Fluid Dynamics (CFD) software or Finite Element Analysis (FEA) and also optimisation tools. Analysis tools are under constant development and have reached a point where a confident application to aeronautical design in conjunction with MDO is possible [1–3]. However, there are limitations to their application within MDO at an industrial level, due to the computational expenses involved. A single 3-D high-fidelity Navier-Stokes CFD computation around an aircraft wing, for example, might take several hours on a supercomputer. Therefore, the continuing challenge has been to develop methodologies such as Design of Experiments (DOE), approximation methods and variable fidelity models that combine and use different fidelity analysis tools during the design and optimisation process to minimise the computational expense.

While the area of traditional optimisation tools for a single discipline is quite mature, the area of robust optimisation tools and approaches for MDO is still in the initial stages of development [4–9]. A well-known approach in aeronautical design and MDO has been the use of traditional deterministic optimisers. These optimisers are efficient for finding optimal global solutions if the objective and constraints are differentiable. However, robust alternative numerical tools are required if a broader application of the optimiser is desired, or the problem is multi-modal, involves approximations, is non-differentiable or involves multiple objectives and physics; as is usually the case in the design of complex multi-disciplinary systems in aeronautics.

A relatively new technique for optimisation is the use of Evolutionary Algorithms (EAs). EAs are based on Darwinian theories of evolution, where populations of individuals evolve over a search space and adapt to the environment through the use of different mechanisms such as mutation, crossover and selection. EAs require no derivatives or gradients of the objective function and have the capability of finding global optimum solutions amongst many local optima. They are easily executed using parallel computing techniques and can be adapted to arbitrary analysis codes without major modification. Another major advantage of EAs is that they can tackle multi-objective problems directly. Together these characteristics give EAs substantial advantages over more conventional deterministic approaches. Interest in EAs for problems in engineering and aeronautics has grown substantially in the past fifteen years. These methods have been successfully applied to different aeronautical design problems including airframe, wing, aerofoil and rotor blade design [10–14].

The application of EAs for MDO problems is nevertheless limited. This is mainly because one of the drawbacks of EAs is that they are slow when compared to traditional deterministic methods, as they require a larger number of function evaluations to converge to an optimal solution. Hence the continuing challenge in evolutionary optimisation has been to reduce the number of function evaluations and the computational expense. To achieve this goal, several approaches have been proposed including a combination of variable fidelity models, parallelisation strategies and hybridisation techniques [15–17] and using *Nash or Stackelberg games scenario described in detail in Chap. 3*.

The content of this book develops multi-objective multi-disciplinary evolutionary methods and a framework in which different aeronautical problems are designed, analysed and optimised. The methods are based on a robust evolutionary optimiser, parallel computing, asynchronous evaluation and a hierarchical topology of fidelity solvers that reduce the computational expense for multi-objective and MDO problems. The methods are applicable to single and multi-objective, inverse or optimization of complex engineering problems that can be multi-modal, involve approximations, non-differentiable, with convex, non-convex or discontinuous Pareto optimal fronts.

These evolutionary methods simplify significantly the task of a designer or a design team by integrating several components so that they can focus on the engineering problem itself. The methods are developed in a sequence of steps consisting of: defining the requirements, formulating the methods, identifying several promising robust analysis and optimisation methods, creating algorithms and testing with mathematical functions and evaluating performances before solving practical real-world problems in aeronautics.

The methods are implemented in a single framework that integrates a Graphical User Interface (GUI), a robust multi-objective evolutionary algorithm, a series of modules for mathematical test functions, aerofoil, nozzle, wing and aircraft/airframe design and optimisation, parallel computing, post-processing and a Design of Experiments (DOE) capability. The framework was developed in a way that different applications and modules could be easily developed and implemented. These modules include real-world applications using analysis tools with variable fidelities, benchmarking of mathematical test functions, DOE studies, and parallel computing implementations.

It should also be noted that although the development of software for MDO has been an active field of research, the common limitation is that these architectures are proprietary or developed by universities and industries with restrictions on their use, and are sometimes specialised and difficult to expand for other applications. The framework and methods described in this book have been developed from scratch by progressively selecting, evaluating and assembling different software components in order to obtain detailed insight into the complexities involved in multi-objective optimisation and the process of MDO.

As detailed further in this book, the methods and framework have the following characteristics:

- *Modularity:* designed with purpose, the system uses object-oriented principles in a modular approach. When a new mathematical test case or a new aeronautical design and optimisation problem needs to be studied, the user only has to develop a few input/output files and lines of code.
- *Scalability and Parallelisation*: new and more complex analysis tools and design modules can be implemented. Also, studies in parallel computation can be performed.
- *Post-processing*: simple intuitive tools for visualisation of intermediate or final results.
- *Robustness*: the framework uses a robust evolutionary tool for optimisation. The framework and its methods have been shown to be robust and have been successfully validated for different single and multi-objective mathematical test functions as well as to a series of design and analysis tools.
- *Equivalence of formulations*: different multi-objective and multi-disciplinary analysis formulations are evaluated directly with evolutionary methods. There are no derivatives or transformations on the mathematical formulation of the problem that may introduce noise or require fine-tuning of the solution.

Of particular interest in this book is the application to Unmanned Aerial Vehicles (UAVs) and Unmanned Combat Aerial Vehicles (UCAV) systems. Robust tools for multi-disciplinary and multi-criteria analysis and optimisation are required as the need and complexity of the missions that these vehicles can achieve increases. UAVs are usually designed for cruising in single flight conditions while meeting other design constraints. The penalty for operating in off-design conditions can be significant. UAV aerodynamic performance might be improved if a multi-criteria multi-disciplinary optimisation can be developed which considers several design points.

1.2 Motivation

The methodology and software described in this book demonstrates the practical use of robust and efficient evolutionary algorithms (EAs) for multi-objective and multi-disciplinary design problems in aeronautics, and describes several applications for *validating* this task. The intention is not to exclude human interaction in the process, but rather to facilitate the other cumbersome exploration of the search space by exploiting these methods to find optimal solutions. Adequate engineering definitions and judgement of the problem are fundamental conditions provided by engineers; the aid and the benefit of computer evaluation present engineers with a broader exploration of design space and non-intuitive designs that could be out of their scope.

At an academic research and industrial level, the methods and framework provide scientists and engineers with a *platform* on which numerical analysis and optimisation tools can be implemented and complex engineering problems can be solved.

1.3 Summary of Chapters

This book considers several evolutionary methods, algorithms, games strategies and framework developed, implemented and tested, which can be used to evaluate, optimise and solve single, multi-objective, multi-disciplinary conceptual, preliminary and detailed design optimisation problems relation to aeronautics manned and unmanned aircraft systems.

The backbone of the book is organised as follows: Chapter 2 Provides an overview of Evolutionary Algorithms in multi-objective EAs and Game theory are considered Chap. 3. Chapter 4 describes a framework for numerical design and optimization algorithms this provides an indication of the complexities, coupling and performance of the methods with parallel computing evolutionary techniques and analysis tools. Real-world applications in aeronautics with increasing levels of complexity are discussed in Chaps. 5–10. Chapter 5 applies evolutionary methods to single objective aerodynamic shape optimisation problems. Chapter 6 is concerned with the application of methods for multi-objective model numerical test cases. Chapters 7, 8 and 9 presents robust multi-objective and multidisiplinary numerical model test cases. Finally Chap. 10 introduces design optimization with Morphing techniques. An Appendix includes and discusses two (2) "hands-on" examples to help the reader to reproduce step-by-step numerical computations with optimization software.

References

1. Agarwal RK (1999) Computational fluid dynamics for whole body aircraft. Annu Rev Fluid Mech 31:125–169
2. Alexandrov N, Kodyalam S (1998) Initial results of an MDO Methods Evaluation Study. In Proceedings of the Seventh AIAA/USAF/NASA/ISSMO Symposium on Multidisciplinary Analysis and Optimization, St. Louis, Missouri, September 1998
3. Alexandrov NM, Lewis RM (2000) Analytical and computational properties of distributed approaches to MDO. In AIAA Paper 2000-4718. AIAA, September 2000
4. Alexandrov NM, Lewis RM, Gumbert CR, Green LL, Newman PA (2000) Optimization with variable-fidelity models applied to wing design. In Proceedings of the 38th Aerospace Sciences Meeting and Exhibit, Reno, Nevada, January 2000. AIAA
5. Bartholomew P (1998) The role of MDO within Aerospace Design and Progress towards an MDO Capability. In Proceedings of Seventh AIAA/USAF/NASA/ISSMO Symposium on Multidisciplinary Analysis and Optimisation, St. Louis, Missouri, September 1998
6. Coello-Coello C, Van Veldhuizen DA, Lamont GB (2002) Evolutionary algorithms for solving multiobjective problems. Kluwer Academic Publishers, New York

7. Deb K (2003) Multi-objective optimization using evolutionary algorithms. Wiley, Greece
8. Kim HJ, Rho OH (1997) Dual-point design of transonic airfoils using the hybrid inverse optimization method. J Aircr 34(5):612–618
9. Mäkinen R, Neittaanmäki P, Périaux P, Toivanen J (1998) A genetic algorithm for multiobjective design optimization in aerodynamics and electromagnetics. In: Papailiou KD et al (ed) Computational fluid dynamics '98, Proceedings of the ECCOMAS 98 Conference, vol 2, Wiley, Greece, pp 418–422
10. Mason WH, Knill DL, Giunta AA, Grossman B, Haftka RT, Watson LT (1998) Getting the full benefits of CFD in conceptual design. In Proceedings of the AIAA 16th Applied Aerodynamics Conference, Albuquerque, New Mexico
11. Obayashi S (1998) Multidisciplinary design optimization of aircraft wing planform based on evolutionary algorithms. In Proceedings of the 1998 IEEE International Conference on Systems, Man, and Cybernetics, La Jolla, California, October 1998. IEEE
12. Oyama A, Liou M, Obayashi S (2002) Transonic axial-flow blade shape optimization using evolutionary algorithm and three-dimensional Navier-Stokes solver. In Proceedings of the 9th AIAA/USAF/NASA/ISSMO Symposium on Multidisciplinary Analysis and Optimization, Atlanta, Georgia, September 2002
13. Parmee IC, Watson AH (1999) Preliminary airframe design Using co-evolutionary multiobjective genetic algorithms. In: Proceedings of GECCO 1999, vol 2, pp 1657–1665. Morgan Kaufmann, San Francisco
14. Raymer D (2002) Enhancing aircraft conceptual design using multidisciplinary optimization. PhD thesis, KTH, Department of Aeronautics, FLYG 2002-2
15. Sobieszczanski-Sobieski J, Haftka RT (1996) Multidisciplinary aerospace design optimization survey of recent developments. Technical Report 96-0711, NASA/American Institute of Aeronautics and Astronautics
16. Sobieszczanski-Sobieski J, Haftka RT (1997) Multidisciplinary aerospace design optimization: survey of recent developments. Struct Optim 14:1–23
17. Thomas Z, Green A (1999) Multidisciplinary design optimization techniques: implications and opportunities for fluid dynamics research. In Proceedings of the 30th AIAA Fluid Dynamics Conference, Norfolk, VA, June 1999. AIAA Paper 1999-3798

Chapter 2
Evolutionary Methods

2.1 Overview

This chapter provides the reader with an overview of Evolutionary Algorithms. The idea of mimicking Evolution as an optimisation tool for engineering problems appeared in the late 50s and early 60s. The concept was to evolve a population of candidate solutions to solve a problem, using operators inspired by natural selection. In the 60s, Rechenberg introduced "Evolution Strategies" (ESs) for airfoil design: this approach being continued by H. Schweffel [2]. Other computer scientists developed evolution inspired algorithms for optimisation and machine learning at the same period when electronic computers appeared for the first time. Genetic Algorithms (GAs) were invented by J. Holland in the late 60s and developed by Holland and his students (D. Goldberg among many others) at the University of Michigan to study within the computer the phenomenon of adaptation as it occurs in nature. Holland's book [3] on "Adaptation in Natural and Artificial Systems" presented the genetic algorithm as an abstraction of biological Evolution, which was a major innovation due to the biological concept of population, selection, crossover and mutation. The theoretical foundation of Genetic Algorithms (GAs) was built on the notion of "*schemas*" and "*building blocks*" which is explained in detail in many books devoted to Genetic Algorithms (c.f. D. Goldberg for instance [4]. In the last decade there has been widespread interaction among researchers studying Evolutionary Computation methods, and the GAs, Evolution Strategies, Evolutionary Programming and other evolutionary approaches were finally unified in the late 90s under the umbrella named Evolutionary Algorithms (EAs).

J. Periaux et al., *Evolutionary Optimization and Game Strategies for Advanced Multi-Disciplinary Design*, Intelligent Systems, Control and Automation: Science and Engineering 75, DOI 10.1007/978-94-017-9520-3_2

2.2 Fundamentals of Evolutionary Algorithms (EAs)

One of the emerging techniques for solving Multi-disciplinary Design Optimization (MDO) and Multi-disciplinary Optimization (MO) problems is Evolutionary Algorithms (EAs). EAs are decision maker algorithms that mimic the natural principle "survival of the fittest" introduced by Charles Darwin's famous book "*Origin of Species*" in 1859. Broadly speaking they operate simply through the iterated mapping of one population of solutions to another population of solutions. This *Information Technologies* (*IT*) based approach contrasted with existing conventional deterministic search techniques such as the simplex method, conjugate gradient method and others, which proceed from one given sub-optimal solution to another, until an optimum solution is reached. Evolutionary algorithms are not deterministic, so that for identical problems and identical starting conditions, the evolution of the solution will not follow the same path on repeated simulations. It is for this reason that EAs fall into the category of stochastic (randomized) optimization methods. Some other stochastic methods that are used are the Monte-Carlo approach (MC), the directed random walk and simulated annealing (SA) However the process of evolution in EAs is of course not completely random (in particular during the selection procedure) because in this case the performance of the algorithm would be no better than simple guessing, and at worst would be equivalent to complete enumeration of the parameter search space. Evolutionary algorithms work by exploiting population statistics to some greater or lesser extent, so that when newer individual solutions or offspring are generated from parents, some will have inferior *genetic characteristics* while others will have superior genetic characteristics. The general working Darwin principle of the iterated mapping then reduces to generate an offspring population, removing a certain number of "below average" evaluated individuals, and obtaining the subsequent population. This can be summarized as the repeated application of two GAs operators on the population, the variation operator (the generation of offspring) and the selection operator (the survival of the fittest) [4]. Some variant approaches to EAs in the literature only differ in the operation of these two operators. The origin of Evolutionary Algorithms for parameter optimization seems to have appeared independently in two separate streams, Genetic Algorithms (GAs) and Evolution Strategies (ESs).

Some of the advantages of EAs are that they require no derivatives (or gradients) of the objective function, have the capability of finding globally optimum solutions amongst many local optima, are easily run in parallel computers and can be adapted to arbitrary solver codes without major modification. Another major advantage of EAs is that they can tackle simultaneously multi-objective problems (by considering fitness vector and not the traditional weighted aggregation of several fitness criteria). It is shown in the sequel how this new feature is used intensively for the capture of Pareto, Nash or Stackelberg solutions using game strategies in multi-criteria optimization problems.

2.3 Evolutionary Algorithms (EAs)

Evolutionary Algorithms (EAs) are search procedures based on the mechanics of natural selection and Darwin's principle: "*survival of the fittest*". GAs, a particular class of EAs, were introduced by J. Holland [3] who explained the adaptive procedure of natural systems and laid down the two main principles of GAs: the ability of a simple bit-string representation to encode complicated structures and the power of simple transformations to improve such structures. A few years later, D. Goldberg brought simple GAs to nonconvex optimization theory for a quantitative study of optima and introduced a decisive thrust in the GAs research field.

The simple Genetic Algorithm (GA) was founded on principles developed by J. Holland [3] in the late 60s, and a number of research topics both in theory and applications were developed. It is generally accepted, however, that modern GAs were placed on a strong foundation in optimization research by Goldberg [4]. Goldberg's initial applications of the GAs were in real-world topics such as gas pipeline control. The original GAs technique revolved around a single binary string (in base 2) encoding of the DNA of chromosomes, which is the genetic material that each individual carries. The binary coded GAs' variation operator is comprised of two parts, *crossover* and *mutation.* Crossover interchanges genetic portions of parental chromosomes while mutation involves the random switching of DNA letters in the chromosome. The *selection* operator has taken many forms, the most basic being the *stochastic/deterministic* fitness-proportionate (or *roulette wheel*) method. Genetic Algorithms have developed significantly in the past decade, and these developments are considered further throughout this chapter.

2.4 Benefits of EAs

One of the main advantages of EAs and GAs in particular is *robustness*: they are computationally simple and powerful in their search for improvement and are not limited by restrictive assumptions about the search space (continuity, existence of derivatives, uni-modality). Furthermore, they accommodate well to discontinuous environments (see in Chap. 3 the capture of discontinuous Pareto fronts). GAs are search procedures that use semi-random search but allow a wider exploration in the search space compared to classical optimization methods which are not so robust but work nicely in a narrow search domain problem.

How are EA different from traditional numerical optimization tools?

- EAs are indifferent to problem specifics: an example in shape airfoil optimization is the value of the drag of an airfoil which represents the so-called fitness function;
- EA use codes of decision variables by adapting artificial DNA chromosomes or individuals rather than adapting the parameters themselves. In practice, the designer will encode the candidate solutions using binary coding GAs with finite–length string *genotype*;

- GAs process populations via *evolving generations* compared to point by point conventional methods which use local information and can be trapped frequently in a local minimum;
- GAs use *randomized operators* instead of strictly deterministic gradient information.

2.4.1 General Presentation of EAs Using Binary Coding

GAs are different from the conventional search procedures encountered in engineering optimisation. To understand the mechanism of GAs, consider a minimization problem with a *fitness function* index $J=f(x)$, where the design parameter is x. The first step of the optimisation process is to encode x as a finite-length string. The length of the binary string is chosen according to the required accuracy. For a binary string of length $l=8$ bits, the lower bound x_{min} is mapped to 00000000 and the upper bound x_{max} is mapped to 11111111, with a linear mapping in between. Then for any given string the corresponding value x can be calculated according to: $x=x_{min} + 1/(2^*l-1).(x_{max}—x_{min})$.With this coding, the initial population is constituted of N individuals, and each of them is a potential solution to the optimization problem.

We must define now a set of EA operators that use the initial population and then create a new population at every generation. There are many EA operators, but the most frequently used are *selection, crossover and mutation.*

Selection consists in choosing solutions that are going to form a new generation. The main idea is that selection should depend on the value of the fitness function: the higher the fitness is, the higher the probability is for the individual to be chosen (akin to the concept of survival of the fittest). But his selection remains a probability, which means not being a deterministic choice: even solutions with a comparative low fitness may be chosen, and they may prove to be very good in the course of events (e.g. if the optimisation is trapped in a local minimum). Two well-known selection techniques are Roulette Wheel and Tournament (c.f. Goldberg [4]).

Reproduction is a process by which a string is copied in the following generation. It may be copied with no change, but it may also undergo a mutation, according to a fixed mutation probability *Pm*. However filling up the next generation is achieved through the operator called *crossover.*

A 001 / **01110**		*A*' 001**10010**
	→	
B 111/**10010**		*B*' 111**01110**

where/denotes the cutting site

First, two strings are randomly selected and sent to the mating pool. Second, a position along the two strings is selected according to a uniform random law. Finally, based on the crossover probability Pc, the paired strings exchange all genetic characters following the cross site. Clearly the *crossover* randomly exchanges structured information between parents A and B to produce two offspring A' and B' which are expected to combine the best characters of their parents.

The third operator, called *mutation,* is a random alteration of a binary (0–1) bit at a string position, and is based on a mutation probability Pm. In the present case, a mutation means flipping a bit 0 into 1 and vice versa. The *mutation* operator enhances population diversity and enables the optimisation to get out of local minima of the search space.

2.4.2 Description of a Simple EA

For the optimization problem *P* dealing with the minimisation of a fitness function *f (x)*, a simple binary coded GA can be run in the computer according to the following step by step procedure:

- Step 1. Generate randomly a population of N individuals,
- Step 2. Evaluate the fitness function of each individual phenotype,
- Step 3. Select a pair of parents with a probability of selection depending of the value of the fitness function. One individual can be selected several times,
- Step 4.Crossover the selected pair at a randomly selected cutting point with probability Pc to form new children,
- Step 5. Mutate the two offspring by flipping a bit with probability Pm,
- Step 6. Repeat steps 3,4,5 until a new population has been generated,
- Step 7. Go to step 2 until convergence.

Algorithm 1 Step by step procedure of a simple EA software

After several generations, one or several highly fitted individuals in the population represent the solution of the problem *P*. The main parameters to adjust for convergence are the size *N* of the population, the length *L* of the bit string, the probabilities *Pc* and *Pm* of crossover and mutation respectively.

In the sequel, examples of optimisation and inverse problems with the use of GAs are implemented step by step and results presented in the appendix of the book.

2.5 Mechanics of EAs

EAs share common elements: representation of individuals, fitness function, and an iterative selection based on fitness, recombination, mutation, elitism and the dilemma between the *exploration* and *exploitation* of the search space.

2.5.1 Representation of Individuals

There are many types of representations, the most common being binary and floating point. The binary representation uses a bit string to represent an individual. With this representation, the real design variables are transformed into binary numbers that are concatenated to form a chromosome. This chromosome encodes the total number of design variables of the problem. In floating-point representation, a vector of real numbers characterizes an individual. In the case of an airfoil shape design, for example, design variables are *control points* for a Béziers or Spline curve that generates the aerodynamic shape. It has been reported by different researchers that real-coded EAs have outperformed binary-coded EAs in many applications. The explanation for this is that in a binary representation the variables are concatenated to represent an individual and this results in a big string length which is difficult to handle. Another problem is that the binary representation of real design parameters presents a difficulty with what is called *"hamming cliffs"* which is the discrepancy between the representation space and the problem space. As a consequence, it is difficult for a binary-coded EA to exploit the search in the vicinity of the current population. On the other hand, a real-code representation is conceptually closer to the real design space and the length of the string vector is equal to the number of design variables.

Fitness Function Similar to the concept of survival of the fittest in nature, EAs use a fitness function to evaluate the performance in order to determine the quality of the vector string and to define whether the individual is a suitable candidate for the next generation. The fitness function is a critical aspect of EAs: a general rule is that it should reflect as closely as possible the desired physical behaviour of the solution. Examples of fitness functions in aeronautics can be, among others, drag minimization, aerodynamic performance or gross weight.

Evaluation of Fitness Evaluation is the means by which each individual in the population is evaluated. This fitness could be in Aerospace Engineering an analytical function or a complex functional depending of linear or non-linear CFD or FEA analysis. For an optimization problem in aeronautics a standard way for evaluation can be an Euler or Navier-Stokes CFD analysis that evaluates the flow around the airfoil and provides a numerical estimate of lift and drag coefficients, or an aero-structural analysis that computes simultaneously aerodynamic performance and structural weight used to evaluate the fitness function of a candidate solution.

Selection Selection is a procedure during which individuals compete and are selected to produce offspring for the next generation allowing candidates' solutions to be selected by comparison of their fitness values. Several parent selection techniques have been proposed, but their application is usually problem dependent [4, 5]. A selection strategy currently used is the "fitness proportional selection". In this case the selection probability of individuals is calculated by dividing their fitness values by the sum of all the other finesse values of individuals.

Parents can also be selected by *Roulette Wheel* selection [4] or Stochastic Universal Sampling [5]. An individual is selected by spinning the wheel, which is divided according to the selection probability. *Tournament selection* operates by choosing some individuals randomly from a population and selecting the best from this group to survive in the next generation. Its simplest form is binary selection, whereby two random pairs of individuals are selected from the population and the pair with higher fitness is copied to the mating pool or population. Another method for selection is *ranking,* whereby individuals are ranked by their fitness values. The best individual receives rank 1; the second receives rank 2 and so on. A selection probability is reassigned in accordance with the ranking order.

An appropriate level of selection pressure is critical for the success of the evolution. Too much pressure applied can generate a loss of diversity, and lead to premature convergence. This inadequate situation can be explained since the population is not infinite and some individuals who are comparatively highly fit but not optimal rapidly dominate the population. The basic idea is then to control the number of reproductive opportunities that each individual has in order to prevent highly fit individuals taking over the population. On the other hand, if a low selection pressure is imposed, the search is ineffective and will require excessive time for convergence.

Recombination Recombination, also known as crossover, is the procedure in which two or more parent individuals (or chromosomes) are combined to produce an offspring chromosome (individual). Recombination is necessary in cases when an offspring is to have multiple parents, since mutation by itself provides no mixing of the chromosomes.

Mutation and Adaptation The importance of mutation is to keep diversity in the population and to expand the search to areas that cannot be represented by the current population. Different mutation operators have been proposed. A common method is uniform mutation, whereby a random number with probability p is added to each component of the individuals [4, 5]. In Gaussian mutation, a number from a Gaussian distribution with zero mean is added to each component of the individual vector.

Another approach developed by Hansen and Ostermeier [6, 7] uses Covariance Matrix Adaption (CMA) and a mutative strategy parameter control (MSC) that is applied to the adaptation of all parameters of an N-dimensional normal mutation distribution and provides a second-order estimation of the problem topology. Some of their results in [6, 7] confirm the efficiency of this approach in stochastic optimization problems.

Elitism Another important aspect of EAs is the use of an elitist strategy. As the evolutionary procedure with EAs depends on stochastic operators, there is no guarantee that there would be a monotonic improvement in the fitness function value. With an elitist strategy best individuals are copied to the next generation without applying any evolutionary operators.

The Exploration—Exploitation Dilemma Using EAs, a critical aspect is the balance between the *exploration* the search space areas and the *exploitation* of the learned knowledge to progress in the evolution. As these are conflicting objectives, EA researchers have developed different alternatives to balance these trade-offs. Therefore, when developing or selecting an EA it is important to test the algorithm to ensure a good balance between these two criteria and that it possesses capabilities for benchmarking different mathematical test functions to test its robustness and performance. An area that has shown promising results is using the concept of subpopulations that explores and exploits different regions of the search space and is refined as the evolutions progresses.

2.6 Evolution Strategies (ESs)

Evolution Strategies (ESs) are used and explained in different chapters of the book. The first version of an ES software introduced by Rechenberg [8] used only two individuals, one parent and one offspring. Each individual was real coded; in optimization problems all design variables of a chromosome were assigned floating point values. The variation operator was applying a random mutation to each floating point value in the parental chromosome to generate the offspring individual. The selection operator was entirely deterministic, and was simply the result of a competition between parent and offspring to determine which of the both remained. In the standard nomenclature this strategy is denoted as the $(1+1)$ ES, the first digit indicating the number of parents, the '+' indicating competition between parents and offspring and the final digit indicating the number of offspring. From the very beginning ESs have been designed almost exclusively with real coding, as opposed to original GAs variants where real parameter optimisation is a piecewise interpretation of the binary chromosome string associated with each individual. An evolution strategy was therefore seen as a logical starting point for an evolutionary optimisation using real coded problem variables.

Further developments in ESs introduced multi-membered populations for both parents and offspring. The first algorithm of this type was the $(\mu+1)$ ES software introduced by Schweffel [2]. This generalization was achieved by applying some variation operator to the parent population to produce a single offspring. The offspring is selected by determining whether it is better than the worst member of μ; in such a case it replaces the worst member μ. Both the $(1+1)$ ES and the $(\mu+1)$ ES approaches used deterministic control of the mutation size (variations applied to design variables) which were normally distributed when applied to real coded problems. Recent developments in both GAs and ESs have greatly modified both their variation and selection operators such that it is not clear whether such a nomenclature division is nowadays particularly justified. The main difference that exists between them today is still the predominance of adaptive mutations in ESs, which have made them very attractive for real coded optimisation, although GAs research has produced some related concepts.

A representative code of a canonical Evolution Strategy is illustrated in algorithm 1. A population (μ_o) is initialized and then evaluated. Then for a number of generations (g) and as long as a stopping condition (maximum number of function evaluation or target fitness value) is not met, off springs (λ^{g+1}) go recursively through the procedure of recombination, mutation, evaluation and selection.

ESs and GAs have distinguishing features according to their evolution modeling and representation, but this classification has been blurred as the features of one method have been incorporated into other evolutionary methods. For instance, some GAs applications and developments gave up the bit string for a floating-point representation, ESs used some form of crossover operators for reproduction and Genetic Programming (GP) introduced by Koza [9] were also extended and is not only limited to evolution of finite state machines. As there is no longer clear separation between these methods, these evolutionary approaches were defined as "Evolutionary Computation" (EC) or "Evolution Algorithms" (EAs). Considering this general denomination, an EA software (including ESs, GAs and GP) can be defined as indicated below in Algorithm 2:

Initialise: $init(\mu_o)$
Evaluate: $f(\mu_o)$
g=0
while stopping condition not met,
  Recombine: $\lambda_R^{g+1} = reco(\mu^g)$
  Mutate: $\lambda_M^{g+1} = mut(\lambda_R^{g+1})$
  Evaluate: $\lambda^{g+1} = f(\lambda_M^{g+1})$
  Select: $\mu^{g+1} = sel(\mu + \lambda)$ (plus strategy) or,
  $\mu^{g+1} = sel(\lambda)$ (comma strategy)
  g=g+1
loop

Algorithm 2: A canonical Evolution Strategy algorithm

2.7 Application of EAs to Constrained Problems

Engineering problems usually involve a number of constraints due to technical, manufacturing, human resources requirements and limitations. It is necessary to incorporate and satisfy those constraints in the optimization procedure to obtain a realistic design. EAs are unconstrained optimization procedures. Therefore some handling techniques have to be introduced to incorporate constraints into fitness functions.

Different approaches have been developed in order to satisfy design constraints [10]. The use of the penalty function is the most common approach and is based on adding penalties to the objective function [11]. When applying a penalty to an infeasible individual it is important to determine if it is penalized for simply being infeasible or penalized also by some amount due to its infeasibility and the number of constraints violated. As reported by different researchers [11, 12], penalties that are functions of the distance from feasibility perform better than those that are only a function of the number of violated constraints. Joines and Houck [13] describe *static penalties* and *dynamic penalties*. In *static penalties*, the user defines several levels of violation and a penalty coefficient is chosen for each so that the penalty coefficient increases as a higher level of violation is reached. The drawback of this approach is that it requires a high number of parameters. In *dynamic penalties*, the dynamic function increases as the optimization progresses through generations.

Other methods include *annealing penalties* that are similar to *simulated annealing* in which penalty coefficients are changed once the algorithm is trapped in a local optimum. The main problem with this approach is that it is sensitive to the values of its parameters and it is difficult to choose an appropriate cooling scheme. *Adapting penalties* work by modifying the penalty based on a feedback from the last k generations. The inconvenience with this approach is the selection of the number of generations to wait before it can be applied.

The *Death penalty* is the easiest way to handle constraints and works by rejecting infeasible individuals. The main drawback of this approach is the loss of genetic information that can be contained in the discarded individual. It can also be lengthy, especially in cases where it is difficult to approach the feasible region.

Within an EA these constraints and their treatment are specified by the designer and treated by the optimizer. They may take the form of simple upper and lower bounds on the object variables, but many more complicated constraints exist and then must be satisfied during the optimization procedure. Problems are often posed so that only certain combinations of object variables can be considered or their bounds are not simply 'upper' and 'lower' but also 'not between' and 'not if'. Object variables merely represent the *genotype* (numerical representation) of the individual, and further constraints will probably exist on the *phenotype* (physical representation) of the individual as well. Such constraints may be imposed on a particular solution such as weight, geometry or some other physical characteristics which are undesirable. Quite often whether there has been an excursion from the phenotypic problem constraints or not, this situation can only be determined after the fitness function has been applied, and this may result in slowing overall performance of the optimizer.

Two basic methods of handling constraints are considered in this study: the 'rejection' method and the 'penalty' method. The rejection method simply involves rejecting any individual that is not compliant with the constraints, by not allowing it an opportunity to contest insertion into the main population. The merit of the rejection method is that no penalty scheme needs to be devised for handling individuals that are out of bounds, and therefore only solutions that satisfy the constraints fully are admitted. The disadvantage with this approach is that individuals that are close to the boundary but not within it are rejected out of hand, even though they may contain useful genetic information.

The penalty method involves adding some penalty fitness to f which (in the context of a minimization problem) reduces its fitness with respect to other individuals in the population, reducing the likelihood that it will be selected next time. For example, if a certain solution-dependent value (s) must be less than a given value (v) a penalty function can be constructed as follows:

$$f' = f + h(s - v).(s - v)^2 \tag{2.1}$$

where f' is the (possibly) penalized fitness, f is the original fitness and h (…) is the Heaviside function. The advantage of the penalty method is that individuals with good genetic material can be allowed to converge from outside the boundary to inside the boundary, if possible. In this volume, multi-objective fitnesses are penalised by adding equal values of the penalty to each fitness value. This ensures that between two otherwise equal solutions, one which is penalized can never dominate one which is not. The disadvantage with the penalty method is of course that the penalty function needs to be devised with some care, especially considering the case there may be many functions to devise. In the example given, a question of weight to choose arises: should a more severe term such as $10^5(s—v)^3$ or a less severe term such as $\frac{1}{4}|s—v|$ have been used instead. In these cases problem specific knowledge is required, so the user must make a 'best guess' of the penalty to apply, or run a number of cases to gain some experience with the particular case involved. Possibly the best compromise will be the use of both rejection and penalty methods together, so that rejection is used on solutions that are obviously not feasible and will not lead to further improvement, while penalties are applied to solutions that show promise, but exceed allowable limits by a small margin. These two methods are used throughout in chapters with applications, and they will be referred to as 'hard' (rejection) and 'soft' (penalty) bounds respectively.

2.8 Summary of Chapter

This chapter reviews different basic concepts of EAs. It can be noticed that there are different methods, architectures and applications of optimization and multi-disciplinary design optimization methods for solving aeronautical design problems. However, further research and investigations into innovative methods such as hybridized EAs with games are still required to master the industrial and academic challenges and future needs in Aeronautics and UAV Design. EAs are an alternative option to satisfy some of these needs, as EAs with their flexibility (no gradient of the objective function(s) required), their particularly adaptability and their easy parallelization, have been used for several multi-objective optimization and successfully applied to different aeronautical *conceptual* design problems. Nonetheless, EAs face challenges in an industrial level in *detailed* design (complex and expensive mathematical modeling of the environment) due to the computational expense involved in optimization procedures and the fact that they require a larger number of function evaluations, compared to traditional deterministic techniques. Efficiency and quality design of such evolutionary optimizers coupled with game strategies are the two major targets developed in the following chapters.

References

1. Mitchell M (1998) An Introduction to genetic algorithms. MIT Press, Cambridge. ISBN 978-0-262-63185-3 pp. I–VIII, 1–208
2. HP S (1975) Evolution strategie and numerische optimierung, PhD. Thesis. Technische Universität, Berlin
3. Holland J. (1975) Adaption in natural and artificial systems. The University of Michigan Press, Michigan
4. Goldberg D. (1989) Genetic algorithms in search, optimization and machine learning. Addison-Wesley
5. Michalewicz Z. (1992) Genetic algorithms + data structures=evolution programs. Artifical intelligence. Springer-Verlag
6. Hansen N, Ostermeier A. (1996) adapting arbitrary normal mutation distributions in evolution strategies: The covariance matrix adaptation. In: Proceedings of the 1996 IEEE International Conference on Evolutionary Computation, pp 312-317
7. Hansen N, Ostermeier A (2001) Completely derandomised self-adaption in evolution strategies. Evolut Comput 9(2):159–195
8. Rechenberg I (1973) Evolution Strategie: Optimierung technisher Systeme nach Principen der biologischen Evolution. Frommann-Holsboog, Stuttgart
9. Koza JR (1992) Genetic programming: on the programming of computers by means of natural selection. The MIT Press, Cambridge
10. Kumar V (1992) Algorithms for constraint-satisfaction problems: a survey. AI Magazine 13(1):32–44
11. Dasgupta D, Michalewicz Z. (1997) Evolutionary algorithms in engineering applications. Springer-Verlag
12. Richardson JT, Palmer MR, Liepins G, Hilliard M. (1989) Some guidelines for genetic algorithms with penalty functions-incorporating problem specific knowledge into genetic algorithms. In: Schaffer JD (ed) Proceedings of the Third International Conference on Genetic Algorithms, pp 191–197, George Mason University. Morgan Kaufmann.
13. Joines JA, Houck CR. (1994) On the use of non-stationary penalty functions to solve nonlinear constrained optimization problems with GA's. In: International Conference on Evolutionary Computation, pp 579–584

Chapter 3
Multi-Objective EAs And Game Theory

3.1 Generalities

The traditional way to address the problem of multiple objective optimization is to associate a scalar objective, generally obtained through some linear combination of weighted objectives. Such an approach may be of interest in some cases—particularly if the weight of each criterion is known beforehand—but besides its ad hoc character, it has several drawbacks since there is a loss of information and a need to define the weights associated to each objective. Moreover, the behavior of the algorithm is very sensitive and is biased by the values of these weights. Schaffer was the first to propose a Genetic Algorithm approach in 1985 for multiple objectives through his Vector Evaluated Genetic Algorithms (VEGA [1]), but it was biased towards the extrema of each objective. Goldberg proposed a solution to this particular problem with both non-dominance Pareto-ranking and sharing, in order to distribute the solutions over the entire Pareto front [4]. All of these approaches are based on Pareto ranking and use either sharing or mating restrictions to ensure diversity. In the following, section we discuss a Pareto-based multi -objective algorithm inspired which is a cooperative approach which gives a whole set of non-dominated solutions—the Pareto front.

Another non-cooperative multiple objective algorithm based on the notion of Nash equilibrium and hierarchical Stackelberg games. We also describe the concept of hybridized games with Genetic Algorithms and presents some mathematical functions to evaluate performances of hybridized games.

J. Periaux et al., *Evolutionary Optimization and Game Strategies for Advanced Multi-Disciplinary Design,* Intelligent Systems, Control and Automation: Science and Engineering 75, DOI 10.1007/978-94-017-9520-3_3

3.2 Definition of A Mult- Objective Problem

Engineering design problems often require a simultaneous optimization of inseparable objectives and associated number of constraints. A multi-criteria optimization problem can be formulated as:

$$Minimise: \quad f_i(x) \quad i=1\ldots N \tag{3.1}$$

Subject to constraints:

$$g_j(x)=0 \quad j=1\ldots M \quad h_k(x)\leq 0 \quad k=1\ldots K, \tag{3.2}$$

where f_i are the objective functions, N is the number of objectives and x is an N-dimensional vector where its arguments are the decision variables and g_j, h_k are equality and inequality constraints respectively.

There are many variants and developments of multi-objective approaches: these include the lexicographic approach, traditional aggregating functions, Pareto, Nash and Stackelberg game approaches [2, 3]. These concepts are extended and applied in combination with EAs in the following sections.

3.3 Cooperative Games: Pareto Optimality

The main interest of a Pareto based evolutionary algorithm is that the optimization procedure does not have to combine its objectives. It derives the fitness values of the candidate solutions directly from the comparison of their respective objective vectors.

A common way to represent the solution to a multi-objective problem is by the use of the concept of Pareto optimality or non-dominated individuals [4]. Figure 3.1 shows the Pareto optimality concept for a problem with two conflicting objectives. A solution to a given multi-objective problem is the Pareto optimal set, found using a cooperative game which computes the set of non-dominated solutions. This spans the complete range of compromised designs between the two objectives. Most real world problems involve a number of inseparable objectives where there is no unique optimum, but a set of compromised individuals known as Pareto optimal (or non-dominated) solutions. We use the Pareto optimality principle where a solution to a multi-objective problem is considered Pareto optimal if there are no other solutions that better satisfy all the objectives simultaneously. The objective of the optimization is then to provide a set of Pareto optimal solutions that represent a trade-off of information amongst the objectives.

For a minimization problem, a vector x_1 is said to be partially less than vector x_2 if and only if:

$$\forall i: \quad f_i(x_1)\leq f_i(x_2) \quad and \quad \exists_i: \; f_i(x_1)< f_i(x_2). \tag{3.3}$$

In this case the solution x_1 *dominates* the solution x_2.

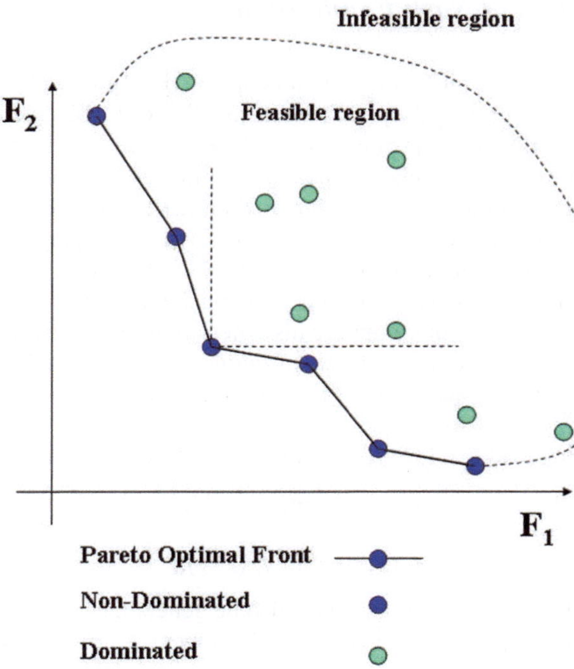

Fig. 3.1 Pareto optimality

As EAs consider multiple points simultaneously, they are capable of finding a number of solutions in a Pareto set. Pareto selection ranks the population and selects the non-dominated individuals for the Pareto front. A comprehensive theory, literature review and implementation of Multi-objective EAs (MOEAs) including the NSGAII and VEGA algorithms is given by Deb in reference [3].

3.4 Competitive Games: Nash Equilibrium

When dealing with multiple objective optimization, Pareto GAs have now become a sort of standard in design environments. With the introduction of non-dominance Pareto-ranking and sharing (in order to distribute the solutions over the entire Pareto front) the Pareto GAs are a very efficient way to find a wide range of solutions to a given problem. However, another multiple objective optimization approach, this time a non-cooperative one, was introduced by John F. Nash in the early 50s [5] this approach introduced the notion of Nash players and aimed at solving multiple objective optimization problems originating from Game Theory in Politics and Economics.

3.4.1 Definition of Nash Equilibrium

Nash optima define the solution of a non-cooperative multiple objective optimization approach first proposed by J. F. Nash [5]. Since it originated from Game Theory, the notion of a player is often used and we will use it in the sequel. During a Nash game, the two players choose their best strategies according to the one decided by the other player to improve their own criteria (their gain). For an optimization problem with n objectives, a Nash strategy is a symmetric game which consists in having n players, each optimizing its own criterion. However, each player has to optimize his criterion given that all the other criteria are fixed by the rest of the players. When no player can further improve his criterion, it means that the system has reached a state of equilibrium called *Nash Equilibrium*. For a set of two variables x and y, let E be the search space for the first criterion and F the search space for the second criterion.

A strategy pair $(\overline{x}, \overline{y}) \in E \times F$ is said to be a Nash equilibrium if and only if:

$$f_E(\overline{x}, \overline{y}) = \inf_{x \in E} f_E(x, \overline{y}), \tag{3.4}$$

$$f_F(\overline{x}, \overline{y}) = \inf_{y \in F} f_F(\overline{x}, y). \tag{3.5}$$

The definition of a Nash equilibrium may also be generalized to N players. In this case, a solution $u = (u_1, \ldots u_m)$ with, u the total number of variables, is a Nash equilibrium if and only if: $\forall i, \forall v_i$,

$$f_i(u_1, \ldots u_{i-1}, u_i, u_{i+1}, \ldots, u_m) \leq f_i(u_1, \ldots u_{i-1}, v_i, v_{i+1}, \ldots, v_m)$$

where f_i is the optimization criterion for player i.

With classical optimization approaches, Nash Equilibria are very difficult to find. It is generally easier to prove that a given solution is a Nash Equilibrium, but exhibiting such a solution may prove to be very hard. And it becomes almost impossible if the criteria are non-differentiable functions. The next section will show that GAs offer an elegant alternative.

3.4.2 Coupling Nash Games and GAs

The approach that combines GAs with a Nash strategy was first presented in [6]. The merger can be achieved with two (2) players trying to optimize two (2) different objectives. Of course, it is possible to consider n players optimizing n criteria, as presented in the previous section. But for the sake of clarity, we will restrict ourselves to $n=2$.

Let $s=XY$ be the string representing the potential solution for a bi-objective optimization problem. X corresponds to the subset of variables handled by Player-1, and optimized using criterion 1. Y corresponds to the subset of variables handled by *Player-2* and optimized using criterion 2. Thus, as introduced by Nash theory,

Player-1 optimizes *s* with respect to the first criterion by modifying *X*, while *Y* is fixed by *Player-2*. Symmetrically, Player-2 optimizes s with respect to the second criterion by modifying *Y* while *X* is fixed by *Player-1*.

Then the next step consists in creating two different populations, one for each player. *Player-1's* optimization task is performed by population 1 whereas *Player-2's* optimization task is performed by population 2. Let X^{k-1} be the best value found by Player-1 at generation *k– 1*, and Y^{k-1} the best value found by *Player-2 at* generation *k – 1*. At generation *k*, Player ~ 1 optimizes X^k while using Y^{k-1} in order to evaluate s (in this case, $s = X^k Y^{k-1}$). At the same time, *Player-2* optimizes *Yk* while using *Xk – 1* ($s = X^{k-1} Y^k$). After running the optimization procedure, Player-1 sends the best value X^k to Player-2 who will use it at generation *k+1*. Similarly, *Player-2* sends the best value X^k to Player-1 who will use it at generation *k+1*. Nash equilibrium is reached when neither *Player-1 nor Player-2* can further improve their criteria.

The repartition of design variables between the players (i.e. which player should optimize which variable?), depends on the structure of the problem. If the problem has n criteria and n variables, it is straightforward that each player should optimize one different variable. However, for problems dealing with more optimization variables than criteria, players must share among themselves the variables. The repartition may be arbitrary, but in most real-life problems, the physics of the problem is likely to suggest a way to split those variables.

As described in Chap. 2, Genetic Algorithms possess robustness for the capture of the global solution of multi-modal optimization problems. On the other hand Nash games can be used for design under conflict and Stackelberg games for hierarchical design. Therefore it is quite natural to combine the two approaches, a Nash Game with GAs or a Stackelberg game with GAs in order to solve multi-criteria design optimization problems either under conflict or with a hierarchy between them. The resulting decentralized optimization can be considered as a set of decision maker algorithms for real design in aerospace engineering well suited to distributed parallel computing environments.

In a Nash game, each player uses a GAs to improve his own criterion along generations constrained by strategies of the other player. In applications, design variables—with their territory—are geometrically split between players who exchange symmetrically their best strategies (best chromosomes) at each generation. Such a procedure is continued until no player can further improve its criterion. At this stage the system has reached the Nash equilibrium. One of the evident properties of Nash/GAs is their inherent parallel structure during evolution.

A flow chart of a Nash-GAs is shown in Fig. 3.2. Let *S=XY* be the string representing the potential solution for a dual objective optimization, where *X* corresponds to the first criterion and *Y* to the second criterion. Player-1 optimizes *X* (*Y* is fixed by Player 2) and Player 2 optimizes *Y* (*X* is fixed by Player-1). Each player has his own GAs with population. Nash equilibrium is reached when neither player can further improve its criterion (see [6] for more details).

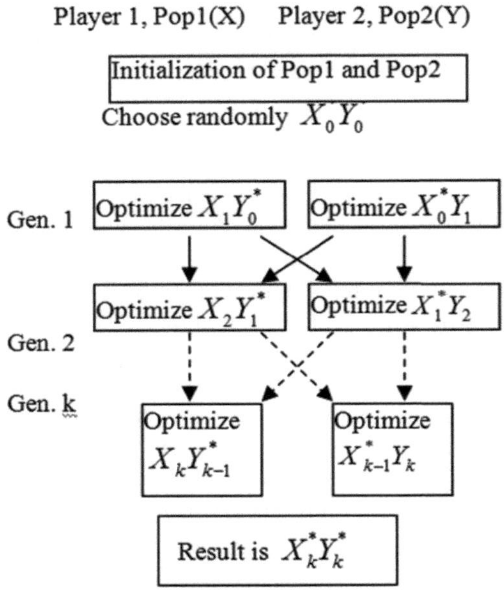

Fig. 3.2 Nash/ GA flowchart

3.4.3 Generalization to N Nash Players

The generalization of this approach to players derives naturally from the case with two (2) players. Let us consider players optimizing a set of function objectives. The optimization variables are distributed among the players, in such a way that each player handles a subset of the set of optimization variables. Let {AU????—X and Y??} be the optimization variables (each can consist of a set of several variables). The Nash GAs will then work by using different populations. The first population will optimize using criterion 1 while all the other variables will be fixed by players. Population 2 will optimize using criterion 2 while all the other variables are fixed by players, and so on. The different players still have to send each other information about their best (elitist) result after each generation (Fig. 3.3).

This setting may be compared to the so-called Island Model in Parallel Genetic Algorithms (PGAs) introduced by H. Muhlenbein et al. [7]) mimicking the ability of bird populations to survive on islands in comparison with mainlands. However, there is a fundamental difference in the sense that PGAs use the same criterion for each sub-population whereas Nash GAs use different criteria (thus introducing the notion of equilibrium).

The authors first developed Nash GAs with binary-coded GAs to solve combinatorial discrete problems. But for the examples presented, a version based on real-coded GAs was used. Therefore in the following sections, each player evolves by a real-coded GA, using a non-uniform mutation scheme. A distance-dependent mutation is also used, a technique allowing diversity to be maintained in small

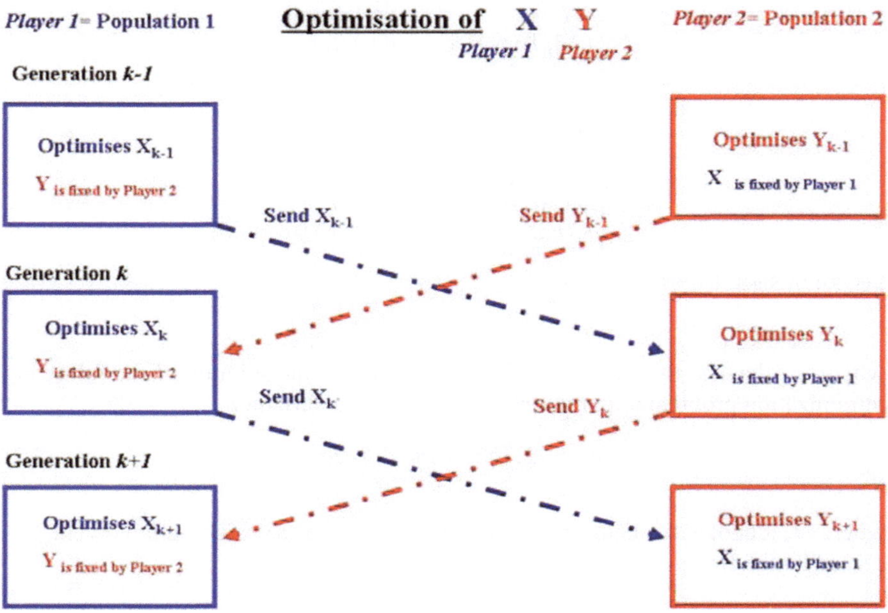

Fig. 3.3 A coupled Nash- EAs algorithm

populations. Instead of a fixed mutation rate, each offspring has its mutation rate adjusted after each mating. This mutation rate depends on the distance between the two parents (for more details, see [8]).

3.5 Hierarchical Game: Stackelberg

A Stackelberg equilibrium is the solution of a hierarchical game (also competitive game), which can be found in Loridan P. and Morgan J. [9]. A Stackelberg game has a non-symmetrical structure with completely different roles of players. For instance in the case of an optimization problem with two criteria, Nash and Stackelberg games are implemented with two players, each player is in charge of one criterion and chooses his best decision in a rational reaction set. The players associated to a Nash game have a symmetric role, while for a Stackelberg game, the leader-follower roles of players are hierarchically defined.

A two-player Stackelberg equilibrium can be characterized as follows: Suppose A denotes the search space of first player—*the leader*, and B the search space of second player—*the follower*, then a strategy pair $(x^*, y^*) \in A \times B$ is a Stackelberg equilibrium if and only if:

$$f_A(x^*, y^*) = \inf_{x \in A} f_A(x, \overline{y}^*), \tag{3.6}$$

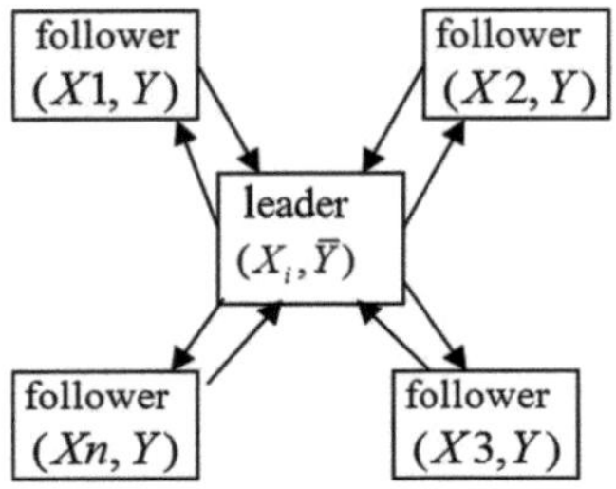

Fig. 3.4 A Stackelberg—GAs flowchart

where f_A denotes the gain of the first player, $\bar{y}^*$ is the solution of the following minimization problem with respect to the y decision variable:

$$f_B(\bar{x}, y) = \inf_{y \in B} f_B(\bar{x}, y) \quad \textit{with } \bar{x} \textit{ frozen from leader}, \tag{3.7}$$

where f_B denotes the gain of second player, and $\bar{x}$ the design variable value received by the first player.

3.5.1 *Coupling a Stackelberg Game with GAs*

The GA implementation of a Stackelberg game with two players is described in Fig. 3.4. Let denotes by $S = XY$, {AU: IT IS NOT CLEAR WHAT IS DENOTED BY WHAT.} X and Y are the strategy set of first player and second player respectively (the first player being the leader and the second player the follower. For each individual $\bar{x}$ frozen of the leader's decision set, the follower searches the corresponding $\bar{Y}*$ to improve his gain. Once all individuals of the leader's decision set have received the corresponding $\bar{Y}*$ values, then the leader changes X to improve his gain. This numerical procedure is repeated until the leader can no longer improve his gain, this means that such a system reaches a Stackelberg equilibrium. It will be noticed that the two GAs of the follower and the leader are respectively embedded.

3.6 Comparison of Analytical Solutions and Numerical Game Solutions for Solving a Two Mathematical Functions Minimisation Problem

Let us consider a game with 2 players A and player B, with the following objective functions:

$$\begin{Bmatrix} f_A(x, y) = (x-1)^2 + (x-y)^2 \\ f_B(x, y) = (y-3)^2 + (x-y)^2 \end{Bmatrix}. \tag{3.8}$$

3.6.1 Analytical Solution

The mathematical definition of a Nash equilibrium described above is not very practical to use. One possible way to find a Nash equilibrium is to use the notion of *rational reaction set*. Let D_A be the rational reaction set for A, and D_B the rational reaction set for B.

$$D_A = \left(\vec{x}, y\right) \in \bar{A} \times \bar{B} \, such \, as \, f_A\left(\vec{x}, y\right) \leq f_A(x, y),$$

$$D_B = \left(x, \vec{y}\right) \in \bar{A} \times \bar{B} \; such \; as \; f_B\left(x, \vec{y}\right) \leq f_B(x, y).$$

An intuitive insight on rational reaction sets is that they are the set of the best solutions a player can achieve in the face of different strategies of his opponent.

D_A and D_B can be built by finding the x and y values that satisfy the following equations:

$$\begin{cases} D_A = \left\{ x \mid \dfrac{\partial f_A(x, y)}{\partial x} = 0 \right\} \\ D_B = \left\{ y \mid \dfrac{\partial f_B(x, y)}{\partial y} = 0 \right\} \end{cases}.$$

The Nash Equilibrium is the intersection of the two rational reaction sets D_A and D_B.

If we go back to the example we have presented, the rational reaction set D_A is the solution of the equation:

$$\frac{\partial f_A(x, y)}{\partial x} = 0$$

which gives

$$\frac{\partial f_A(x, y)}{\partial x} = 0 \Leftrightarrow 2(x-1) + 2(x-y) = 0 \Leftrightarrow y = 2x - 1.$$

It follows that the rational reaction set D_A is the line $y = 2x - 1$. The second rational reaction set D_B is defined by the solution of the equation:

$$\frac{\partial f_B(x, y)}{\partial y} = 0,$$

$$\frac{\partial f_B(x, y)}{\partial y} = 0 \Leftrightarrow 2(y-3) - 2(x-y) = 0 \Leftrightarrow y = \frac{x+3}{2}.$$

Hence, the rational reaction D_B is the line $y = \dfrac{x+3}{2}$.

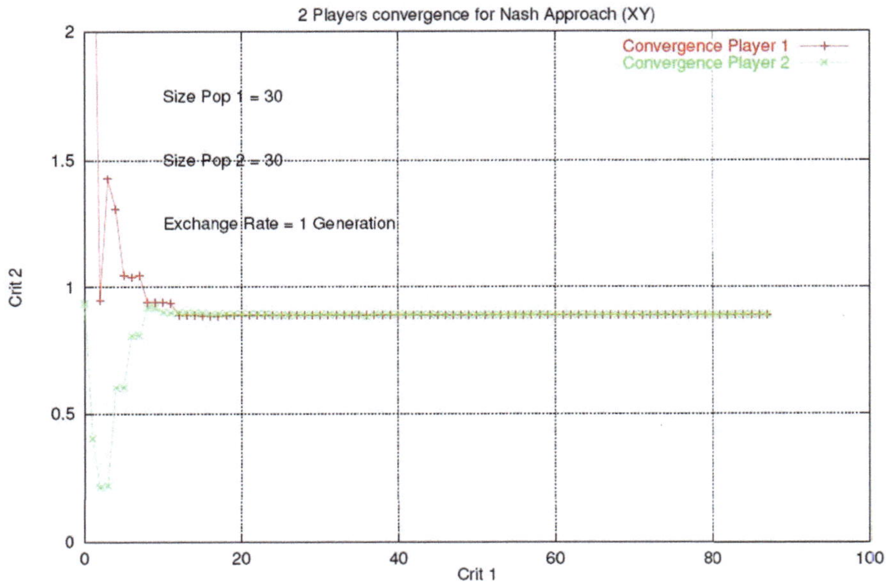

Fig. 3.5 Computed Nash equilibrium

Since the Nash Equilibrium is the intersection of the two rational sets, it can be determined by solving the system:

$$\begin{cases} y = 2x - 1 \\ y = \dfrac{x+3}{2} \end{cases} \Rightarrow \begin{cases} y = 2x - 1 \\ 3y = 7 \end{cases} \Rightarrow \begin{cases} x = \dfrac{5}{3} \\ y = \dfrac{7}{3} \end{cases},$$

which means that the Nash Equilibrium is the point

$$\begin{pmatrix} \dfrac{5}{3} \\ \dfrac{7}{3} \end{pmatrix} = \begin{pmatrix} 1.66 \\ 2.33 \end{pmatrix}.$$

3.6.2 *Nash/Gas and Stackleberg/Gas Numerical Solutions*

A similar numerical two-objective optimization is performed using the Nash/GAs approach.

Figure 3.5 shows the evolution of the best chromosome for the two populations. Population 1 which optimizes criterion f_A converges towards 0.88. Population 2 which optimizes criterion f_B converges towards 0.88 as well. Both populations converge after only 50 generations.

Those values seem to be different from the ones found analytically, but this is because the figure shows the optimization process on the objective space. We can easily check that:

$$f_A\left(\frac{5}{3},\frac{7}{3}\right) = 0.88 \; and \; f_B\left(\frac{5}{3},\frac{7}{3}\right) = 0.88.$$

The point $\begin{pmatrix} 0.88 \\ 0.88 \end{pmatrix}$ is actually the point $\begin{pmatrix} f_A(1.66) \\ f_B(2.33) \end{pmatrix}$ in the plan of the criteria, which means that we actually find the right point $\begin{pmatrix} 0.88 \\ 0.88 \end{pmatrix}$ in the plan (x, y).

Note: if the objective functions f_A and f_B are permuted, with player A still handling x and player B still handling y, then the Nash equilibrium is not a single point. The analytical solution yields the line $x=y$ and the different runs of the algorithm converge towards different points on that line.

The analytical and differentiable functions selected above are still selected to compare solutions of multi-objective optimizations through Nash/GAs and Stackelberg/GAs approaches as follows:

$$\begin{cases} F1: & f_1 = (x-1)^2 + (x-y)^2 \\ F2: & f_2 = (y-3)^2 + (x-y)^2 \end{cases} \quad -5 \le x, y \le 5. \tag{3.9}$$

The optimization problems are to minimize F1 and F2 in search space $x, y \in [-5, 5]$ using Nash/GAs and Stackelberg/GAs in a two-player game with which the global search space will be decomposed into two parts and each player takes one. With the Nash/GAs approach, the two players have a symmetrical role during the game, but a non-symmetrical nested role in the Stackelberg/GAs approach.

The analytical solutions for Nash and Stackelberg strategies could be calculated according to the mathematical definition of the game, and are shown in Table 3.1.

The numerical solutions are obtained using Nash/ GAs and Stackelberg/ GAs approaches, and shown on Table 3.1. It can be noticed that numerical solutions with GAs agree quite well with analytical solutions.

Figure 3.6 shows the solution trace during the GAs optimization procedure. In the case of a Stackelberg strategy, different leader definitions provide different optimal solutions, and these solutions are different from the Nash equilibrium. Finally Fig. 3.7 shows a comparison of convergence history of different optimization strategies reached after twenty five (25) generations.

Table 3.1 Analytical and numerical results: a comparison

	Nash	Stackelberg	
		F1 leader	F2 leader
X	1.666625	1.399945	1.799998
Y	2.333309	2.200030	2.600001
F1	0.888857	0.800092	1.280001
F2	0.888943	1.280002	0.800004
Analytical result	X=5/3	X=1.4	X=1.8
	Y=7/3	Y=2.2	Y=2.6
	F1=F2=8/9	F1=0.8	F1=1.28
		F2=1.28	F2=0.8

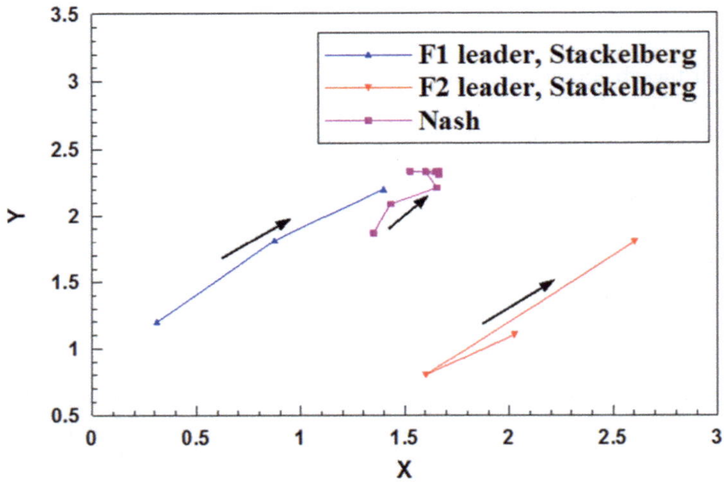

Fig. 3.6 Nash and Stackelberg Game solutions reaching equilibrium

3.7 Hybridized Games

The Hybrid-Game uses the concepts of Nash-game and Pareto optimality and hence it can simultaneously produce Nash-equilibrium and a set of Pareto non-dominated solutions. The reason for implementation of Nash-game is to speed up the search for one global solution or the capture of a Pareto Front. The elite design variables from a Nash-game are seeded to a Pareto-game at every generation. Each Nash-Player has its design criterion using his own optimisation strategy. The shape of hybrid Nash-HAPEA topology (Hierarchical Asynchronous Parallel Evolution Algorithm) is a top view of a trigonal pyramid as shown in Fig. 3.8. More details in this section can be found in Lee [10].

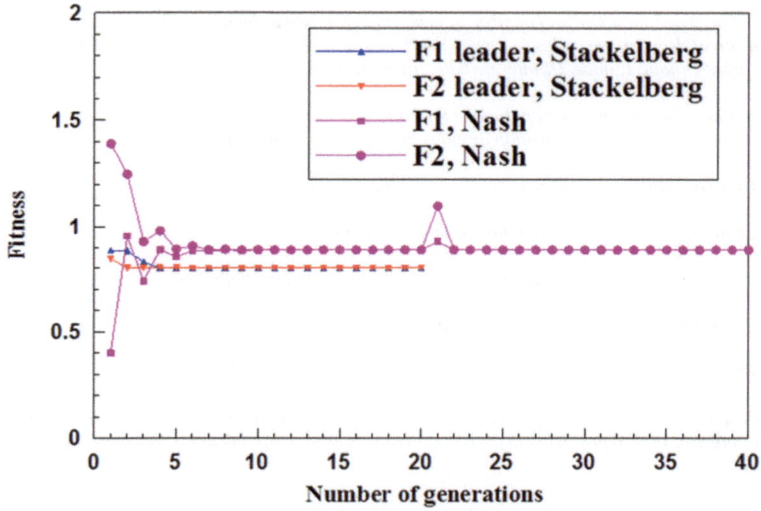

Fig. 3.7 Comparison of convergence

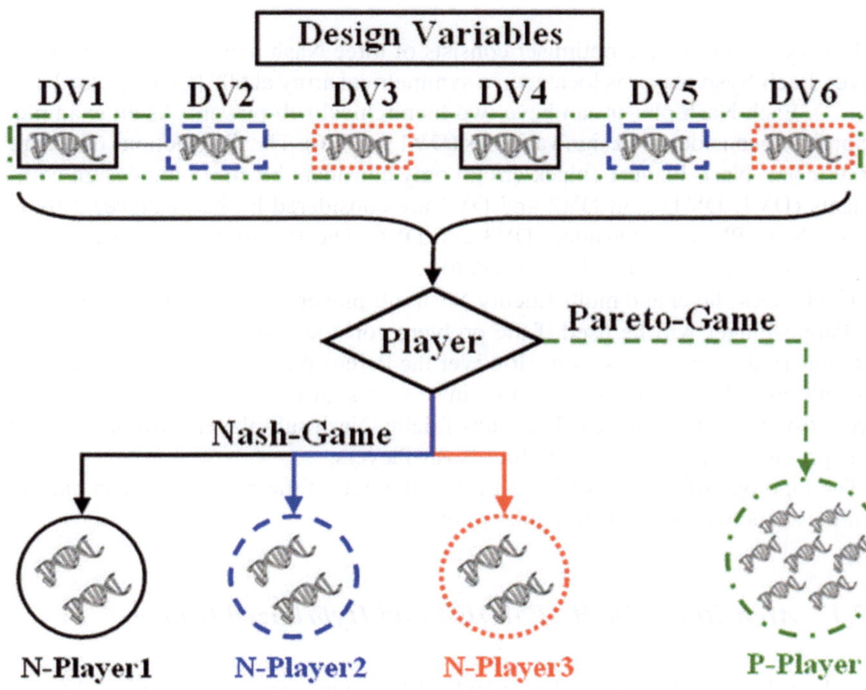

Fig. 3.8 Hybridized games

Step 1: Define population size and number of generation for hierarchical topo logy (node 0 to node 6), dimension of decision variables (x_1, x_2, x_3) and their design bounds, model quality (layer 1 (node 0): high, layer 2 (node 1, node 2): intermediate, layer3 (node3 to node 6): low)

Step 2: Initialize seven random populations for node 0 to node 6.
while termination condition (generation or elapsed time or pre-defined fitness value)

Step 3: Generate offspring using mutation or recombination operations.

Step 4: Evaluate offspring corresponding to fitness functions.

Step 4-1: Evaluate offspring for each node in terms of high, intermediate, low.

Step 5: Sort each population for each node based on its fitness.

Step 6: Replace best individual into non-dominated population of each node. End

Step 7: Designate results: Pareto optimal front obtained at node 0 on the first layer (precise model) for multi-objective design problem otherwise plot convergence of optimization based on best-so-far individual.

Step 8: Post-optimization procedure: for instance an optimization problem considering aerodynamic wing design, then Mach sweep will be plotted corresponding to objective (C_D, C_L or L/D).

Fig. 3.9 Pseudo-code of HAPMOEA algorithm

It can be seen that the optimiser consists of three Nash players with one Pareto player. Each Nash player is located in a symmetrical array at 60° (Line 1, Line2 and Line 3). Each Nash player can have two hierarchical sub-players. As an example, the problem considers 6 design variables (DV1 to DV6). The distributions of design variable are; Nash-Player1 (black circle) only considers black square design components (DV1, DV4), and DV2 and DV5 are considered by Nash-Player 2 (blue circle). Nash-Player 3 considers DV3 and DV6. The Pareto player considers the whole design variable span (DV1 to DV6).

The Pareto-Player and multi-fidelity Nash sub-players are optional. For instance, the Pareto-Player will be used if the problem considers multi-objective or multi-disciplinary design optimisation. However the Pareto-Player will not be used if the problem considers a reconstruction or inverse design optimisation problem since Pareto fronts are not required. The multi-fidelity Nash sub-players will be not used if the problem requires more than four Nash-Players.

The topology of hybrid Nash-HAPEA is flexible; if there are four Nash players then the shape will be a quadrangular pyramid.

3.7.1 Algorithms for HAPMOEA and Hybridized Games

The algorithms for HAPMOEA and Hybrid-Game are described on Figs. 3.9 and 3.10 where it is assumed that the problem considers the fitness function $f=min\ (x1, x2, x3)$. The validation of Hybrid-Game and HAPMOEA can be found in [10]. The reader will consult Chap. 4 for the definition of HAPMOEA.

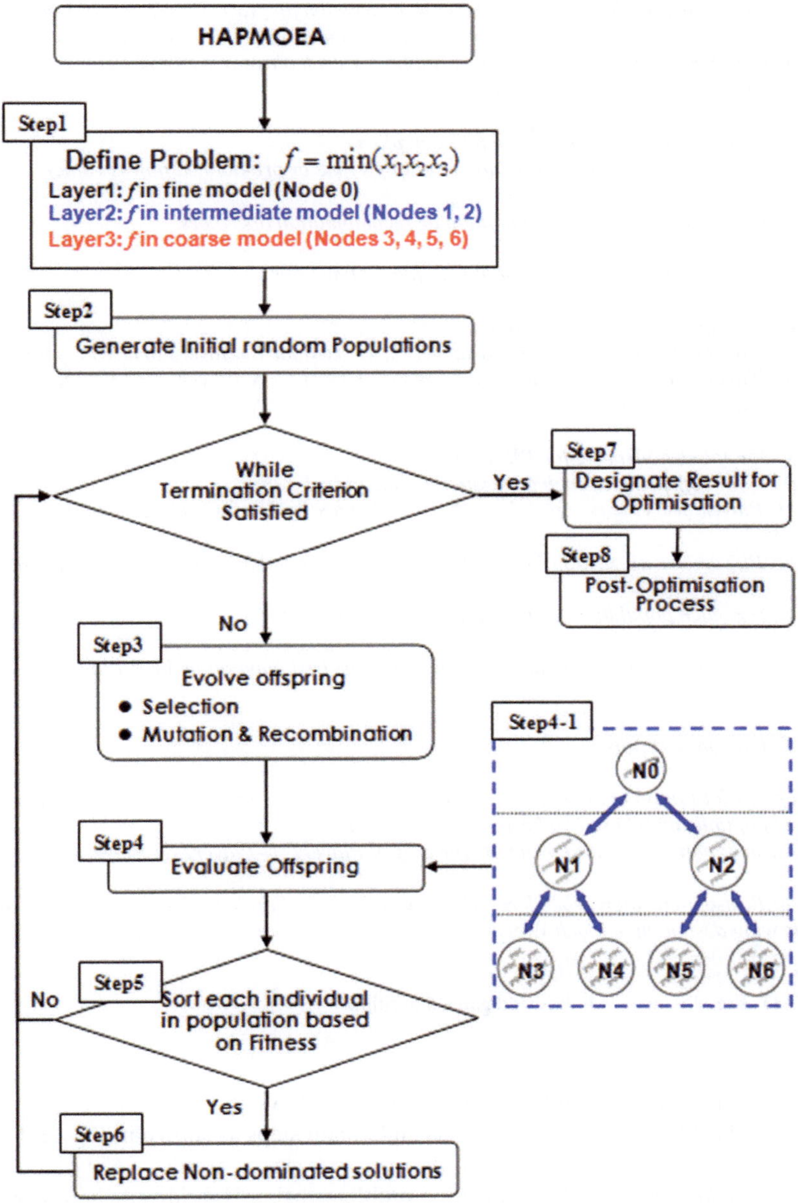

Fig. 3.10 Flow chart code of HAPMOEA algorithm

Step 1: Define population size and number of generation for Nash-Players (N-Player1, N-Player2, N-Player3) and Pareto Player (P-Player), dimension of decision variables (x_1, x_2, x_3) and their design bounds. Split decision variables for each player (N-Player 1: x_1, N-Player 2: x_2, N-Player 3: x_3, P-Player: x_1, x_2, x_3).

Step 2: Initialize random population for each player.
while termination condition (generation or elapsed time or pre-defined fitness value)

Step 3: Generate offspring using mutation or recombination operators.

Step 4: Evaluate offspring corresponding to fitness functions.

Step4-1: Evaluate offspring in Nash-Game.
N-Player1: use x1 with design variables x2, x3 fixed for N-Player2 and N-Player3.
N-Player2: use x2 with design variables x1, x3 fixed for N-Player1 and N-Player3.
N-Player3: use x3 with design variables x1, x2 fixed for N-Player1 and N-Player2.

Step 4-2: Evaluate offspring for P-Player.
if the first offspring at each generation is considered
P-Player: use elite design (x1, x2*, x3*) obtained by Nash-Game at Step4-1.*
else
P-Player: use x1, x2, x3 obtained by mutation or recombination operator as default.

Step 5: Sort each population for each player based on its fitness.

Step 6: Replace the non-dominated individual into the best population for P-Player.
end

Step 7: Designate results :

P-Player: Plot Pareto optimal front for multi-objective design problem otherwise plot convergence of optimization based on best-so-far individual
Nash-Game: plot Nash-equilibrium obtained by N-Player1, N-Player2, N-Player3

Step 8: Post-optimization procedure: if for instance optimization problem considers aerodynamic wing design, then Mach sweep will be plotted corresponding to objective (CD, CL or L/D).

Fig. 3.11 Pseudo-code for hybridized game algorithm

The HAPMOEA has eight (8) main steps as follows:

The hybrid Pareto-Nash method has eight main steps as shown in Figs. 3.11 and 3.12:

Numerical examples of optimization problems using Hybridized Games are implemented and discussed in Chap. 4.

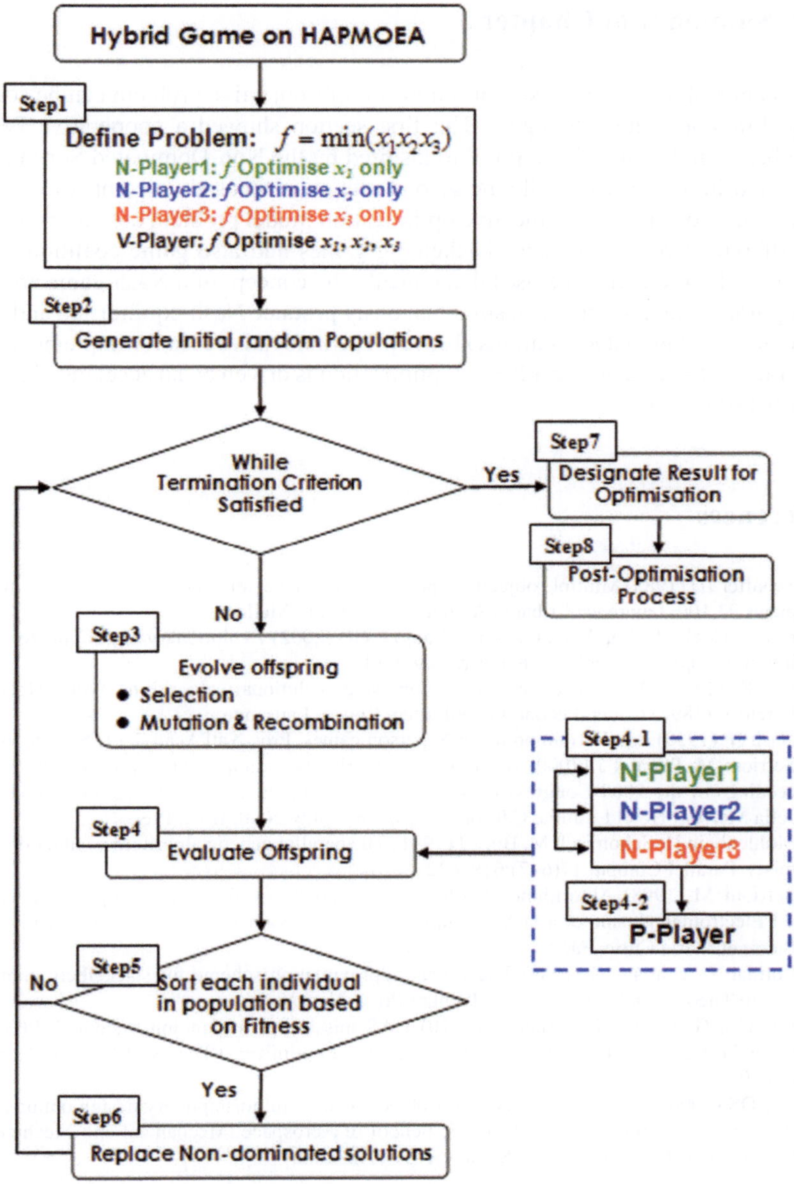

Fig. 3.12 Flow chart for hybridized game algorithm

3.8 Summary of Chapter 3

This chapter shows simple examples how a multi-objective problem can be solved using different Game Strategies. The first section showed a cooperative Pareto game based multi-objective algorithm inspired by the Non-Dominated Sorting Genetic Algorithm (NSGA) while the second section is focused on competitive Nash games. Solutions of a two-objective optimization model problem are computed using different Pareto, Nash and Stackelberg games and also game coalitions. The concept of hybridized game uses dynamically the concept of a Nash-game and Pareto optimality and hence it can simultaneously produce Nash-equilibrium and a set of Pareto non-dominated solutions [10, 11]. The innovative idea for implementing a dynamic Nash-game during a Pareto optimization is driven by an accelerated search of global solutions.

References

1. Schaffer JD (1985) Multiple objective optimization with vector evaluated genetic algorithms, pages 93.100. Lawrence Erlbaum Associates, Carnegie-Mellon
2. Coello-Coello C, Van Veldhuizen Da, Lamont GB (2002) Evolutionary algorithms for solving multiobjective problems. Kluwer, New York
3. Deb K. (2003) Multi-objective optimization using evolutionary algorithms. Wiley, Hoboken
4. Pareto V (1896) Cours d'economie politique. Rouge, Lausanne
5. Nash JF (1950) Equilibrium points in N-person games. Proc Natl Acad Sci U S A 36:46–49
6. Sefrioui M, Periaux J (2000) Nash genetic algorithms: examples and applications. In Proceedings of the 2000 Congress on Evolutionary Computation CEC00, pages 509-516, La Jolla Marriott Hotel La Jolla, California, USA, 6–9 July 2000. IEEE Press
7. Mühlenbein H, Schomisch M, Born J (1991) The parallel genetic algorithm as function optimiser. Parallel Comput 17(6–7):619–632
8. Sefrioui M (1998) Algorithmes Evolutionnaires pour le Calcul Scientifique. Application à l'Electromagnétisme et à la Mécanique des Fluides Numériques. PhD thesis, University Pierre et Marie Curie, Paris
9. Loridan P, Morgan JA (1989) Theoretical approximation scheme for stackelberg games. J Optim Theory Appl 61(1):95–110. Plenum Press, New York
10. Lee DS, Gonzalez LF, Periaux J (2010) UAS mission path planning system (MPPS) using hybrid-game coupled to multi-objective design optimizer. J Dyn Syst Measure Control. doi:10.1115/1.4001336
11. Lee DS (2006) Uncertainty based multi-objective and multidisciplinary design optimisation in aerospace engineering. PhD Thesis, School of Aerospace, Mechanical and Mechatronic Engineering, J07 University of Sydney, NSW, Australia

Chapter 4
Advanced Techniques for Evolutionary Algorithms (EAs)

4.1 Generalities

One of the main drawbacks of EAs is the CPU time needed due to many expensive fitness or objective function evaluations of candidate solutions. For this reason it is necessary to introduce advanced techniques. This chapter describes techniques including distributed and parallel EAs, hierarchical EAs, asynchronous EAs, advanced mutation operators as well as game strategies and hybridized games. All these techniques aim to increase diversity and speed up the search for optimal solutions.

4.2 Distributed and Parallel EAS

EAs are particularly well suited to parallel computing, as candidate individuals or populations are sent to remote machines for evaluation. The most common approach for parallelisation is global Parallel EAs (PEAs). Such an approach consists of a well known master- slave implementation whereby the master controls the process and sends individuals to solver nodes where their fitness is evaluated by processors (slaves). The master collects the results and applies evolutionary operators to produce the next generation of EAs.

Instead of relying on a single large population, PEAs use a network of interconnected small populations, thus defining a paradigm called the *Island model* introduced by H. Muhlenbein [1]. Figure 4.1 shows such a model, with the yellow area corresponding to a node (or sub-population) and its connections to its neighbours. The idea is that each of these sub- Populations will evolve *independently* for a given period of time (or epoch). After each epoch, a period of migration and information exchange takes place, before the isolated computation resumes.

Figure 4.2 shows that this approach allows different subpopulations to explore different regions of the search space during the isolation stage. Yet, the most prom-

J. Periaux et al., *Evolutionary Optimization and Game Strategies for Advanced Multi-Disciplinary Design,* Intelligent Systems, Control and Automation: Science and Engineering 75, DOI 10.1007/978-94-017-9520-3_4

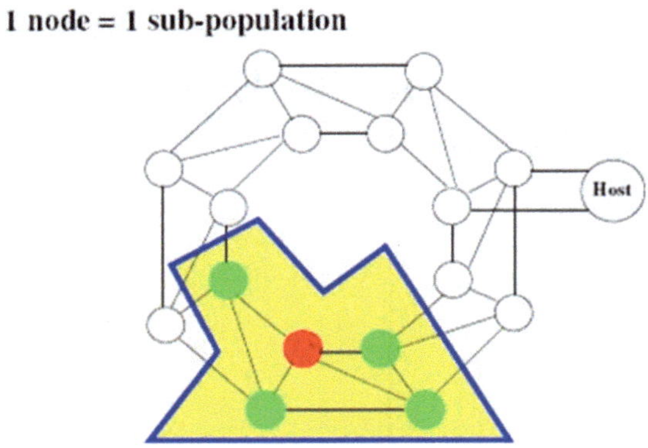

Fig. 4.1 Parallel EAs

ising solutions are shared by the whole set of sub-populations in sending them to neighbours.

The main idea considers a number of small, interconnected, sub-populations instead of a single large population; these sub-populations evolve independently on each node for a time or period called an "epoch". After each epoch a period of migration and information can be exchanged between nodes, after which successive new periods of isolation occur.

With this approach sub-populations can explore different regions of the search space; by doing this the robustness of the PEAs improves and it is easier doing so to escape from local minima. Another common approach is to preserve the global population while parallelising the EA operators that are restricted to neighbouring individuals. This is considered an extension to the second approach and is categorised as a cellular EA. Details on PEAs methods can be found in Veldhuizen et al. [2] and Cantu-Paz et al. [3, 4].

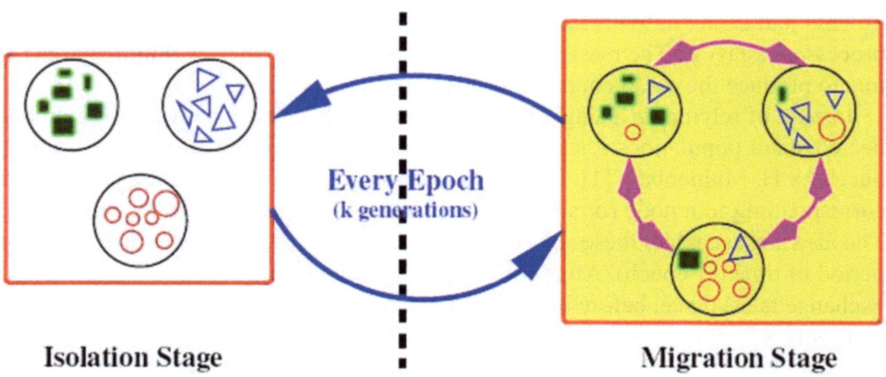

Fig. 4.2 Isolation and migration stages

4.3 Hierarchical EAS (HEAs)

Hierarchical Genetic Algorithms (HGAs) or more generally, Hierarchical EAs (HEAs), are a particular approach developed by Sefrioui and Périaux [5]. This approach uses a hierarchical topology for the layout of the sub-populations. Figure 4.3 illustrates this concept. The bottom layer can be entirely devoted to exploration, the intermediate layer is a compromise between exploitation and exploration and the top layer exploits refinement of solutions.

The main feature is the interaction between the multi-layers. The best solutions progress from the bottom layer to the top layer where they are refined. This circulation of solutions up and down the "tree" works as a speed up procedure if each node can be handled by a different EAs where specific parameters can be tuned. In other words, the nodes of each layer have different tasks, defined by their associated EAs:

1. The top layer concentrates on refining solutions. This can be achieved by tuning the EAs in a way that proceeds with very small steps between successive crossover and mutation operations.
2. The intermediate layer is a compromise between exploitation and exploration.
3. The bottom layer is entirely devoted to exploration. That means that the EAs can use large mutations in the search space.

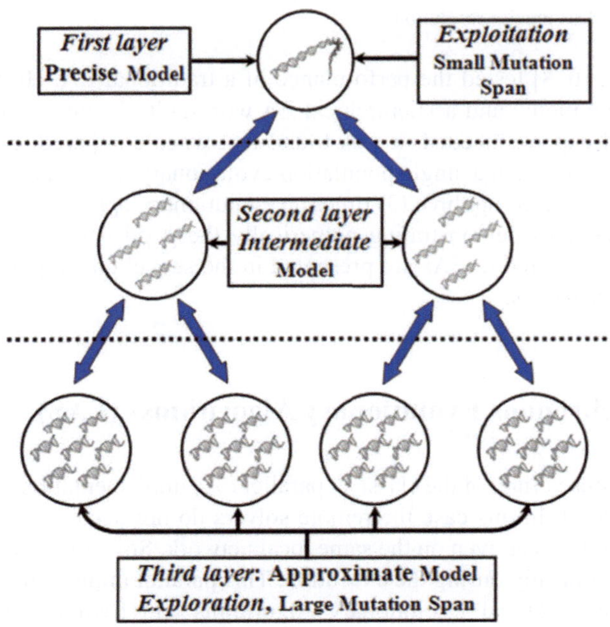

Fig. 4.3 Hierarchical multi-population/multi-fidelity topology

All the nodes of the bottom layer 3 can use a less accurate but fast model to compute the fitness function of the individuals of the sub-populations. Even though these solutions may be evaluated rather roughly, the hierarchical topology allows their genetic material information content to be used. As these solutions are sent up to the intermediate layer during the migration phase, they are re-evaluated using a more precise model to give a more accurate representation of the actual quality of the solution. However, the model used on layer 2 is also an approximate model, being chosen deliberately not too precise for the sake of speed. The procedure is repeated again by sending the solutions up to the top layer during the migration process. These solutions are re-evaluated with the model of the top layer 1, the most precise model that provides a genuinely accurate value for the fitness function.

For a practical numerical implementation, this hierarchical model is described below:

Layers	Description
First layer	Concentrates on refining solutions, which can be achieved by tuning the EA in a way that makes mutation take very small steps. This layer incorporates the most precise formulation of the high fidelity solver or the finest resolution
Second layer	Is a compromise between exploitation and exploration. This layer incorporates an intermediate approximation of the solver or the resolution
Third layer	Entirely devoted to exploration which means that the EAs can make a large steps in the search space. This layer includes a middle/low fidelity solver or a less precise resolution

Whitney et al. [6–8] tested the performance of a traditional EA, Hierarchical EA with one single model and a Hierarchical EA with multiple models, based on the computational expense needed. It was found that when compared to a traditional EA implementation with a single population evolutionary algorithm, the hierarchical approach can speed up three (3) times an evolutionary optimisation procedure.

Several examples demonstrating numerically the good performances of Hierarchical EAs compared to EAs are presented in the sequel on simple optimization aerodynamic problems.

4.4 Asynchronous Evolutionary Algorithmss (EAS)

An asynchronous variant of the classical parallel EAs implementation was proposed by Whitney [6–8]. In this case the remote solvers do not need to run at the same speed (synchronous) or even on the same local network. Solver nodes can be added or deleted dynamically during the execution. This parallel implementation requires modifying the canonical Evolutionary Algorithms and Evolutionary Strategies which ordinarily evaluate entire populations simultaneously before proceeding with a new population. The distinctive method of an asynchronous approach is that it generates, based on an individual approach, only one candidate solution at a time

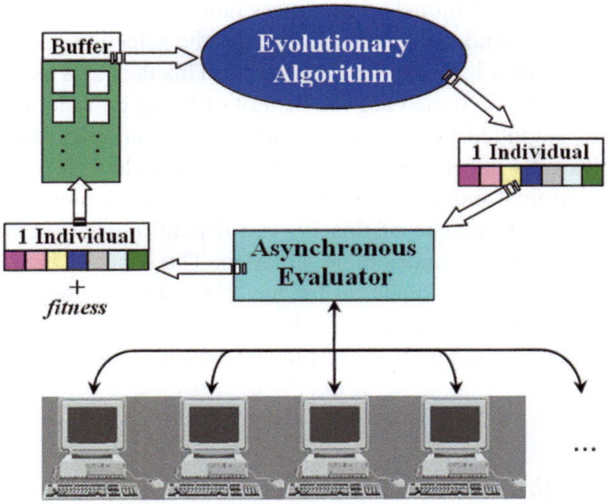

Fig. 4.4 Parallel computing and asynchronous evaluation

and only re-incorporates one individual at a time, rather than an entire population at every generation as is usual with traditional EAs. Consequently, solutions can be generated and returned *out of order*. This allows the implementation of an asynchronous fitness evaluation, giving the method its name. This is an extension of the work by Wakunda and Zell [9]. With an asynchronous approach there is no waiting time (or bottleneck) for individuals to return compared to the classical notion of generations since the buffer of the EAs or ESs is continuously updated by evaluated individuals. As soon as one solution has been evaluated, its genetic material is incorporated back into the buffer of the optimization procedure. Figure 4.4 illustrates this concept.

4.5 Advanced Operators

There are a number of advanced operators been developed. In this book, two techniques are described but others are suggested [10, 1].

4.5.1 Covariance Matrix Adaptation (CMA)

One extension to the canonical Evolution Strategy described in Chapter 2 is the incorporation of the Covariance Matrix Adaptation (CMA) mutation model developed and extensively studied by Hansen and Ostermeier [12, 13]. This mutation model takes its inspiration in conventional optimisation techniques which work by developing a second-order model of the fitness function. Through information gathered about the fitness landscape, learned through the selection procedure, a model of the covariance matrix is built up progressively. This then acts as a scaling metric on subsequent mutations. It also has the benefit of being de-randomized: the same mutation vector is applied to both object and strategy variables, and this has been shown to significantly accelerate the optimisation procedure. Hansen [12, 13] has shown that the original mutation mechanism used by the canonical ES has some drawbacks as the performance of this ES can depend on the permutation of the object variables, clearly an undesirable situation. The basis of Hansen and Ostermeier's method is that individuals no longer 'carry' the strategy variables themselves, and these are stored as a single set by the evolution algorithm itself [12, 13]. Details on the CMA method implementation are not provided here but can be found in Reference 17.

4.5.2 Pareto Tournament Selection

One desirable feature of optimization is the use of a suitable selection operator capable of handling single or multi-objective(s). One approach is to implement the on-the-fly selection operator by means of a Pareto tournament selection operator [10, 11, 14, 15].

Most Evolutionary Algorithms configured for multi-objective optimisation currently use the *non-dominated sorting* approach [11]. This is a straightforward way to adapt an algorithm that is designed as a single objective optimiser into a multi-objective optimiser, and is used by many researchers [11]. The problem with sorting approaches is that the method is not a fully integrated one. Briefly, a sorting method works by computing the set of non-dominated solutions amongst a large statistical sampling (either a large population or previous data), and assigning these solutions a Rank1. Then ignoring these points, the procedure is repeated until a 'second' Pareto front is found, and this is assigned Rank2. This procedure continues until all points are ranked, and then the value of the rank is assigned to the individual as a new single objective fitness. An example of the ranking procedure is shown in Fig. 4.5.

A problem arises now on whether it is fair to assign individuals in the second rank numerically half the fitness of the first, and whether the third rank deserves a third of the fitness of the first. This poses a dilemma regarding the level of equality present amongst the solutions, as often solutions with excellent information may lie adjacent to, but not in, Rank1. To solve this 'artificial scaling' problem, it is possible to introduce scaling, sharing and niching schemes, however all of these require problem-specific parameters or knowledge, even in adaptive approaches. It is of course always better dealing with an algorithm that does not introduce such unnecessary parameters.

In the Pareto tournament approach, the operator requires no additional 'tuning' parameters, works seamlessly with the asynchronous selection buffer (B), and is very

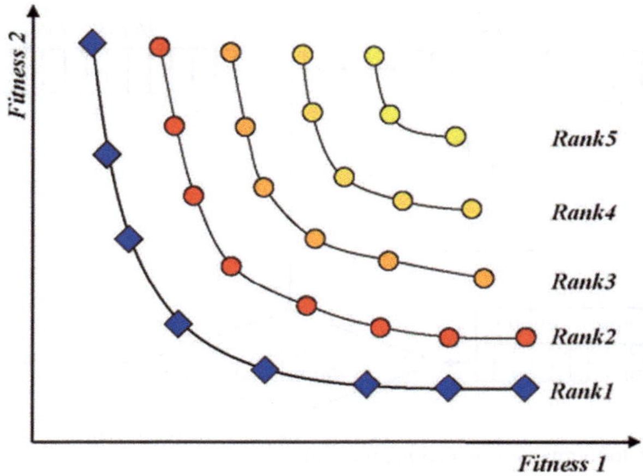

Fig. 4.5 Pareto ranking process

easy to encode. Simply, to determine whether a new individual x is to be accepted into the main population, it is compared with the selection buffer by assembling a small subset of the buffer called the t tournament functions $Q = [q_1, q_2, \ldots, q_n]$. Q is assembled by selecting individuals from the buffer, exclusively at random, until it is full. The next step is to ensure that the new individual is not dominated by any in the tournament. If this is the case, then it is immediately accepted, and is inserted according to the replacement rules. The only tuning parameter that needs to be determined in advance is the tournament size, a parameter that would exist in a single objective optimisation anyway. Selection of this parameter requires a small amount of problem specific knowledge, and should vary between $Q = \frac{1}{2}B$ (strong selective pressure) and $Q = \frac{1}{6}B$ (weak selective pressure). EAs optimisers are not overly sensitive to this value, provided the user errs on the side of weak selective pressure (smaller tournaments) in the absence of better information. The egalitarian approach to the tournament (by selecting individuals at random) ensures good diversity amongst the selected individuals: no niching or forced separation of individuals has been found necessary. It can also be seen that in the event the fitness vectors have only one element (a single objective optimisation), this operator is simplified to a standard tournament selection operator. The above Pareto tournament concept is illustrated in Fig. 4.6.

4.6 Advanced Games

Complementary to the cooperative and non-cooperative games discussed in Chap. 3, other advanced approaches have also been considered and developed to provide the EA optimiser with a better efficiency.

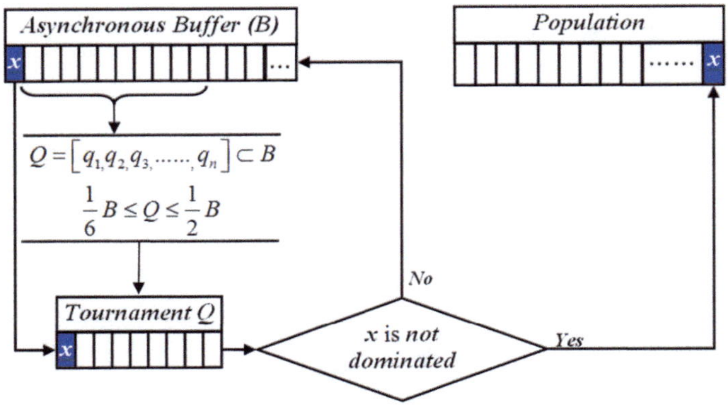

Fig. 4.6 The Pareto tournament selection procedure

4.6.1 Virtual And Real Nash-Games

There exist two Nash-Game strategies: the first strategy is a *virtual* Nash-Game which considers a single global objective. The second strategy is a *real* Nash-Game which considers multi-objectives in terms of two or more objectives according to the physics of the problem.

In a virtual Nash-Game, a split of a single objective (one for each discipline) design problem into several simpler design problems is introduced. This results in a multi-objective design problem using several Nash-Players. Each Nash-Player aims to optimize its own design criterion and the sum of design domains for each Nash-Player is equal to the search space of the original design problem. For instance, if the design problem considers a search space C which is split into A and B, i.e., $C = A \cup B$. The global solution can be captured when a Nash-equilibrium occurs. The global solution consists of the elite design string of each Nash-Player. As it will be shown in Chap. 7, Virtual Nash-Games are very efficient for solving inverse design optimisation problems.

The concept of virtual Nash-Game can be integrated with an Evolutionary Algorithm to accelerate an optimisation procedure [16]. The Nash-Equilibrium is a result of a game based on symmetric information exchanged between game players. Each player is in charge of one objective, has its own strategy set and its own criterion.

The *real* Nash-Game is an extended concept of the *virtual* Nash-Game. *A Real* Nash-Game aims to solve multi-objective/multi-disciplinary objectives: each Nash-Player optimises one of the objectives or disciplines in a multi-disciplinary design optimisation (MDO) problem by using the same design space or part of search space C, i.e., $C = A \cap B$ or $C = A \cup B$. If the problem for instance is an aero-structural multi-disciplinary design problem with two objective functions considering the design space C, i.e., $f_{Aerodynamics} = \min(C)$, $f_{Structures} = \min(C)$, two Nash-Players 1 (P_1) and 2 (P_2) can be employed to optimise $f_{Aerodynamics} = \min(C, f_{Structures}(C^*))$ and $f_{Structures} = \min(f_{Aerodynamics}(C^*), C)$ using strategy profiles $f_{Aerodynamics}(C)$ and $S_{Structures}(C)$.

A Nash-Game splits one multi-disciplinary design optimisation to two single-disciplinary design optimisations.

Both *virtual* and *real* Nash-Games have a common characteristic: the decomposition of design problems with a split of the territory of design variables. They can also be coupled with other Game Strategies such as a combination of Global and Pareto-Games (Hybridized Games or Games coalition). Examples of application of a *real* Nash-Game can be found in Chaps. 7–9.

4.6.2 Nash-Game and Hierarchical Asynchronous Parallel EAs (NASH-HAPEA)

The competitive Nash-Game approach described in the previous chapter can also be combined with the HEAs or HAPEAs approach described earlier in this chapter. This concept is implemented using one hierarchical EAs for each player, as illustrated in Fig. 4.7, whereby information is exchanged between the EAs after a migration period has occurred.

There are two migrations present when using the hierarchical EA-Nash scheme. The first one is a circulation of solutions up and down; the best solutions progress from the bottom layer to the top layer where they are refined and then a second Nash migration where information between players is exchanged after an epoch. The new variables for each player are updated on each node and on each hierarchical tree.

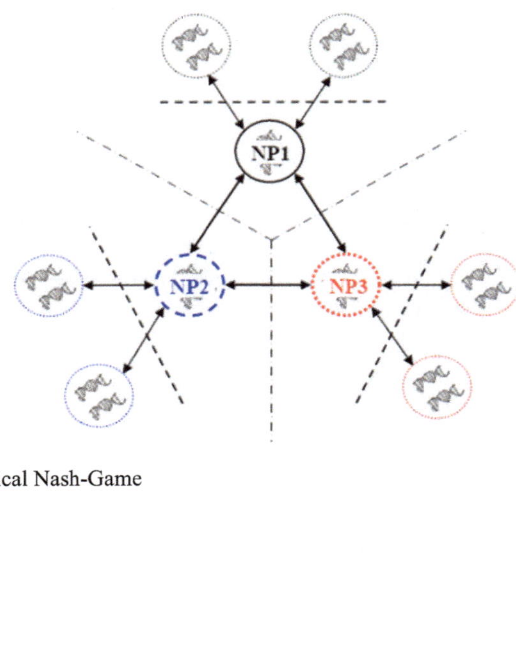

Fig. 4.7 A hierarchical Nash-Game

4.6.3 Hybrid-Game Coupled with Single-Objective or Multi-Objective Evolutionary Algorithms

A Hybrid-Game is an advanced optimization method that uses the concepts of Nash-game and Pareto optimality *dynamically* and hence it can simultaneously produce Nash-equilibrium and a set of Pareto non-dominated solutions [17]. The reason for implementing the Hybridized Nash/GAs strategy is to speed up the search of one or several global solutions when solving single-objective or multi-objective design problems. The elitist solution from each Nash-Player will be seeded to a single Player for single objective or to a Pareto-Player at every generation. This coalition based mechanism increases diversity of Pareto-Player search during the optimisation procedure. Each Nash-Player has its own design criterion and uses its own optimisation strategy.

Figure 4.8 shows an example of a possible topology for Hybrid-Games which consists of three Nash-Players and one Global/Pareto-Player. The Nash-Players choose their own strategy to improve their own objective. The Nash-Players seed/update their elite designs to the buffer of the Global/Pareto-Player (Node 0). The topology of Hybrid-Game is flexible; if there are four (4) Nash players then the shape will be a quadrangular pyramid.

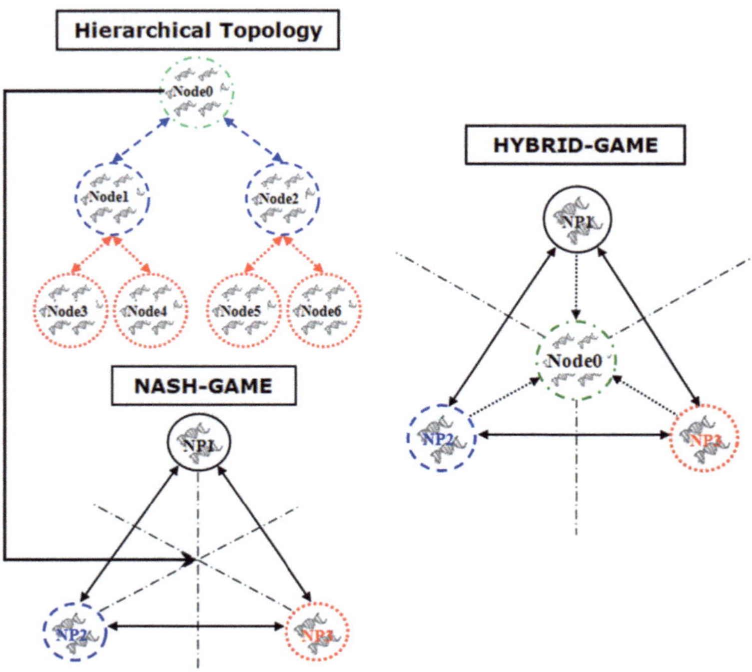

Fig. 4.8 A Topological example of Hybrid -Game

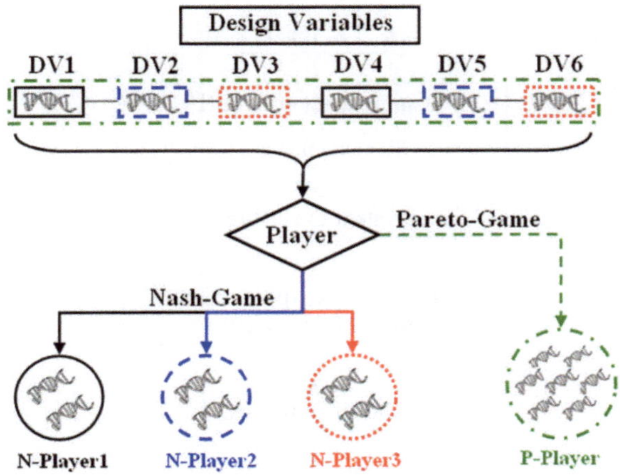

Fig. 4.9 Topology of a Pareto-Nash Game coalition

Figure 4.9 shows an example with an optimization problem considering six (6) design variables (DV1 to DV6). The distribution of design variables in this example are: the Nash-Player1 (black circle) only considers black square design components (DV1, DV4), DV2 and DV5 are considered by Nash-Player 2 (blue circle) while Nash-Player 3 considers DV3 and DV6. The Pareto-Player considers the whole design variable span (DV1 to DV6). It can be noticed that the sum of Nash-Players design variables is the same as the number of design variables for the Pareto-Player. This is because a set of elite designs (DV1 ~ DV6) obtained by Nash-Game will be seeded to the population of Pareto-Player. In this example, Nash-Game decomposes the problem into 3 simpler problems corresponding to Nash-Player 1, Nash-Player 2 and Nash-Player 3 which play the roles of a speed-up maker of the Pareto-Player to capture the Pareto Front.

If the problem is a multi-objective design optimization, then the Hybridized Nash-Game will split the problem into several single-objective design problems. If the problem is a multi-disciplinary/multi-physics design problem then the Hybridized Nash-Game will split the problem into single discipline design problems.

4.7 Meta Model Assisted EAS

Meta model assisted EAs is another technique to accelerate the convergence of an EA. The use of high-fidelity analysis CFD and FEA software for aeronautical system design is now a common practice at scientific and industrial levels. However, their use for optimisation and MDO is limited by the time and cost of performing a number of function evaluations during the optimization procedure. One alternative is to construct an approximate model. When compared to the full analysis using high-fidelity tools, the approximate model requires very little time to evaluate the

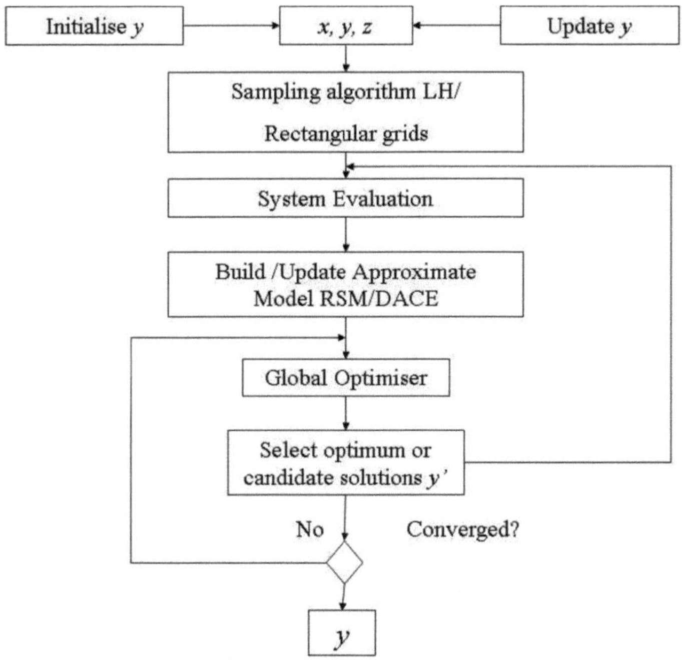

Fig. 4.10 MDO using surrogate/approximate models

objective function. Therefore, there are important savings in computational time. In order to build an approximation for optimisation, a series of steps have to be followed:

1. Define the experimental design sample using Latin Hypercube sampling or rectangular grids;
2. Define a model to represent the data (RSM, Kriging);
3. Fit the model to the observed data.

This optimization procedure is illustrated in Fig. 4.10. In this case the global optimiser is one of the Canonical GAs described in Chaps. 2, 3 or an efficient algorithm that incorporates advanced techniques described in this chapter.

Meta models with EAs are not described in detail in this chapter but can be found in References [18–22]

4.8 Summary of Chapter

EAs are computationally expensive therefore researchers have been developing advanced techniques to accelerate convergence and improve diversity. This chapter described briefly several techniques coupled to EAs including distributed and parallel computing, hierarchical topology, asynchronous computing, game strategies, hybridized games coalition and meta models that assist an EA during an optimization procedure.

References

1. Muhlenbein M (1991) Evolution in time and space—the parallel genetic algorithm. Found Genet Algorithm 1:316–337
2. Veldhuizen DA van, Zydallis JB, Lamont GB (2003) Considerations in engineering parallel multiobjective evolutionary algorithms. IEEE Trans Evolut Comput 7(2):144–173
3. Cantú-Paz E (1995) A summary of research on parallel genetic algorithms. Technical report 95007, illinois genetic algorithms laboratory. University of Illinois, Urbana-Champaign
4. Cantú-Paz E (2000) Efficient and accurate parallel genetic algorithms. Kluwer Academic Publisher, New York
5. Sefrioui M, Périaux J (2000)A hierarchical genetic algorithm using multiple models for optimization. In: Proceedings of the Sixth International Conference Parallel Problem Solving from Nature (PPSN-VI), pp 879–888. Springer
6. Whitney EJ, González L, Srinivas K, Périaux J (2002, July) Multi-criteria aerodynamic shape design problems in CFD using a modern evolutionary algorithm on distributed computers. In: Armfield S, Morgan P, Srinivas K (eds) Proceedings of the Second International Conference on Computational Fluid Dynamics (ICCFD2). Springer, Sydney, pp 597–602
7. Whitney EJ, Sefrioui M, Srinivas K, Périaux J (2002, Feb) Advances in hierarchical, parallel evolutionary algorithms for aerodynamic shape optimisation. JSME Int J 45(1):23–28
8. Whitney EJ, González LF, Srinivas K, Périaux J (2003) Adaptive evolution design without problem specific knowledge. In: Proceedings of Evolutionary Algorithms in Engineering and Computer Science (EUROGEN'03)
9. Wakunda J, Zell A (2000) Median selection for parallel steady-state evolution strategies. In: Proceedings of the Sixth International Conference Parallel Problem Solving from Nature (PPSN-VI), Springer, pp 405–414
10. Deb K, (2003) Multi-objective optimization using evolutionary algorithms. Wiley West Sussex
11. Srinivas N, Deb K (1995) Multiobjective optimisation using non-dominated sorting in genetic algorithms. Evolut Comput 2(3):221–248
12. Hansen N, Ostermeier A (1996) Adapting arbitrary normal mutation distributions in evolution strategies: the covariance matrix adaptation. In: Proceedings of the 1996 IEEE International Conference on Evolutionary Computation, pp 312-317
13. Hansen N, Ostermeier A (2001) Completely De-randomised self-adaption in evolution strategies. Evolut Comput 9(2):159–195
14. Coello Coello CA, Van Veldhuizen DA, Lamont GB (2002) Evolutionary algorithms for solving multi-objective problems. Kluwer Academic Publishers, New York
15. Horn J, Nafpliotis N, Goldberg D (1994) A niched Pareto genetic algorithm for multiobjective optimization. In: Proceedings First IIIE Conference on Evolutionary Computation Symposium on the theory of Computing
16. Periaux J, Lee DS, Gonzalez LF Fast reconstruction of aerodynamic shapes using evolutionary algorithms and virtual Nash strategies in a CFD design environment. J Comput Appl Mathemat 232(1):61–71
17. Lee DS, Gonzalez LF, Periaux J UAS mission path planning system (mpps) using hybrid-game coupled to multi-objective design optimizer. J Dyn Syst, Meas Control—ASME, DS-09-1135 132(4) 041005-1-11
18. Karakasis KM, Giannakoglou KC (2006) On the use of metamodel-assisted, multi-objective evolutionary algorithm. Optimization 38(8):941–957

19. Emmerich M, Giotis A, Özdemir M, Bäck T, Giannakoglou KC (2002) Metamodel-assisted evolution strategies In: Parallel problem solving from nature VII. Springer, Berlin, pp 361–370
20. Giannakoglou KC, Kampolis IC (2010) Multilevel optimization algorithms based on metamodel-and fitness inheritance-assisted evolutionary algorithms. In: Computational intelligence in expensive optimization problems, adaptation learning and optimization, vol 2. pp 61–84
21. Asouti VG, Kampolis IC, Giannakoglou KC (2009) A grid-enabled asynchronous metamodel-assisted evolutionary algorithm for aerodynamic optimization. Genet Program Evol Mach 10(4):373–389
22. Li M, Li G, Azarm S (2006) A Kriging Metamodel assisted multi-objective genetic algorithm for design optimization In: Proceedings of IDETC/CIE 2006, ASME 2006 International Design Engineering Technical Conferences and Computers and Information in Engineering Conference, DETC2006-99316, Philadelphia, Pennsylvania, September 10–13

Chapter 5
Multidisciplinary Design Optimisation and Robust Design in Aerospace Systems

5.1 Generalities

The evolution of aerospace systems has progressed rapidly and the type of problems that engineers are faced with is becoming more and more complex each year. The traditional phases of conceptual preliminary and detailed design have now been complemented with the need to integrate more than one discipline optimisation method as well as uncertainties in all phases of the design process. This chapter provides an overview of multi-disciplinary and robust design optimization tools.

5.2 Conceptual, Preliminary and Detailed Design

An aircraft design process can be divided into three major phases: *conceptual, preliminary* and *detailed* design. All these phases can benefit from the application of optimisation methods.

In *conceptual design*, basic questions such as size, weight, configuration arrangement and performance are answered. In this phase, a large number of possible alternatives are evaluated together with trade studies. In *conceptual* design, the design requirements are evaluated and studied to guide and evaluate different aircraft configurations. *Conceptual* design is a very fluid process and the configuration layout is changed constantly to incorporate new concepts and re-evaluate potential improvements. Trade studies and sophistication of design are improved as the concept design progresses in time. As the level of complexity increases, the level of detail increases and new phases of the design are considered.

In *preliminary Design*, the level of understanding of the concept defined at the end of the *conceptual* design phase matures and the mathematical modelling of the outside surface of the aircraft is detailed with accuracy. Different experts in aerodynamics, control and structures perform complex analyses in their areas, usually

J. Periaux et al., *Evolutionary Optimization and Game Strategies for Advanced Multi-Disciplinary Design,* Intelligent Systems, Control and Automation: Science and Engineering 75, DOI 10.1007/978-94-017-9520-3_5

with the aid of CFD and FEA codes. The use of sophisticated CAD tools, that accurately model the aircraft geometry, are part of the preliminary design process, allowing different disciplines to work with a general configuration that follows the requirements validated in the *conceptual* design phase and allows for introduction of a new analysis of the current design phase. The final output of the *preliminary* design is an integrated proposal of the model for entering a full-scale development in the *detailed* design process.

Detailed Design includes specific details such as the proper location of holes for fasteners, and design of components such as hinges, brackets and racks. Some considerations on production design, whereby specialists determine how the aircraft will be fabricated, are also part of the *detailed* design. Actual structures are tested and a simulation of the control laws for the system is performed during this phase.

5.3 Multi-Disciplinary Design Analysis (MDA) and Optimisation

5.3.1 Definition

In aerospace engineering the designer is usually presented with a problem which involves not only a single objective but numerous objectives and complex multi-physics. A systematic approach that accounts for the coupling between the disciplines and variables regarded as Multi-disciplinary Design Analysis (MDA) and as Multi-disciplinary Design Optimisation (MDO) when coupled with optimization techniques such as gradient based methods or Evolutionary Algorithms. In aircraft design, for example, multi-physics include aerodynamics, structures, propulsion, control among other disciplines.

These multi-physics are interrelated and interdisciplinary constraints must be satisfied to solve the optimization problem. The area of MDO has matured as a separate discipline now with different journals, specialised technical committees, conferences and publications devoted solely to the topic. The needs for, and benefits of MDO are clear, given that a small improvement in the performance of the aircraft can be significant. An MDO procedure also includes the evaluation of constraints on multiple disciplines from the early stages of the design, thus reducing the expense of re-designing an aircraft system [1].

5.3.2 Challenges and needs for MDO

The aircraft design industry has well understood the benefits of optimization and the use of MDO tools during the design process, but as traditional concepts are revised, new configurations are studied and market pressures and industrial requirements emerge, the need for developing alternative numerical methods tools and their application to current designs is becoming a requirement.

A comprehensive survey of MDO methods, their development and limitations is provided by Sobieski et al. [2]. The research classified the different methods and highlighted some important needs, including a multi-platform operation, the use of parallel computation to reduce computational expense and time design cycle and space visualisation as the designer might be interested in the optimum neighbourhood rather than the optimum itself. In most of the methods described in the survey, the optimisation algorithms for MDO use traditional gradient methods for computing the solution, but as has been discussed, these methods have some limitations. Sobieski and Hafka [3] evaluated some developments in multi-disciplinary aerospace design and optimization and identified several categories of problem formulations and also two other main challenges for MDO, the computational expense and organisational (architectural) complexity.

Another survey by Giesing and Barthel [4] identified several industrial applications and summarized some of the needs for MDO. Most of the applications are related to *detailed* design; few applications are developed for *conceptual* or *preliminary* MDO studies. With respect to the classification of needs, their research also describes how an MDO framework should be flexible to accept whatever function is needed and should address the issue of *low and high -fidelity models*, but not to compromise the optimisation. It also points out the need for efficient models that describe the physics to keep computing time at a reasonable level. A critical aspect mentioned in the paper is the need for accurate realistic design by identification of constraints, mechanisms and underlying physics of the various disciplines involved.

An open issue on MDO studies is the fact that many high-fidelity analysers such as Euler, Navier-Stokes, CFD or FEA codes are too complex to couple to an MDO as many of them are not automated, robust or fast enough. Also, the need for approximation techniques such as Response Surface Modelling (RSM) or Design and Analysis of Computer Experiments (DACE) arises because of the computational expense of using an analysis code for all evaluations during the optimisation procedure and also from the fact that some analysis codes cannot be directly integrated with MDO architecture.

Dealing with MDO framework architectures, there is a requirement for more efficient, robust flexible framework architectures and methods with industrial codes. These codes should be easily coupled and reconfigurable, and adaptable to commercial solvers. One of the problems with current MDO architectures is that these are usually developed at universities and in industries with restrictions on their use, and sometimes they are specialized and difficult to expand for other applications.

On the topic of optimization tools for MDO, it is important to determine the presence of multi-modalities, non-linearities and multiple objectives that might cause a traditional deterministic method to fail. Despite of the use of existing automated differentiation [23] the computational cost of gradients and the presence of multiple disciplines is also an important factor. Therefore, the continuing challenge has been on developing and improving numerical optimization techniques and enhancing their speed and robustness for their efficient use within MDO. One of the emerging optimisation techniques for MDO is EAs, but they have found limited applications in MDO due to the computational burden associated with them.

Applications of EAs for MDO problems include the work by Giesing and Barthelem [4] who presented a short discussion on supporting design space search methods such as Evolutionary Algorithms (EAs), and explained how they are gaining popularity for MDO because they are simple to couple with analysis modules and do not incur the cost of computing derivatives.

5.3.3 MDO Application Using Gradient-Based Methods

There are a number of applications of MDO to aeronautical problems [5, 6]. Some requirements, benefits and drawbacks of MDO are described in several applications to aeronautical problems. Wakayama and Kroo [7] applied MDO for subsonic wing platform design; their results highlighted the importance of considering a flexible architecture coupled with appropriate detailed analysis for structural and aerodynamic analysis to achieve realistic designs.

Bartholomew [1] explains the role of MDO within aerospace design and progress toward an MDO capability, based on lessons learned from research activities within the European community. The study stresses the importance of the incorporation of MDO in the design process at an industrial level. It also points out the requirement for an MDO framework to provide a flexible user interface definition and to monitor the MDO progress, and stresses the fact that the benefits of MDO provide means to avoid fragmentation with current recursive or intuitive methods that are time-consuming and limit the efficiency in the design.

The benefits of MDO, parallel computing and variable fidelity models for aircraft design have been studied by Giunta et al [8]. In these studies the researchers compared the use of RSM and DACE techniques and obtained optimum realistic aircraft design configurations. Other applications of MDO in engineering are related to complex systems, such as ground vehicle design [9–11]. These studies, although not directly related to aircraft or aircraft component design, provide an indication of the complexities of the MDO procedure and the benefit of parallel computing when large space systems are analysed.

5.4 Approaches to MDO

The selection of the appropriate optimization architecture is of great importance for providing an efficient solution to MDO problems, as one MDO architecture may find a feasible solution without high computational expense while other solutions might be too slow or even fail [12, 13]. There are several approaches for MDO, the most common being: the Multi-disciplinary Design Feasible (MDF) method [14], the Individual Discipline Feasible (IDF) method [15], the Collaborative Optimization (CO) method [16], the Concurrent Sub Space Optimisation with Response Surfaces (CSSO/RS) [17] and the Bi-Level Integrated System Synthesis with Response Surfaces (BLISS/RS) method [18]. This book develops algorithms and examples of the first two variants and a modified version of CO.

5.4.1 Multi-Disciplinary Design Feasible (MDF)

This formulation has been extensively used in the field of optimisation and can be considered as the standard approach. It is also known as single-level optimisation or All-in-One (A-i-O). The MDF can be stated as:

$$\text{Minimise:} f\left(z, y(x,y,z)\right) \qquad (5.1)$$

$$\textit{Subject to}: \ g\left(z, y(x,y,z)\right) \leq 0$$

where f is the objective function and g represents all systems and/or disciplinary constraints.

This method is illustrated below in Fig. 5.1.

In this approach the set of design variables x, z is evaluated by an integrated analysis. This analysis evaluates the coupled system of equations, guarantees inter-disciplinary constraints and returns the objective function that is manipulated by the optimizer for the search of the optimal solution. To achieve a design that is physically feasible, this approach usually requires a computationally expensive iterative procedure during each integrated analysis call. As an example, the minimization of the drag of a wing in a two-discipline problem dealing with structures and aerodynamics, may be considered. The design variables can be, for example, wing sweep, wing span and twist or taper. With an integrated approach the optimizer sends the values of these variables to an integral system of equations that represent the aerodynamics and structures analysis coupling.

These analyses are iteratively evaluated to conform and converge on each discipline to a consistent solution. The objective function and constraints are then evaluated and manipulated by the optimizer to improve the design. This approach is conceptually very simple. Once all disciplines are coupled to form a single multidisciplinary analysis module, the same techniques that are used for a single-discipline

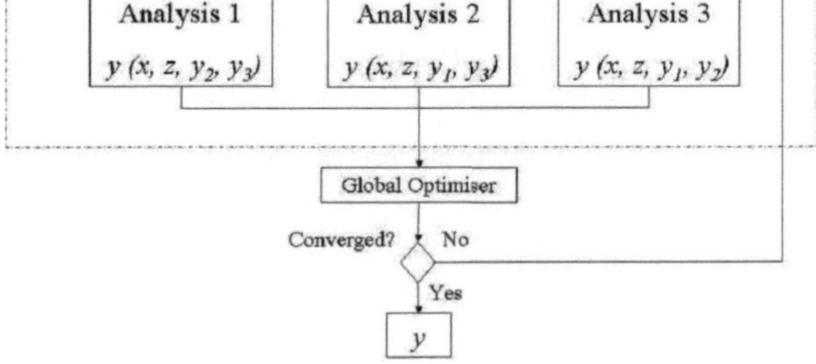

Fig. 5.1 Multi-disciplinary design feasible (*MDF*)

optimisation can be applied. The disadvantage of this approach is that the solution of a single system could be very expensive and does not enable the potential decoupling of the individual disciplines into analysis modules that can be run in parallel.

5.4.2 *Individual Discipline Feasible (IDF)*

This method belongs to the distributed analysis architectures and differs from the standard analysis by the way in which the different disciplines are decoupled. The disciplinary analyses are decoupled but the optimisation at the system level is kept to form a single-level decomposed optimisation problem. The subsystems are individually analysed and the optimization is performed for the system as a whole. Constraints that impose multi-disciplinary feasibility are introduced by using extra coupling variables. In this approach the disciplines are individually feasible but the complete system may not be feasible until the optimization procedure converges.

The IDF optimisation statement is:

$$\begin{aligned} &\text{Minimise:} f\left(z, y\left(x, y', z\right)\right) \\ &\textit{Subject to}:\ g\left(z, y\left(x, y', z\right)\right) \leq 0 \\ &y' - y\left(x, y', z\right) = 0 \end{aligned} \tag{5.2}$$

where y' are the auxiliary disciplinary input variables. The second constraint ensures multi-disciplinary feasibility when y is equal to y'. The IDF approach is illustrated in Fig. 5.2. The IDF approach is illustrated below in Fig. 5.2.

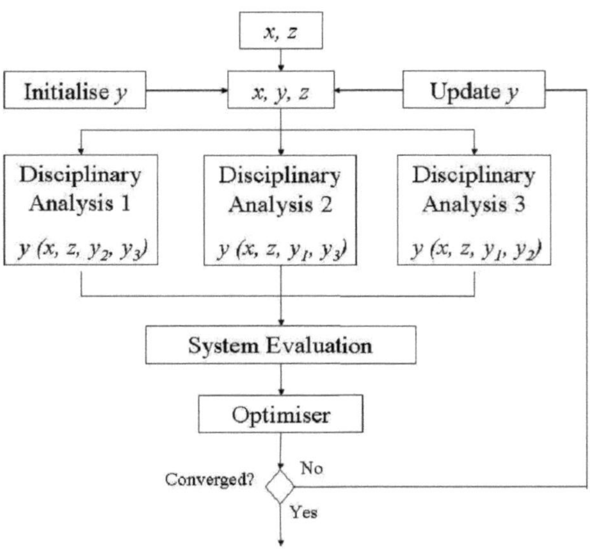

Fig. 5.2 Individual discipline feasible (*IDF*)

In this case the design variables x, y, z are sent to each discipline; these variables are analysed by each discipline block (that is, aerodynamics-structures). These analyses return an evaluated solution that satisfies the governing equations and domain-specific constraints on each discipline, but that might not be interdisciplinary compatible.

The system-level optimizer is responsible for ensuring compatibility of the overall solution, guaranteeing satisfaction of interdisciplinary constraints and minimizing the objective function.

A distributed architecture has several organisational and computational advantages over a traditional integral standard approach.

Computational advantages

- A reduction in re-iteration and communication requirements;
- The use of parallel heterogeneous platforms to evaluate the different disciplinary analyses;
- Removal of iteration loops;
- A reduced level of disciplinary requirements.

Organisational advantages

- Increased autonomy and disciplinary expertise in the design process;
- The benefits of being able to alter a part of the analysis without having to reformulate the whole problem.
- Resembles the division into areas of expertise in most design companies.

5.4.3 Collaborative Optimization (CO)

The CO approach is a two-level hierarchical scheme for MDO which decomposes and re-formulates the problem as a bi-level optimization, with the top level being the system optimizer that forces the multi-disciplinary variables or system level targets z to satisfy interdisciplinary compatibility constraints c while minimising the objective f. The problem is decomposed into the analysis and subspace optimizers that are integrated within each analysis block [19].

The system is optimised at the coordination level by determining the target values for subsystem responses and shared design variables with compatibility constraints that ensure multi-disciplinary feasibility. The optimization objective at each subsystem is to match as closely as possible these target values while satisfying local disciplinary constraints.

Through subspace optimization each block is given control over its own set of design variables and changed to satisfy its own domain-specific constraints. Specific knowledge of each of the other groups or design variables is not required. The objective of each subspace analysis is to agree upon the values of the interdisciplinary variables with the other groups. Again, a system-level optimizer is used to ensure compatibility at the overall solution, guarantee satisfaction of interdisciplinary constraints and minimize the objective function.

Similar to the IDF approach, the multi-disciplinary feasibility in CO is reached at the end of the procedure. If the target values corresponding to the shared and state variables are z and y' respectively, the system level optimisation can be written as:

$$\begin{aligned} &\text{Minimise:} f(z, y') \\ &\textit{Subject to}: \ c\left(z, z^*, y', y\left(x_i^*, y', z_i^*\right)\right) = 0 \end{aligned} \tag{5.3}$$

where c represents the compatibility constraints, one for each subsystem, and takes the form:

$$c_i \left(z - z_i^*\right)^2 + \left(y' - y\left(x_i^*, y', z_i^*\right)\right)^2 \tag{5.4}$$

where asterisks (*) indicate optimal subsystem values.

At the disciplinary level I^{th} the subsystem optimization can be stated as:

$$\begin{aligned} &\text{Minimise: } c_i\left(z, z_i, y', y\left(x_i, y', z_i\right)\right) = 0 \\ &\textit{Subject to}: \ g\left(x, z, y(x, y, z)\right) \leq 0 \end{aligned} \tag{5.5}$$

where the objective c_i is of the same form as the constraints at the global level. The CO approach is illustrated below in Fig. 5.3

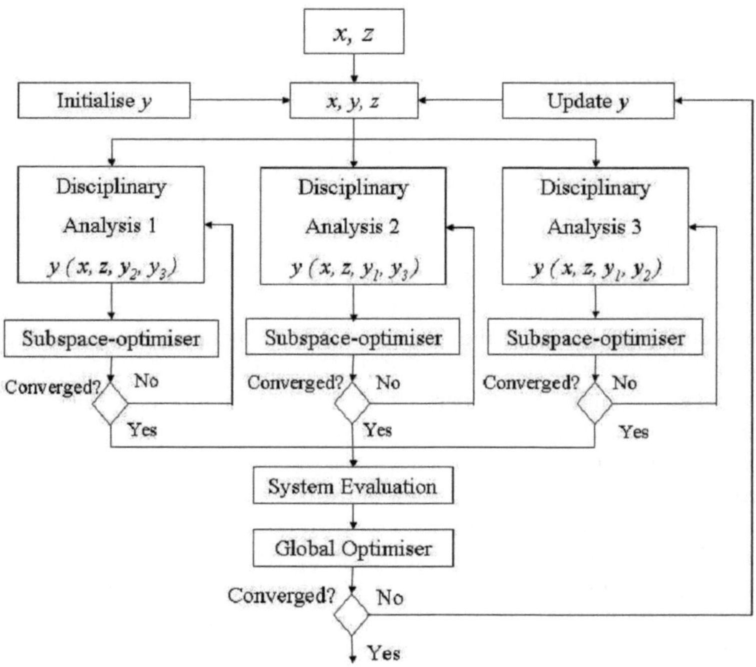

Fig. 5.3 Collaborative optimization (*CO*)

5.4.4 Criteria and Performance of MDO Implementations

Alexandrov and Lewis [20] classified different MDO implementations and identified several criteria to classify, analyse and evaluate approaches to MDO. They stated the distinction between formulating the MDO problem and solving the resulting computational problem. This technique involves two major elements: posing the problem formulation as a set of mathematical statements amenable to a solution and then defining a procedure algorithm to solve the problem once it is posed. As explained in the same paper, this distinction is crucial and is often blurred in presentations of new MDO approaches. According to this research, the main attributes for an MDO formulation are consistency, well-posedness, equivalence of formulations, optimality conditions and sensibility to solutions, while algorithm considerations are on local convergence rates, global convergence properties and iterative cost. Their research also provides guidelines and poses several questions for new formulations and analysis for MDO methods. Examples of these questions are: how is the original MDO problem formulated and does the formulation lead to an optimisation procedure that is not amenable to solution by existing optimisation algorithms

Another concern about the performance and solution of MDO problems is on the use of different fidelity analysis tools for MDO, due to the computational expense involved in the procedure. For this reason, continuing research has been devoted to developing MDO formulation and in combination with variable fidelity analysis, approximation techniques and Design of Experiments (DOE) theory [8]. Results indicate that these are possible avenues for research, as they lower the computational expense while maintaining the robustness of the solution [8].

5.5 Uncertainty Based Robust Design

There are many engineering applications where some values of the design parameters cannot be provided accurately. These are related to geometry manufacturing accuracy (wingspan, length, angles) or operational conditions that vary due to the presence of fluctuations (Mach, angle of attack, air density and temperature). Uncertainty based Robust Design is used to deal with uncertainties present during manufacturing or within the natural environment. It is on this premise that an approach taking account of uncertainties needs to be incorporated into the analysis for design of an aeronautical/aerospace design. In these circumstances, some of the input parameters need to be estimated by mean and variance values.

5.5.1 Robust/Uncertainty Method

The Robust Design method was pioneered by Taguchi and Chowdhury [21]. The method considers possible uncertainties in input parameters, manufacturing or operational constraints to improve quality of engineering productivity. In Robust Design an optimisation problem could be re-defined as:

$$Maximisation\,/\,Minimisation\; f = f(x_1,\ldots,x_n,x_{n+1},\ldots,x_m) \tag{5.6}$$

where $x_1,\ldots,x_n$ represent design parameters and $x_{n+1},\ldots,x_m$ represent uncertainty parameters which can be distributed with mean and standard deviation.

In robust design optimisation the aim is to minimize the variability of the performance under uncertain operating conditions. An optimization with uncertainty parameters can be defined fitness objectives associated with the statistical formulas; mean $(\bar{f})$ and variance of function (δf) or standard deviation $(\sigma f = \sqrt{\delta f})$ for the sensitive analysis.

$$\bar{f} = \frac{1}{K}\sum_{j=1}^{K} f_j \tag{5.7}$$

$$\sigma f = \sqrt{\delta f} = \sqrt{\frac{1}{K-1}\sum_{j=1}^{K}\left(f_j - \bar{f}\right)^2} \tag{5.8}$$

where K denotes the number of subintervals of variation flow conditions.

In aerodynamics for example, uncertainties at desired operating conditions can be described with the fluctuation on lift coefficient (C_l) at different free stream Mach numbers. Figure 5.4 for example shows the fluctuation on C_l and compares the total drag distribution (physical performance) obtained by conventional meth-

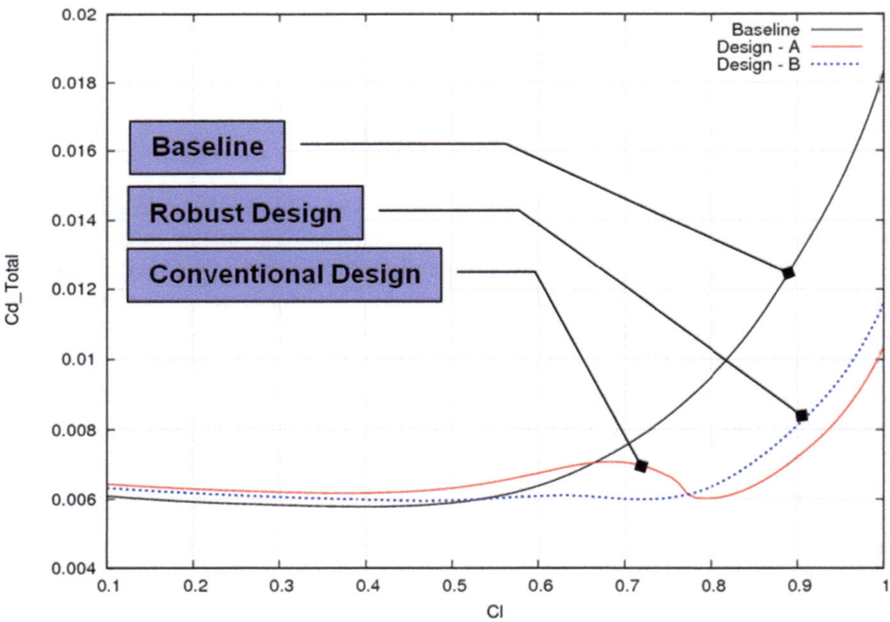

Fig. 5.4 Drag distributions obtained by the conventional method (*Design-A*) and Robust Design Method (*Design-B*)

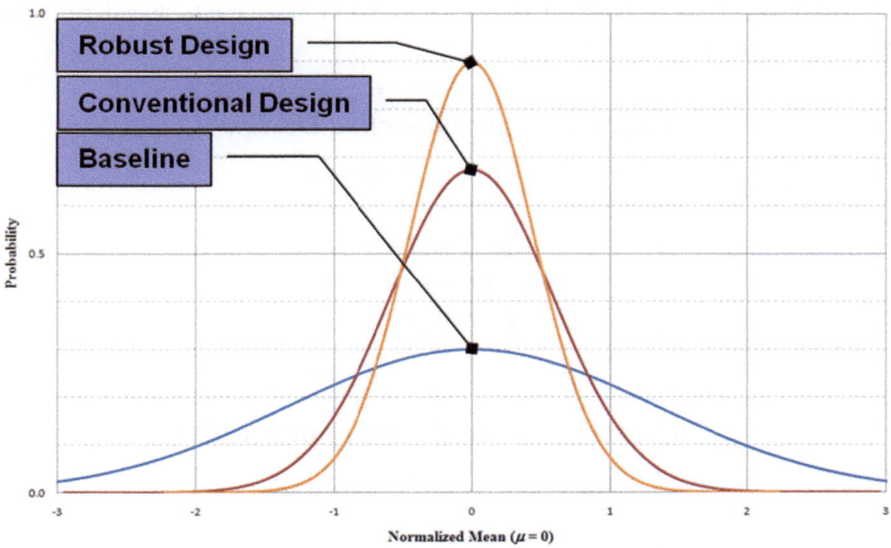

Fig. 5.5 Sensitivity comparison obtained by the conventional method and robust design

ods (Design-A) and by the robust design method (Design-B). Even though Design-A produces lower drag coefficient at the design point (when the lift coefficient is 0.78), Design-A fluctuates along the lift coefficients [C_l=0.5:0.8]. In other words, the conventional design optimization approach produces an over-optimised solution (Design-A) that has good performance at desired flight condition (C_l=0.78) but with poor off-design characteristics (high sensitivity). Design-B obtained by the robust design has a stable performance (low sensitivity) over a range of C_l conditions without fluctuations.

The performance sensitivity obtained by the conventional and the robust design methods can also be compared using a Probability Density Function (PDF) as shown in Fig. 5.5. It can be seen that the robust design method produces a narrower bell curve (lower standard deviation which means lower sensitivity and stable physical performance) when compared to a baseline design and the results obtained by a conventional method.

5.5.2 From Single-Objective to Multi-Objective Design Optimisation Using the Robust Design Method

A single objective problem can be converted to a Multi-Objective Design Optimisation when we consider Robust Design. The original single objective function can be modified using mean and variance to represent the model quality in terms of performance and stability/sensitivity at a set of variable design conditions. For instance, when uncertainty is applied to single-objective problem such as minimisation of drag (*f*=*min (CD)*).

The problem can be modified as robust uncertainty based multi-objective design optimisation problem as follows:

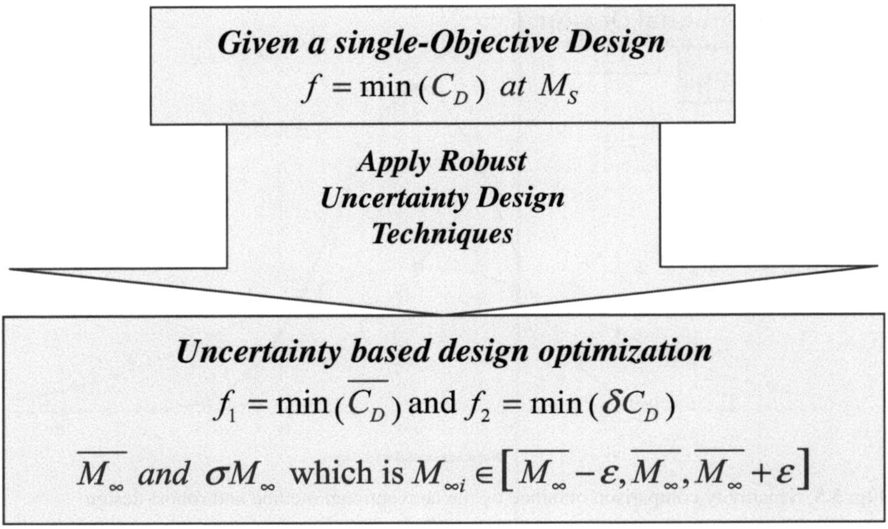

1. Compute K, number of off-design conditions defined by the mean M_∞ and the standard deviation of M_∞ which represents a vector of flight conditions $M_{\infty i} \in \left[\overline{M_\infty} - \varepsilon, \overline{M_\infty}, \overline{M_\infty} + \varepsilon\right]$.
2. Split the single-objective/fitness function into the mean and variance or standard deviation of drag coefficient ($\overline{C_D}$: Eq. 5.9) and (δC_D : Eq. 5.10).

$$\overline{C_D} = \frac{1}{K}\sum_{i=1}^{K} C_{Di} \tag{5.9}$$

$$\sigma C_D = \sqrt{\delta C_D} = \sqrt{\frac{1}{K-1}\sum_{i=1}^{K}\left(C_{Di} - \overline{C_D}\right)^2} \tag{5.10}$$

where K represents the number of uncertainty conditions.

The main role of robust design in this example is to improve C_D quality; reduction of both magnitude of drag coefficient and drag coefficient *sensitivity* at uncertain flight conditions by computing mean and variance of criteria. Additional details on the uncertainty based technique can be found in Chap. 6.

5.5.3 Robust Multi-Objective/Multi-Disciplinary Design Optimization

Multi-objective and Multi-disciplinary Design Optimisation problems can also be analysed using Robust Design optimisation techniques using variance to mean ratio (VMR). Considering such an approach the objective functions can take into account for instance both aerodynamic quality and structural weight quality if the design problem considers interdisciplinary aero-structural design problems−==v with uncertainty parameters. For instance, the aerodynamic quality fitness function can be defined as shown Eq. (5.11) which combine Eqs. 5.9 and 5.10;

$$f_{AerodynamicQuality} = \min\left(\frac{\delta(L/D)}{\overline{L/D}}\right) \quad (5.11)$$

and the structural quality function can be written as;

$$f_{StructuralQuality} = \min(\overline{W_{Wing}}\,\delta(W_{Wing})) \quad (5.12)$$

where W_{Wing} represents the weight of wing.

5.6 Limitations of Traditional Optimisation Techniques for MDO and Robust Design

5.6.1 MDO Using Traditional Method and Evolutionary Algorithms

As discussed in the previous sections, most of the optimization methods used for Multi-disciplinary Design Optimisation (MDO) rely on traditional deterministic techniques. While the use of these methods is largely successful and efficient at finding optimal global solutions, problems still do arise. Gradient-based methods usually work best with uni-modal functions, but their effectiveness decreases with the presence of local optima or ridges in the fitness landscape. Moreover, the presence of numerical noise inhibits the application of many gradient-based optimization techniques. The numerical noise causes an inaccurate calculation of gradients, which in turn slows or prevents convergence during optimisation [8, 22]. In 3-D aircraft design, for example, the problem of numerical noise is of special concern when an accurate solution is sought through a high-fidelity analysis but the computation of gradients is complex and a single aerodynamic or structural analysis might take several CPU hours on a supercomputer.

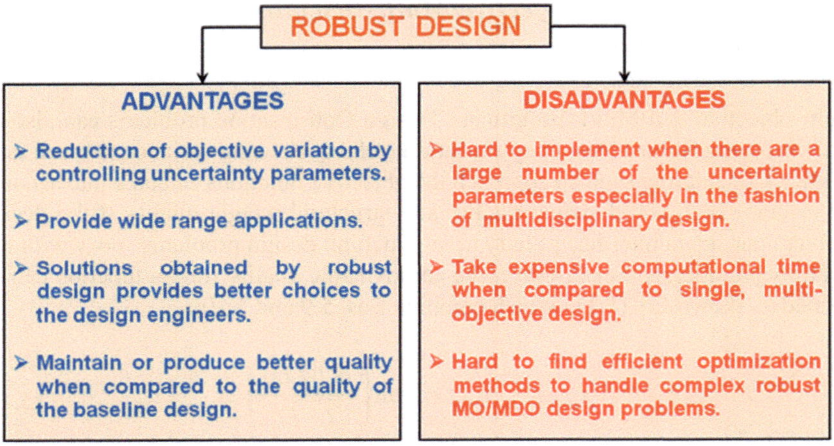

Fig. 5.6 Advantages and disadvantages of using Robust design

An emerging optimization technique for MDO is Evolutionary Algorithms (EAs); these techniques are robust in finding optimal solutions for single- and multi-objective problems, but have found limited applications to MDO due to the computational burden associated with them. The challenge is then to study, develop, apply and improve the speed and robustness of these methods so that confident applications and use within MDO is possible.

It is important to highlight that in this book the use of EAs will be restricted to *conceptual* and *preliminary* MDO studies where the number of variables is still relatively small, less than a hundred, and where the use of EAs is still of potential benefit. On a larger scale, the use of EAs can be extended for an increased number of variables and coupled with other techniques such as Games Strategies or Design of Experiments (DOE).

5.6.2 Advantages and Drawbacks of Robust Design

The advantages and drawbacks of using Robust Design are compared in Fig. 5.6. The Robust Design method significantly affects solution quality in terms of performance and sensitivity. In other words, optimal solutions obtained by Robust Design have capabilities to produce higher performances and also maintain/improve its performance along the uncertain variation when compared to the baseline design. Design engineers will have available a better choice to solve real-world design after introducing Robust Design in their optimization procedures.

Even though the Robust Design method possesses significant benefits in several aspects, its counterpart balances considerable disadvantages: difficulty to define/implement uncertainty design into complex multi-objective/multi-disciplinary optimisation, high computational cost due to the further iteration caused by a set of uncertain variation. The major drawback of Robust Design is difficult to find an efficient optimization method to overcome the above first and second disadvantages.

Subsequent chapters address the issue of computational expense when using EAs, MDO and Robust Design with coupled advanced techniques described in Chap. 4.

5.7 Summary of Chapter

This chapter described some of the concepts and techniques used in multi-disciplinary design optimization (MDO) and Robust/uncertainty based Design. Some of these techniques are used in subsequent chapters.

References

1. Bartholomew P (1998) The role of MDO within Aerospace Design and progress towards an MDO Ccapability. In Proceedings of Seventh AIAA/USAF/NASA/ISSMO Symposium on Multidisciplinary Analysis and Optimisation, St. Louis, Missouri. September 1998
2. Sobieszczanski-Sobieski J, Haftka RT (1996) Multidisciplinary aerospace design optimization survey of recent developments. Technical Report 96–0711, NASA/American Institute of Aeronautics and Astronautics, 1996
3. Sobieszczanski-Sobieski J, Haftka RT (1997) Multidisciplinary aerospace design optimization: survey of recent developments. Struct Optim 14:1–23
4. Giesing JP, Barthelem JF (1998) A summary of industry MDO applications and needs. In Proceedings of Seventh AIAA/USAF/NASA/ISSMO Symposium on Multidisciplinary Analysis and Optimisation. AIAA, St. Louis, Missouri, September 1998.
5. Blair M, LeClair S, Zweber J, Chemaly A (1997) Multidisciplinary design for uninhabited air vehicles. In 6th Workshop on Enabling Technologies Infrastructure for Collaborative Enterprises (WET-ICE '97). AIAA, 1997
6. Gantois K, Morris AJ (2004) The multi-disciplinary design of a large-scale civil aircraft wing taking account of manufacturing costs. Struct Multidiscip Optim 28(1):31–46
7. Wakayama S, Kroo I (1995) Subsonic wing planform design using multidisciplinary optimization. AIAA J Aircr 12(4):746–753
8. Giunta AA, Balabanov V, Haim D, Grossman B, Mason WH, Haftka RT, Watson LT (1997) Multidisciplinary optimization of a supersonic transport using design of experiments theory and response surface modelling. Aeronautic J:347–356 query for volue det
9. Bennett J, Fenyes P, Haering W, Neal M (1998) Issues in industrial multidisciplinary optimization. In Proceedings of Seventh AIAA/USAF/NASA/ISSMO Symposium on Multidisciplinary Analysis and Optimization, St. Louis, Missouri, September 1998
10. Kodiyalam K, Yang RJ, Gu L, Tho C-H (2004) Large-scale, multidisciplinary optimization of a vehicle system in a scalable, high performance computing environment. Struct Multidiscip Optim 26:256–263
11. McAllister C, Simpson T, Hacker K, Lewis K (2002) Application of multidisciplinary design optimization to racecar design analysis. In Proceedings of the 9th AIAA/USAF/NASA/ISSMO Symposium on Multidisciplinary Analysis and Optimization, Atlanta, Georgia, September 2002
12. Braum R, Gage P, Kroo I, Sobieski I (1996) Implementation and performance issues in c.o. In Proceedings of Sixth AIAA/USAF/NASA/ISSMO Symposium on Multidisciplinary Analysis and Optimization, Bellevue, Washington, September 4–6 1996. AIAA Paper No. 96–4017

13. Dulikravich GS, Dennis BH, Martin TJ, Egorov IN (2001) Multi-disciplinary analysis and design optimization. in invited lecture, mini-symposium on inverse problems—state of art and future trends, XXIV Brazilian Congress on Applied and Computational Mathematics, Belo Horizonte, Brazil, September 10–13, 2001
14. Alexandrov N, Kodyalam S (1998) Initial results of an MDO methods evaluation study. In Proceedings of the Seventh AIAA/USAF/NASA/ISSMO Symposium on Multidisciplinary Analysis and Optimization, St. Louis, Missouri, September 1998
15. Cramer EJ, Dennis JE, Frank PD, Lewis RM, Shubin GR (1994) Problem formulation for multidisciplinary optimization. SIAM J Optim 4:754–776
16. Braum R, Gage P, Kroo I, Sobieski I (1996) Implementation and performance issues in c.o. In Proceedings of Sixth AIAA/USAF/NASA/ISSMO Symposium on Multidisciplinary Analysis and Optimization, Bellevue, Washington, September 4–6 1996. AIAA Paper No. 96–4017
17. Renaud G, Shi G (2002) Evaluation and implementation of multidisciplinary design optimization (MDO) strategies. In Proceedings of the 23rd ICAS Congress, Toronto, Canada, September 2002
18. Kodiyalam K, Yang RJ, Gu L, C-H Tho (2004) Large-scale, multidisciplinary optimization of a vehicle system in a scalable, high performance computing environment. Struct Multidiscip Optim 26:256–263
19. Kroo I, Altus S, Braun R, Gage P, Sobieski I (1994) Multi-disciplinary optimization methods for aircraft preliminary design, AIAA 94–4325, Fifth AIAA/USAF/NASA/ISSMO Symposium on Multidisciplinary Analysis and Optimization, September 7–9, Panama City, Florida
20. Alexandrov NM, Lewis RM (2000) Analytical and computational properties of distributed approaches to MDO. In AIAA Paper 2000–4718. AIAA, September 2000
21. Taguchi G, Chowdhury S (2000) Robust engineering. McGraw-Hill, New York
22. Giunta A, Balabanov V, Burgee S, Grossman B, Haftka RT, Mason WH, Watson LT (1995) Variable-complexity multidisciplinary design optimization using parallel computers. In: Alturi SN, Yagawa G, Cruse TA (eds) Computational mechanics 95–theory and applications, Proc. of ICES '95, International Conference on Computational Engineering Science, pp 489–494, Mauna Lani, Big Island, Hawaii, July 1995. Springer
23. Bartholomew-Biggs M, Brown S, Christianson B, Dixon L (2000) Automatic differentiation of algorithms. J Comput Appl Math 124(1–2):171–190

Chapter 6
A Framework for Numerical Design and Optimization Algorithms

6.1 Overview

As described in previous chapters design and optimization in aeronautics are complex tasks due to nonlinearity, multi-objective, multi-disciplinary considerations involved in the optimization procedure. In order to handle the level of complexity of the design problem it is desirable to develop a framework. A framework facilitates integration of a series of design tools, graphical user interfaces, post-processing capabilities and other means to solve the design problem.

This chapter focuses on the requirements, development and implementation of a framework that uses evolutionary techniques and analysis tools in which different multi-disciplinary and multi-objective problems in aeronautics can be analysed. Detailed pseudo-code and algorithms for the advanced techniques described in previous chapters are also presented.

6.2 An Optimization Framework

A framework for robust, practical and efficient optimization procedures needs to satisfy a series of requirements which can be subdivided into problem formulation, optimization methods, problem execution, architectural design and information access [1–5].

Problem Formulation

- The framework should allow the user to configure and reconfigure different MO and MDO formulations easily with low level programming.
- The framework should handle problem formulations that can be multi-objective, multi-modal, discontinuous, or dealing with noisy search spaces.

J. Periaux et al., *Evolutionary Optimization and Game Strategies for Advanced Multi-Disciplinary Design,* Intelligent Systems, Control and Automation: Science and Engineering 75, DOI 10.1007/978-94-017-9520-3_6

Optimization Methods

- The framework should allow ease of integration of robust optimization methods.
- It should also allow integrating different discipline analysis with different optimization methods and should provide schemes which involve sub-optimization within each design module.

Problem Execution

- The framework should allow the execution and movement of data in an automated fashion;
- The framework should be able to execute multiple processes in parallel and on heterogeneous computers;
- A batch mode should be also implemented.

Implementation and Architectural Design

- The framework should be developed using object-oriented principles;
- The framework should allow the user to incorporate legacy codes, which can be written in different programming languages, and proprietary software where the source code is not available;
- The framework should provide an easy to use and intuitive Graphics User Interfaces (GUI);
- The framework should be easily extended by developing new interfaces required to integrate new processes into a system;
- The framework should not impose unreasonable overhead on the optimization procedure;
- The framework should be based on standards.

Information Access

- The framework should provide facilities for database management;
- The framework should provide capabilities to visualize intermediate and final result from the analysis or optimization;
- The framework should also allow capabilities for monitoring and viewing the status of an execution and its system status;
- It should be also a mechanism for fault tolerance.

With these requirements in mind a general scope for the framework can be identified. The framework explained in this chapter address these requirements to some extent. Figure 6.1 shows a representation of different components to satisfy these requirements. The framework has seven major constituents: a robust optimization tool, a problem formulation capability within each analysis module, some architectural design considerations such as a Graphics User Interface, a Design of Experiments (DOE) module, some analysis modules, and capabilities for parallel computing and post-processing. In the following sections and subsections each of these constituents is detailed.

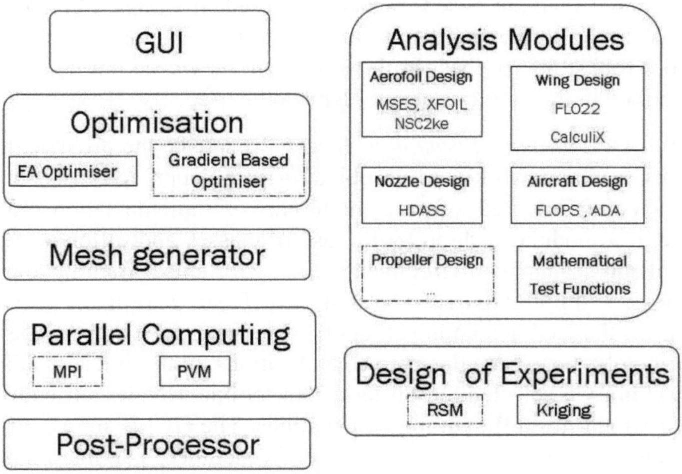

Fig. 6.1 A MDO framework

6.3 Implementation of the Framework

Integrating all these components remains a complex task. This section considers a general formulation for MDO and multi-criteria problems with selection and description of an optimization tool, development of the architecture, a GUI and capabilities for pre- and post-processing. The DOE capability has been accounted for, but has been evaluated only for simple mathematical test cases. The following sub-sections detail how the requirements are satisfied.

Problem Formulation and Solution There are mathematical and engineering optimization problems that give difficulties or cannot be solved by traditional gradient based techniques. When these problems are formulated the search space can be multi-modal, non-convex or discontinuous, with multiple local minima and noise or problems where multiple solutions simultaneously, a Nash equilibrium point or a set of non-dominated solutions, are sought. Some problems in aeronautical and Unmanned Aerial Vehicle (UAV) design fall into this category.

Population based techniques are now well known as they can be suited to solve these type of problems. However one major drawback of Evolutionary Algorithm based optimizers is that they are slow in converging as they require a large number of function evaluations to find optimal solutions and have relatively poor performance with an increasing number of discipline specific and interdisciplinary variables. Hence a continuing effort has been dedicated to developing robust and faster numerical techniques to solve complex problem formulations, overcome these challenges and facilitate the complex task of design and optimization in aeronautics.

Selection of a Robust Optimization Tool A second consideration is the selection of an optimization tool that is appropriate for the problem to be solved. For many problems a traditional gradient based method, a canonical simple Genetic Algorithm (GAs) or Evolution Strategy (ES), will work nicely but more complex problems require more robust approaches. In this direction, *advanced* optimizers and algorithms described in the previous chapter have to be used.

Architectural Design and Information Access To satisfy the architectural design requirements the framework uses an object-oriented approach in C++. The benefits of using object-oriented software are the ease of implementation and extension of software in a modular fashion by the use of classes and different methods. In an industrial and academic environment the need for a user-friendly application is required hence a simple GUI may be designed. There were many considerations and options for the GUI development, but knowledge in C++ and the use of object-oriented principles have been the main considerations. In this book, the Fast Toolkit (FLTK.org) library is selected for this task. This toolkit provides a friendly and easy to use environment for different implementations. The GUI is simple and modular on its implementation and consists of five main modules. The main modules are: Design and Analysis, Design of Experiments, Post-processing and Parallel Processing. A GUI facilitates development, extension and modifications of modules in a rather simple manner. The user has to create only a few subroutines within the corresponding module.

Design and Analysis Module A design module allows the user to conduct a single design and optimization for different aeronautical applications and mathematical test cases. So far this module contains five sub-modules for aerofoil, multi-element aerofoil, nozzle, wing, aircraft and mathematical functions design or optimization. As designed the framework is flexible and provides for ease of implementation of other design modules. Modules currently under development are those for propeller, cascade aerofoils and rotor blade design.

Development of Aeronautical Design Modules Before implementing a sub-module it is necessary to develop a design module interface. This comprises a series of files written in C++ that allow communication between the GUI, analysis codes, the optimizer and the parallel processing capability. In the design of the interface a choice has to be made depending on whether the source code for the analysis tool was available or not. In the current implementations minimal modification to the source code was required. Ideally it is desirable to operate only through the input/output files of the analysis tool. In the implementations considered, a design template was used in conjunction with one or two additional files which contain the necessary linking subroutines allowing a rather fast implementation of the design modules. There are subroutines for aircraft, nozzle, wing and full aircraft configuration design. Each of these options allows the designer to perform a single design analysis or a full optimization.

Aerofoil Design and Optimization Module This module allows the user to perform a single analysis or a full airfoil optimization routine. Three different CFD codes and a combination of them can be used: A panel method (XFOIL) [6], an Euler + boundary layer (MSES) [7], an Euler Finite Element Analysis software (PUMI) [8] or Navier -Stokes flow analysis solver (NSC2KE [9]).

Wing Design and Optimization Module This module allows the user to conduct a single analysis on a wing or an optimization study. These could be studies in one or several objectives or with multiple disciplines.

Aircraft Design and Optimization Module This module allows the user to analyse and optimize different problems related to aircraft external configuration design. It can be used to design and optimize different subsonic, Unmanned Aerial Vehicles, transport or supersonic aircraft. Single or multi-criteria optimization studies can be performed. Comparison of different multi-criteria analysis such as Pareto optimality and the Nash equilibrium approach are possible. The user can select from two different analysis codes: an object-oriented Aircraft Design and Analysis Software (ADA) or the Flight Optimization System (FLOPS) software developed by A. McCullers at NASA Langley. ADA is conceptual design and analysis software written using object-oriented principles and is based on the formulation described in Raymer [10, 11]. FLOPS [12], a more robust solver, are a workstation-based code which has capabilities for *conceptual* and *preliminary* design and evaluation of advanced concepts. The sizing and synthesis analysis in FLOPS are multi-disciplinary in nature. It has a numerous modules and analysis capabilities for take-off, performance, structural, control, aerodynamics, noise and electromagnetic Radar Cross Section (RCS) freeware. This code is used in some universities as well as in aerospace firms and governmental institutions for MDO development and it allows an integral multi-disciplinary analysis for the entire aircraft mission and calculation of performance parameters such as range, endurance, take-off field length and landing field length.

Multi-Element Airfoil Design and Optimization Module Similar to the airfoil design module, it allows the designer to perform a single analysis or a full optimization, which the user can choose from an Euler or Navier-Stokes analysis.

Mathematical Test Functions Module This module allows the user to design, and evaluate single, or multi-criteria mathematical test functions which give confidence in the robustness and performance of the optimization method before deciding on its application to real world problems. The current implementation includes mathematical test functions for single or multiple criteria, constrained optimization, DOE and non-linear goal programming problems.

Robust Design Module A robust design module allows the user to evaluate aerodynamic quality, structural quality, and electromagnetic quality for a design model in terms of mean and standard deviation or variance in the MDO system. The module contains four sub-modules for airfoil, multi-element airfoils, wing, and aircraft optimization. The module can be easily coupled to any optimizer for robust aeronautical, aero-structural or aero-structure-electromagnetic applications (cf. applications in the following chapters).

Parallel Computing Module One of the drawbacks of EAs is slow convergence but this module allows the users to dynamically create, add or delete nodes on the parallel implementation. Recent work on multi-criteria parallel evolutionary algorithms has allowed significant performance and robustness gains in global and parallel optimization [13–15]. The framework considers the implementation of a cluster of PCs, wherein the master carries on the optimization process while remote nodes compute the solver code. The message-passing model used is the Parallel Virtual Machine (PVM) [16].

Post Processing The approach considered for post-processing is to use a combination of visualisation capabilities within each analysis software. Common to all design modules is the visualisation of the evolution progress of the fitness function, Nash Equilibria and Pareto fronts for multi-criteria problems. Post-processing tools on each analysis module include a top view of the wing plan forms and a general 3D view of the resulting aircraft configurations. Visualisation tools within each analysis software module include the surface pressure coefficients distribution on the airfoil using an Euler with Boundary Layer effects flow solver or pressure or Mach contours using a Navier-Stokes flow solver.

Implementation of Different Legacy Codes The framework also implements legacy codes in different programming languages C, C++, Fortran 90, and Fortran 77. The optimizer has been successfully coupled with the following aerodynamic and analysis software: FLO22 [17] FLOPS, ADA, XFOIL, MSES, NSC2ke and PUMI. One of the benefits of using an Evolutionary optimizer is that EAs require no derivatives of the objective function. The coupling of the algorithm with different analysis codes is achieved by simple function calls and input and output data files.

6.4 Optimization Methodology

The design optimization methodology used in the examples presented in the following chapters can be summarised as:

- *Problem Definition* (Includes problem formulation, definition of the EA strategy, fitness function and constraints),
- *Design Variables* (includes definition and encoding of design variables),
- *Implementation* (includes the definition of analysis tools-software, design and optimization rationale and evaluation with EA Method),
- *Numerical Results* (includes optimization results and post-processing of compromised optimal solutions).

Problem Definition A first step consists of problem definition or formulation. When the scientist or the technologist are considering the solution and optimization of a real -world engineering problem, usually they try to create a well-posed mathematical formulation which is representative of the problem at hand. The problem in fact is that there are several complexities involved in the procedure. Therefore some assumptions have to be made. This can be illustrated with a simple example: the shape of the design and optimization of an aircraft wing or a wing section; during the entire aircraft mission the wing is subject to numerous flight conditions which are characterised by different Mach numbers, Reynolds numbers, angles of attack and other flight parameters. To obtain the optimum shape for this wing, the design has to exhibit good characteristics with regard to the aircraft payload, total mass, aerodynamic and structural performances. This is a complex task, therefore a sound solution to the problem is to make some assumptions and identify the main flight conditions, design parameters, constraints and then construct an optimization problem in order to improve performances.

A second step is the selection of an optimization strategy. As discussed in Chaps. 1 and 2, even though traditional gradient based methods are fast, the complexity, nonlinearity and multi-objective characteristics of some problems may require the use of more robust optimizers. Also, the computational cost of a turbulent Navier-Stokes or Euler solution around a high lift device, 3D wing for example involves high computational expense, therefore it is also desirable to use parallel computations and a multi-fidelity approach. In these cases, an efficient and robust Parallel Multi-Objective EA (PMOEA) is used.

In this step the user also needs to specify a series of constraints, such as maximum and minimum thickness location, pitching moment and lower and upper bounds for the parameterized Bézier or Spline control points.

The objective or *fitness* function is specified by the end user of the optimizer. It can be referred to by many other names, such as the cost function and also payoff function. It may be purely mathematical, the result of some solution procedure (Computational Fluid Dynamics, Finite Element Analysis, etc.) or be the result of an experiment. Formally, the fitness function is simply defined as:

$$f = f(x) \tag{6.2}$$

where x is the vector of object variables and f is the resulting vector of fitness values. Neither f nor x can be empty vectors. One vector of object variables is associated with each individual, so it is also possible to write: $f=f(\mathrm{I})$.

Further, it can be applied to each member of a population, so it can be written as: $\mu_{eval}=f(\mu)$. The distinction between problems in single and multiple objectives occurs here, with the fitness vector only having one element for a single objective problem.

Design Variables The selection of an appropriate definition of design variables and geometric representation that accounts for the complexities of the design space is the third step. In aerofoil shape optimization, parameterizations with Bézier or Spline curves or a PARSEC representation can be used. In the Bézier curves representation, for the design variables, a combination of mean line and thickness distribution control points is used (see Sect. 8.3.1 for more details). Similar to the NACA series airfoils, the PARSEC parameterises upper and lower surface of the airfoil using polynomial in coordinates X, Z as:

$$Z = \sum_{n=1}^{6} a_n . X^{n-1/2} \tag{6.1}$$

where a_n are real coefficients. The PARSEC aerofoils are defined by basic geometric parameters: leading edge radius (r_{le}), upper and lower crest locations including curvatures (X_{us}, Z_{us}, Z_{xxus}, X_{ls}, Z_{ls}, Z_{xxls}), trailing edge ordinate (Z_{te}), thickness (ΔZ_{te}) and directions and wedge angles (α_{te}, β_{te}). These parameters can be expressed by the original coefficients a_n by solving a simple set of simultaneous equations.

Implementation In this step it is recommended to develop an optimization algorithm. An example of an optimization flow chart is illustrated in Fig. 6.2. Initially the design variables, design constraints, flow conditions and fitness functions are defined and then an initial population of wing geometries is generated at random. Then while the termination criteria has not been satisfied, the algorithm generates shapes, evaluates fitness, evolves by recombining and mutating individuals, sorting non-dominated solutions, computes Pareto fronts and updates non-dominated solutions and proceeds with migration process if there are multiple (hierarchical population or hybridized Nash-Player/Pareto Game populations). The termination criterion is satisfied when the prescribed number of function evaluations is reached or when the fitness value goes below a prescribed number. The last stage is designation of the result such as best-so-far individual, Pareto non-dominated solutions (Pareto-front) and/or Nash-Equilibrium and termination of the optimization procedure.

The next step is to decide which type and fidelity of solver to use. A decision has to be made in order to compromise on the use of a higher fidelity solver, such as a Navier-Stokes solver/Finite Element Analysis which are computationally expensive but accurate, the use of lower fidelity models, such as Euler or panel or analytical methods, which are fast but could be unstable, or a combination of both. The structural properties can be computed with a combination of FEA and an analytical method. Another alternative is using a single model but defining different mesh densities.

Numerical Results This refers to the importance of keeping track of the statistics of the evolution procedure, number of generations, Pareto fronts, Nash Equilibria, but without adding a major computational burden to the optimization procedure. These are for example plots for value of fitness function versus evolution procedure or a plot of the current non-dominated solutions and whole population. Equally important is the post-processing of final solutions. This is related to what is interesting to the designer: these can be for example comparing shapes on the Pareto optimal front, comparing the *Cp* distribution over an airfoil, and/or comparison of aerodynamic performance of the baseline geometry and optimal solutions.

6.5 Optimization Algorithms

6.5.1 Overall Optimization Algorithm

Figure 6.2 (*Algorithm 1*) describes the general aerospace optimization problem and initially it defines:

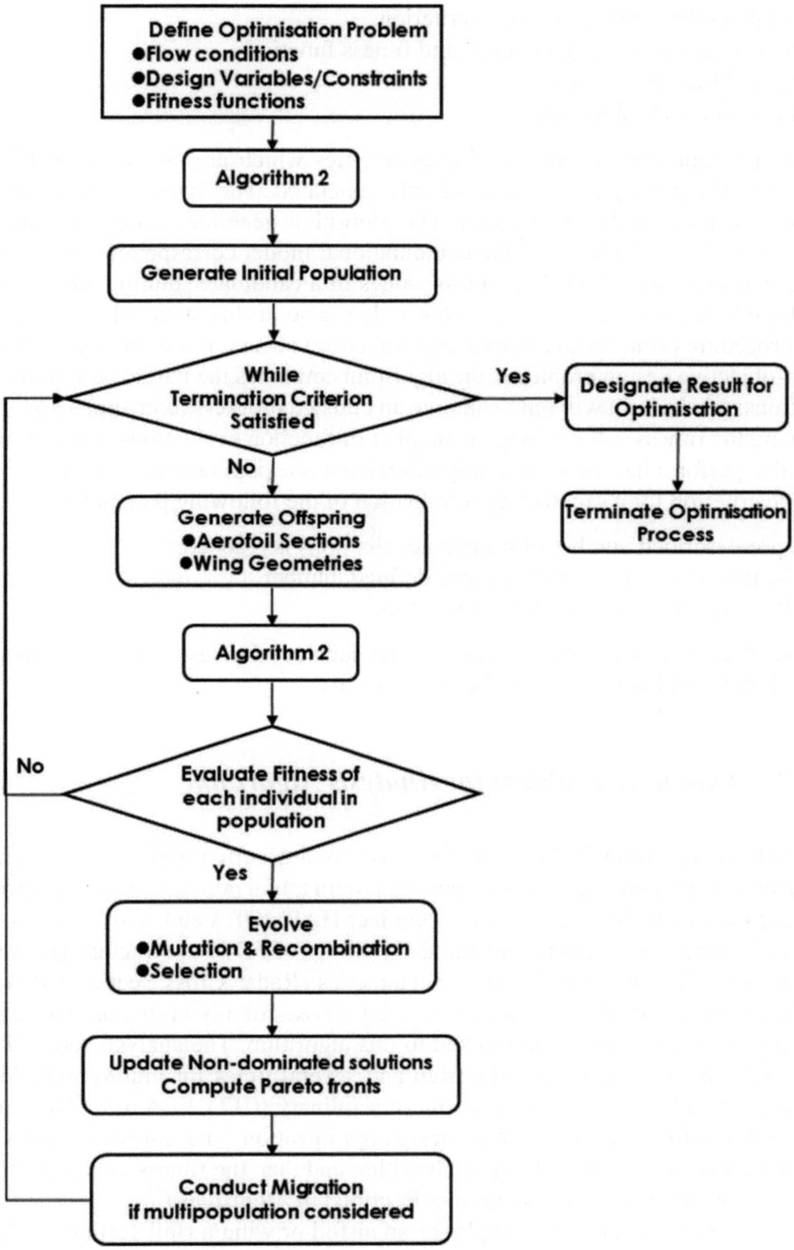

Fig. 6.2 Overall algorithm for design optimization problems

- Number of run and termination criterion,
- Design variables, design bounds and fitness functions,
- Desired flow conditions,
- Hierarchical topology setup.

An initial population of airfoil/wing geometries which are pre -described in the hierarchical topology layers are randomly generated. The optimizer will run until the termination conditions are met. The algorithm generates candidate wing geometries to be evaluated with the computational model corresponding to the hierarchical nodes (Sect. 4.3). The fitness values of a candidate solution are evaluated by *Algorithm 2* and used by *Algorithm 1*. It is also at this stage where the evolution procedure (mutation, crossover and selection) occurs. If the problem considers multi-objective design problems, the algorithm computes the Pareto non-dominated solutions; otherwise it will only compute an updated progress (convergence) history showing the fitness values along the number of function evaluations. The algorithm will also perform best individual migration from one population to another.

The criterion for a satisfactory termination of the following parameters:

- A pre-described number of function evaluations is reached;
- The fitness value goes below a pre-defined number;
- CPU time goes over the pre-defined time.

The final stage is the designation of results such as 'the best-so-far individual' or 'non-dominated individuals' on the Pareto fronts.

6.5.2 A Generic Problem for Analysis Algorithm

Figure 6.3 (*Algorithm 2*) shows the flow chart of a generic problem for an analysis algorithm to compute the fitness values of random initial population and off-springs. The algorithm couples evolutionary optimizer HAPMOEA and analysis tools used in aerodynamics (Computational Fluid Dynamics (CFD)), in structure (Finite Element Analysis (FEA)) and in electromagnetics (Radar Cross Section (RCS)). It can be easily modified to accommodate the purposes of any problems. These tools are chosen by the designer and linked to this algorithm. The analysis tool will run only one time for a single point design and several times for multi-points design problems. In addition, two or three inter-disciplinary (CFD, FEA or RCS) analysis tools will run for multi-disciplinary design optimization. The fitness functions will be formulated and computed in this algorithm and then the fitness values are transferred to the optimizer: these steps are described in *Algorithm 1*.

The figure illustrates an example for an airfoil or wing/airfoil sections and plan form shape analysis. The algorithm starts with collecting information for the individual from *Algorithm 1* shown in Fig. 6.2 . This information includes design variables, flow conditions, size of fitness vectors, and the type of analysis depending on the hierarchical multi-population/multi-fidelity/multi-resolution in which an individual belongs to precise (node 0), intermediate (node 1 and 2) or less precise (node 3 to 6) models.

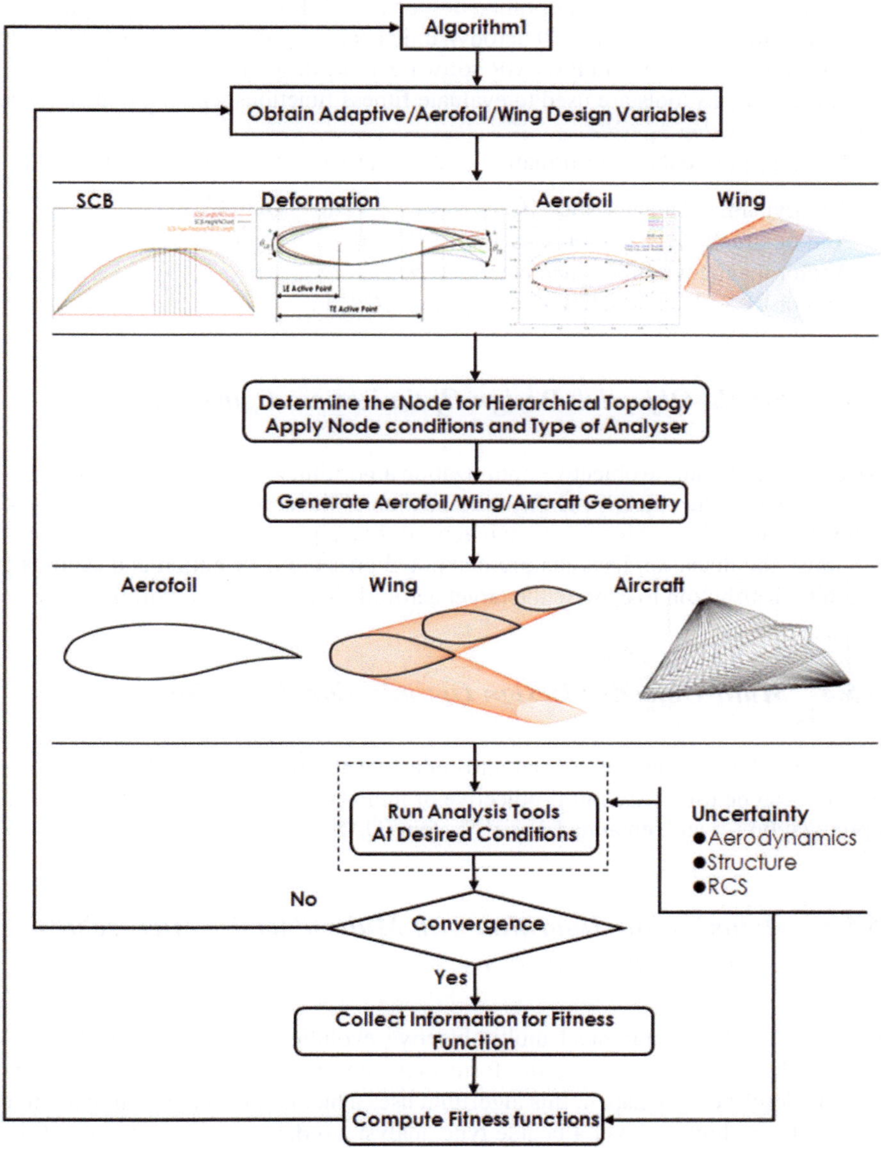

Fig. 6.3 *Algorithm 2*: Generic aeronautical design optimization

The shape of the airfoil is then created by using Bézier or spline curves. The wing geometry is generated based on wing design variables including sweep angles, taper ratios and the break/crank positions.

Analysis tools such as MSES (Euler and boundary layer effects) and NSC2ke (Navier stokes) will be run to compute the aerodynamic properties of airfoil if the problem considers the design optimization of airfoil. For a wing design optimization, the wing is analysed by 3D aerodynamic analysis software: the FLO22 Euler solver with FRICTION and PANAIR software at the desired flow conditions.

These analysis tools are used to evaluate fitness functions which are then transferred to *Algorithm 1*.

The algorithm collects information of the fitness function parameters including

- Aerodynamic characteristics $(C_L, C_D, C_{M_{Pitching}}, C_{M_{Bending}}$ and $C_{M_{Yawing}})$
- And/or structural characteristics $(W_{Wing}, W_{Rib}, W_{Skin}, W_{Spar}$ and $W_{Penalty})$
- And/or manoeuvrability characteristics $(r_{turning}, \omega_{turning})$
- And/or RCS characteristics $(RCS_{Mono}$ and $RCS_{Bi})$.

6.5.3 Single-Objective Design Optimization Algorithm

An example of a single objective optimization algorithm is shown on Fig. 6.4 (*Algorithm 3*) where the objective is the minimisation of the drag coefficient at a desired flight condition. The objective of an inverse design algorithm can be for example to minimise the deviation between pre-computed pressure distribution and candidate pressure distribution to capture the target aerofoil shape or wing aerofoil sections.

6.5.4 Multi-Objective Design Optimization Algorithm

Figure 6.5 shows an example of a multi-objective algorithm (*Algorithm 4*). This algorithm can be used for a multi-objective or multi-point aerodynamic shape design optimization for an Unmanned Aerial System (UAS).

6.5.5 Optimization Algorithm with Multi-objective Hierarchical Evolutionary Algorithms

Integrated with a hierarchical multi-objective evolutionary algorithm (HMOEA) (Sect. 4.3), this analysis takes the form illustrated in Fig. 6.6 (*Algorithm 5*). As already described in Chap. 4, this algorithm uses a hierarchical approach with three levels, on the bottom level a coarse type analysis to direct the exploration, at the top level a more precise model that better describes the physics involved and at an intermediate level, a compromised balance between top and bottom layers is used. Initially the system will specify the design variables $\boldsymbol{x}$, constraints g_i, g_{ij}, and parameters, then it will generate random subpopulations of individuals μ_0 at each layer, then it defines the number of subpopulations (nodes) i and number of hierarchical levels which for simplicity is equal to the number of analysis k. Once these initial populations are generated the algorithm will go through the steps described in the previous algorithms. The scheduler first determines whether given stopping

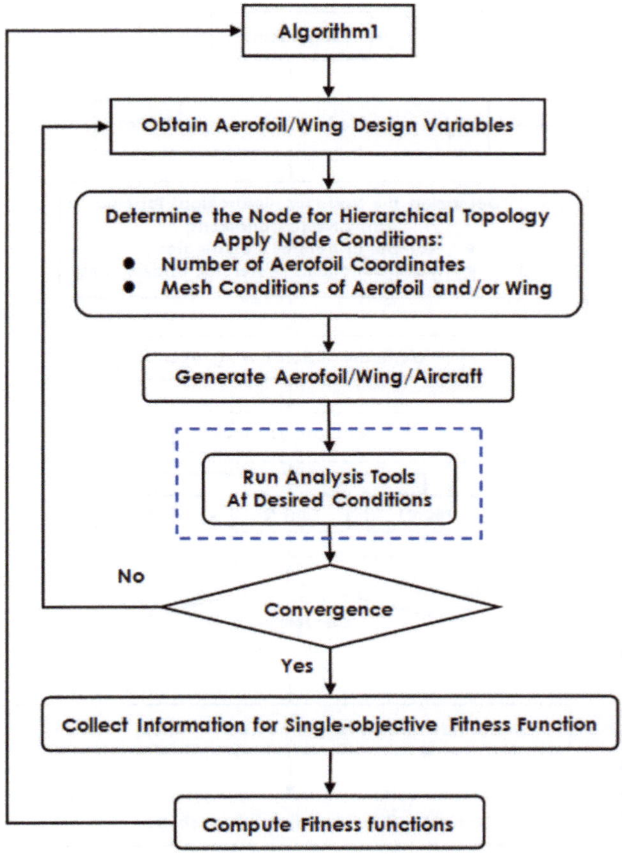

Fig. 6.4 *Algorithm 3*: Single-objective or inverse design optimization flowchart

conditions have been met, and if so the evolutionary loop is exited and the entire procedure is stopped. If no stopping conditions are met, the scheduler updates the asynchronous solver so that further progress can be made. Then the scheduler determines whether or not candidate solutions which have been solved are ready for incorporation into the population. If such solutions exist, the incorporation routine is called and available candidates which now have had a fitness assigned are processed; it receives the individual, ages the population and buffer, performs Pareto tournament selection, deletes the oldest from the buffer and if the acceptance is true it is inserted in the population which I subsequently sorted, it then updates the CMA parameters.

Finally, the scheduler determines whether it is possible to generate more candidates, by polling the asynchronous solver. If this is possible, then the generation routine is called and individuals are generated via the evolutionary operators, by doing recombination, mutation via CMA, checking upper and lower bounds. During

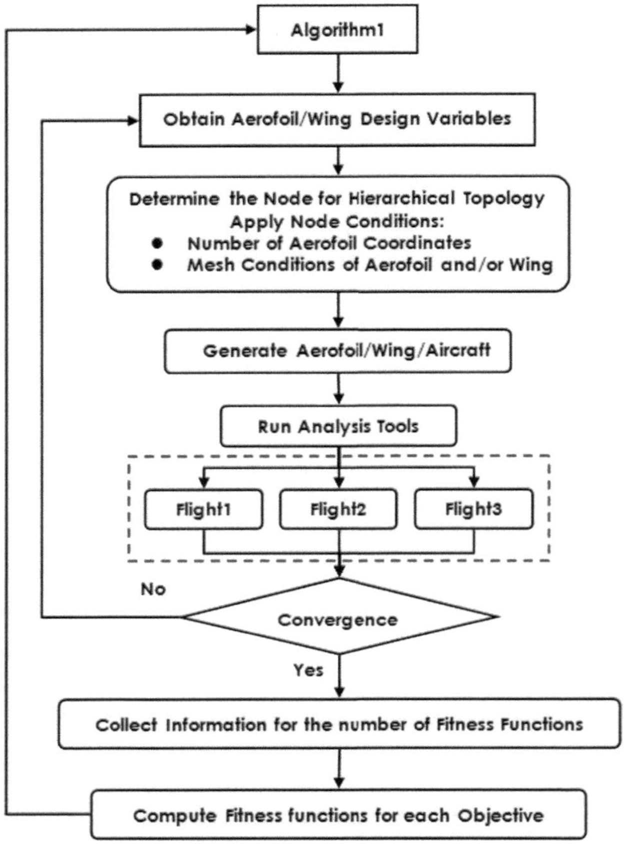

Fig. 6.5 *Algorithm 4*: Multi-point/Multi-objective design optimization

evaluation, the optimizer will take output analysis a_i and parameters p to guarantee satisfaction of constraints and compute the overall fitness function. If the problem is multi-objective the algorithm will find the non-dominated individuals and will calculate the Pareto fronts.

On a hierarchical topology with three levels, when the epoch is finished or the migration criteria is satisfied, the migration phase occurs:

Layer 1 gets best solutions from Layer 2 and re-evaluates them using analysis type one, Layer 2 gets random solutions from Layer 1 gets best solutions from Layer 3 and re-evaluates them using analysis type two, Layer 3 gets random solutions from Layer 2 and re-evaluates them using analysis type three.

This process continues until a stopping condition is reached. These can be for example reaching a limited number of function evaluations, CPU time hours or a prescribed value on the fitness function.

Define design variables $\boldsymbol{x}$ parameters p, and constraints $g : Define(x, p, g_i, g_{ij})$

Define number of subpopulations (nodes) i , hierarchical levels and integrated analysis $k : Define(i, k)$

for all levels initialize subpopulations $(\mu_1^0, \mu_2^0, \mu_3^0, ..., \mu_n^0) : Analysis_k$

Layer 1: Uses Type 1 integrated analysis: $init(\mu_1^0) : Analysis_1$

Layer 2: Uses Type 2 integrated analysis: $init(\mu_2^0, \mu_3^0) : Analysis_2$

Layer 3: Uses Type 3 integrated analysis: $init(\mu_3^0, \mu_4^0, \mu_5^0, \mu_6^0) : Analysis_3$

loop
while stopping condition not met,
Update the asynchronous solver.
if solved individuals are available: incorporate them.
A new individual is received: I_{new}
Age population: For each μ_i in μ; $\mu'_{i\_age} = \mu_{i\_age} + 1$
Age selection buffer: For each b_i in B; $b'_{i\_age} = b_{i\_age} + 1$
Perform tournament selection: $accept = tournament\ (I_{new}, b)$
Delete oldest b_i in selection buffer, insert I_{new}
if accept=true
Perform replacement in μ: *replace* (μ, I_{new})
Sort μ
endif
Update CMA parameters (ref alg 5)
if new individuals can be provided: generate them.
$\boldsymbol{x}_{new} dist(p)) = recombine(\mu)$
Mutate offspring via CMA:
for $i=1,...N$,
$z_i = N(0,1)$
loop
$\mathbf{x}' = \mathbf{x} + \mathbf{BDz}$
Store $\mathbf{z}$ in $\boldsymbol{I}_{new}$
endif
for $i=1,...N$,,
if $x'_i > U_i$ or $x'_i > L_i$ *return* false
loop
return true

Evaluate candidate using specific integrated analysis type: $\lambda^{g+1} = f(\lambda_M^{g+1}) : Analysis_i$

Get output analysis a_i, parameters p, check constraints g and add ...penalty: $\lambda^{g+1} = f(\lambda_M^{g+1}) + penalty$

if Multi-objective: Calculate Pareto fronts: Pareto $Pareto\ Front = Pareto(\lambda_M^{g+1})$

if epoch completed:

Start migration: $\mu_i^{g_k} = mig(\mu_i^{g_k} \rightarrow \mu_{i\pm1}^{g_k}) : Analysis_i$

Layer 1: Receive best solutions from layer 2 reevaluate using Type 1 integrated analysis:
$(\mu_1^{g_k}, \mu_2^{g_k} \rightarrow \mu_0^{g_k}), (\mu_o^{g_k} = f(\mu_o^{g_k})) : Analysis_1$

Layer 2: Receive random solutions from layer 1 and best from layer 3 reevaluate them using type 2 integrated analysis:
$(\mu_0^{g_k} \rightarrow \mu_{1,2}^{g_k}, \mu_{3,45,6}^{g_k} \rightarrow \mu_{1,2}^{g_k}), (\mu_{1,2}^{g_k} = f(\mu_{1,2}^{g_k})) : Analysis_2$

Layer 3: Receive random solutions from layer 2 reevaluates them using type 3 integrated analysis:
$(\mu_{1,2}^{g_k} \rightarrow \mu_{3,45,6}^{g_k}), (\mu_{3,4,5,6}^{g_k} = f(\mu_{1,2}^{g_k})) : Analysis_3$

loop

Fig. 6.6 *Algorithm 5*: Hierarchical multi-objective evolutionary algorithm

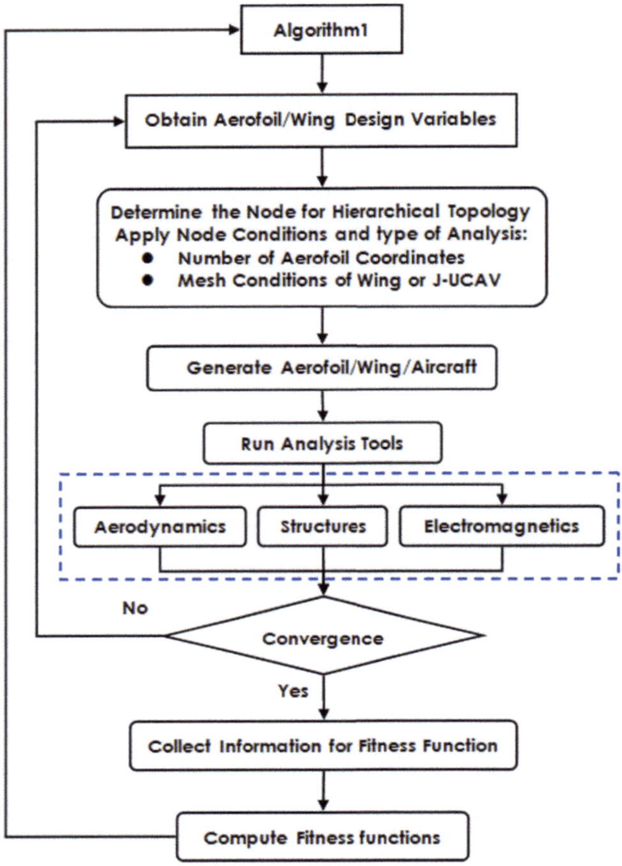

Fig. 6.7 *Algorithm 6*: Multi-disciplinary design optimization

6.5.6 Multi-Disciplinary Design Optimization Algorithm

Chapter 9 considers three main examples of multi-disciplinary design optimization but the reader can expand or use these examples for their own applications:

- Aerodynamic and Structural design optimization.
- Aerodynamic and Electromagnetic design optimization.
- Aerodynamic and Manoeuvrability (structural weight) and Electromagnetic design optimization.

For example, the first test case considers a multi-disciplinary design optimization problem for a Multi mission (MM) -UAV considered (Chaps. 9 and 10). The objectives can be to maximize the aerodynamic performance while minimising the wing structural weight where the wing structure is satisfied for a critical load. More

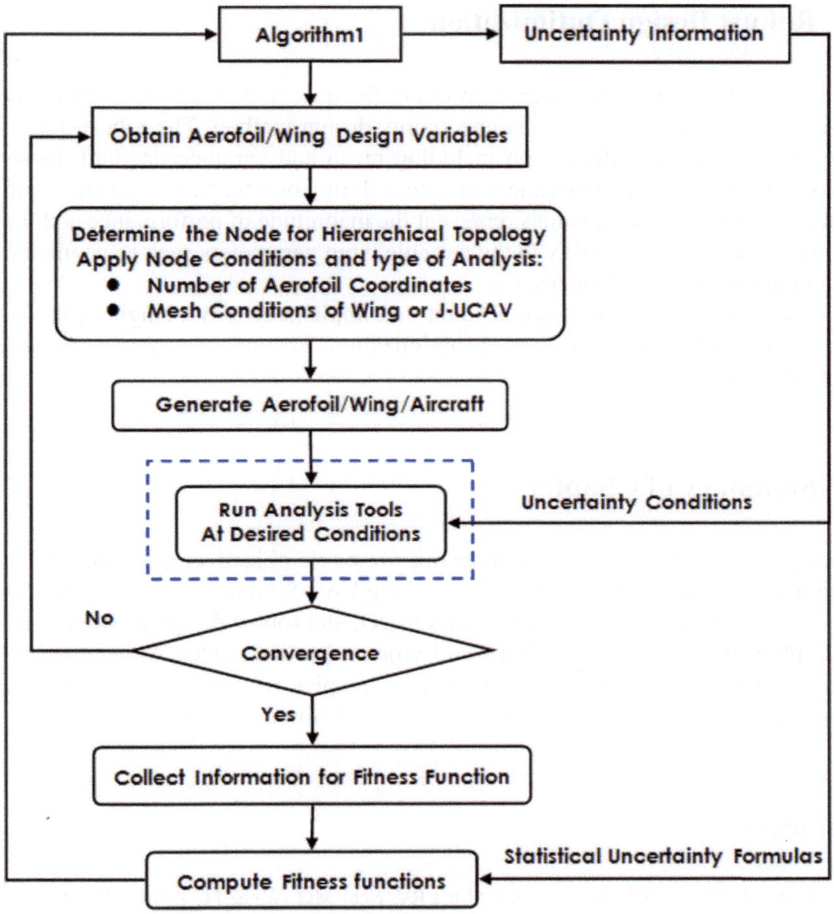

Fig. 6.8 *Algorithm 7*: Uncertainty based design optimization

complex cases are shown in Chap. 10 in which three disciplines including aerodynamics, manoeuvrability and electromagnetism at mono and bi-static radar systems are considered. *Algorithm 6* shown in Fig. 6.7 illustrates the concept: the candidate geometry is analysed by the different disciplines (aerodynamics, structures, electromagnetics) after receiving the information from *Algorithm 1*.

6.6 Robust Design Optimization

Figure 6.8 *(Algorithm 7)* considers improving the quality of a set of solutions by applying the concepts of the robust/uncertainty design method. The robust design method uses statistical information including mean and variance/standard deviation associated to the objective fitness functions during optimization. The mean and variance standard deviation values represent the magnitude of performance and the performance sensitivity stability to the variable input parameters such as geometry, flight conditions, radar frequency, etc.

Uncertainties can be taken into account and implemented in a single or multi-objective or multi-disciplinary design optimization; examples are presented and discussed in Chap. 9.

6.7 Summary of Chapter

This chapter described a generic framework for single objective, multi-objectives and multi-disciplinary design optimization using EAs. Simple example algorithms for a single objective, multi-objectives and MDO, and robust design problems and their implementation were introduced and proposed to the reader. Although quite simplistic for better understanding, these few examples provide an introduction to more complex examples described and detailed in following chapters.

References

1. Bennett JA, Botkin ME, Koromilas C, Lust RV, Neal MO, Wang JT, Zwiers RI (1997) A multidisciplinary framework for preliminary vehicle analysis and design. In: Proceedings of the ICASE/NASA Langley workshop on multidisciplinary design optimization
2. Booker J, Dennis JE, Frank PD, Serafini DB Jr, Torczon V, Trosset MW (1999) A rigorous framework for optimization of expensive functions by surrogates. Struct Optim 17(1):1–13
3. Salas O, Townsend JC (1998) Framework requirements for MDO application development. In: Proceedings of seventh AIAA/USAF/NASA/ISSMO symposium on multidisciplinary analysis and optimisation, St. Louis, Missouri, September 1998
4. Sankar MR, Amitay I, Mujumdar PM, Sudhakar K (2002) MDO framework development-A case study with an elementary model of airborne early warning system optimization. In: Proceedings of the 9th AIAA/USAF/NASA/ISSMO symposium on multidisciplinary analysis and optimization, Atlanta, Georgia, September 2002
5. Weston RP, Townsend JC, Eidson TM, Gates RL (1994) A distributed computing environment for multidisciplinary design. In: Proceedings of fifth AIAA/USAF/NASA/ISSMO symposium on multidisciplinary analysis and optimization, Panama City, Florida, September 7–9 1994
6. Drela M. 6.94 User guide. MIT Aero Astro
7. Drela M A user's guide to MSES V2.3
8. Flores R, Ortega E PUMI: an explicit 3D unstructured finite element solver for the Euler equations. Technical report, CIMNE, Spain

9. Mohammadi B Fluid Dynamics Computation with NSC2KE: an User-Guide Release 1.0, Institut National de Recherche en Informatique et en Automatique, N RT-0164
10. Raymer D Enhancing aircraft conceptual design using multidisciplinary optimization. PhD thesis, KTH, Department of Aeronautics, FLYG 2002-2
11. Raymer D Aircraft design: a conceptual approach, 3rd edn. American Institute of Aeronautics and Astronautics American Institute of Aeronautics and Astronautics
12. McCullers A FLOPS user's guide, release 6.02, NASA Langley Research Center
13. Lee DS, Srinivas K, Gonzalez LF, Periaux J, Obayashi S. Robust multidisciplinary design optimisation using CFD & advanced evolutionary algorithms. In Hafez, MM, Oshima K, Kwak D (eds) Computation fluid dynamics review 2010. World Scientific Publishing Company, Incorporated, Singapore, pp 469–491. ISBN 978-981-4313-36-0
14. Lee DS, Gonzalez LF, Periaux J, Srinivas K Evolutionary optimisation methods with uncertainty for modern multidisciplinary design in aeronautical engineering, notes on numerical fluid mechanics and multidisciplinary design (NNFM 100), 100 volumes NNFM and 40 years numerical fluid mechanics, Chap. 3. Springer-Berlin, Heidelberg, pp 271–284 (ISBN 978-3-540-70804-9)
15. Lee DS, Gonzalez LF, Periaux J UAS Mission Path Planning System (MPPS) using hybrid-game coupled to multi-objective design optimizer. J Dyn Syst Meas Control 132(4). ASME, DS-09-1135 041005-1-11
16. Geist A, Beguelin A, Dongarra J, Jiang W, Manchek R, Sunderam V PVM: parallel virtual machine. A user's guide and tutorial for networked parallel computing. Massachusetts Institute of Technology, USA
17. Jameson A, Caughey D, Newman P, Davis R A brief description of the Jameson Caughey NYU transonic swept-wing computer program FLO22, NASA Technical Memorandum, NASA TM X-73996

Chapter 7
Single Objective Model Test Case Problems

7.1 Overview

There are a number of engineering optimization problems that are single-objective or can be represented by a weighted sum if the weights are known in advance. Inverse design including reconstruction problems are also another type of single-objective problems. This chapter illustrates the application of some of the techniques described in previous chapters to wing airfoil reconstruction design, active flow control bump design, and generic aircraft wing airfoil section design optimizations.

7.2 Wing Reconstruction Using Hierarchial Asynchronous Parallel Multi-Objective Evolutionary Algorthms (HAPMOEA) and Nash-Evolutionary Algorithms

Problem Definition In this example, the pressure distribution of an ONERA M6 wing is reconstructed using two different optimization methods: the first method considers Hierarchical Asynchronous Parallel Multi-Objective Evolutionary Algorithms (HAPMOEA) using three layers with seven populations as shown in Sect. 4.3. Each node (Node 0 ~ Node 6) of HAPMOEA will consider whole airfoil section design variables at root, crank and tip through different resolutions achieved with an Euler flow analysis solver. The second method is Nash-EAs using three Nash-Players with six (6) multi-fidelity sub-players and topology of Nash-EAs is shown in Sect. 4.6.2. Each Nash Player (NP 1 ~ NP 3) considers only airfoil sections at root or crank or tip but Nash players' decisions are coupled via the nonlinear partial differential equations of the flow modelling in the computational domain around the wing.

J. Periaux et al., *Evolutionary Optimization and Game Strategies for Advanced Multi-Disciplinary Design,* Intelligent Systems, Control and Automation: Science and Engineering 75, DOI 10.1007/978-94-017-9520-3_7

Table 7.1 Flow conditions

Description	M_∞	α	Re
Values	0.84	3.06°	11.72×10^6

This problem concerns only one single-objective to minimize the deviation between pre-computed pressures on ONERA M6 wing and candidate solution pressures distribution using Algorithm 3 shown in Sect. 6.5.4.

The fitness function is shown below in (7.1)

$$f = \min\left\{\frac{1}{n+m}\left[abs\left[\sum_{i=1}^{n}\left[\sum_{j=1}^{m}\left(Cp_{\text{Target}_{ij}} - Cp_{\text{Candidate}_{ij}}\right)\right]\right]\right]\right\} \quad (7.1)$$

where n and m represent the number of span wise and chord wise sections respectively.

In this problem, n and m consider twenty one (21) span wise sections and one hundred forty one (141) chord wise points respectively for the precise model in HAPMOEA and Nash-EAs optimizers. The stopping criterion is:

$$\text{Stop } if\,(f \leq 0.0175 \,||\, time_{elapsed} \geq 50h) \quad (7.2)$$

The optimization run will be stopped when the fitness value is lower than 0.0175 and elapsed time is over fifty (50) h. For example, HAPMOEA will be terminated when the first layer (Node 0) reaches pressure deviation less than 0.0175. Nash-EAs will be terminated when all Nash-Players (NP1, NP2 and NP3) have a deviation value less than 0.0175. The flow conditions are shown on Table 7.1.

where Re is abbreviation for Reynolds number

Design Variables The design variables, namely airfoil sections at root, crank and tip are considered with fixed external wing geometry. Each airfoil section has nineteen (19) control points and its design upper and lower bounds are shown in Fig. 7.1.

In total, fifty seven (57) design variables are considered for the reconstruction design problem.

The baseline wing geometry is described on Table 7.2 and Fig. 7.2.

Implementation The Euler FLO22 flow analyser [5] is run to evaluate candidate solutions and the following specific parameters are considered for the HAPMOEA evolutionary optimizer using a hierarchical topology.

1st Layer	Population size of 5 and intermediate recombination with residual 1×10^{-6}
2nd Layer	Population size of 20 and intermediate recombination with residual 1×10^{-5}
3rd Layer	Population size of 30 and intermediate recombination with residual 1×10^{-4}

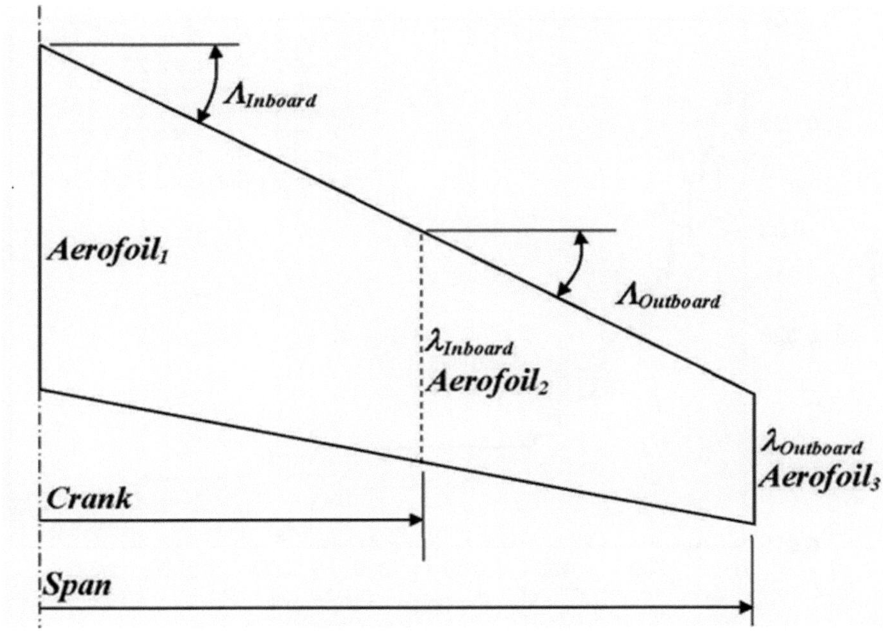

Fig. 7.1 An airfoil section design bounds at root, crank and tip

Table 7.2 The ONERA M6 wing configuration zz

Variables	AR	B	Λ_{In}	Λ_{Out}	λ_{In}	λ_{Out}	$\Gamma_{Overall}$
Values	3.806	2.39	30°	30°	0.781	0.562	0°

Fig. 7.2 The ONERA M6 wing geometry

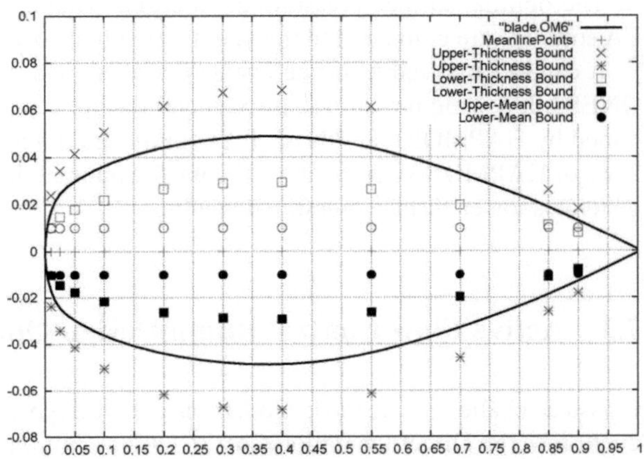

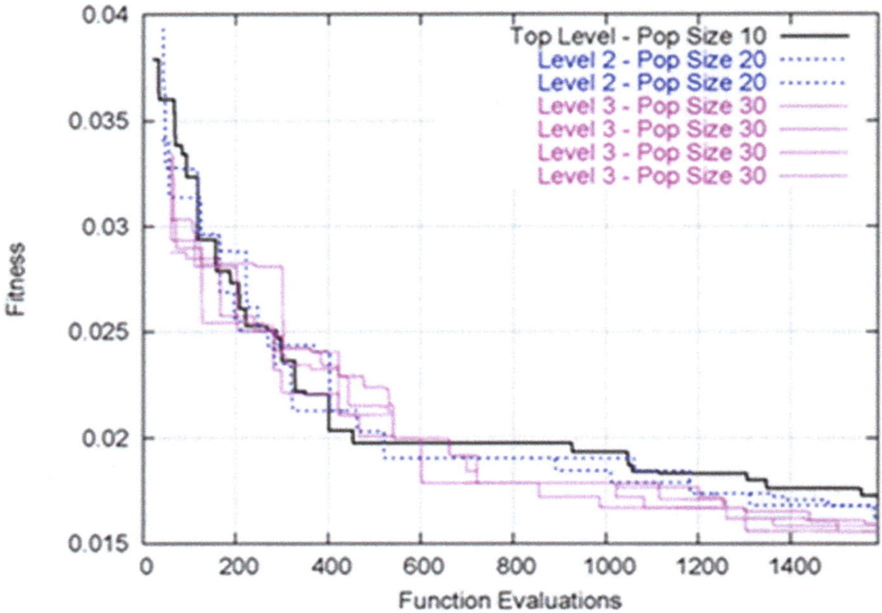

Fig. 7.3 Convergence history obtained with the use of the HAPMOEA software optimizer

Numerical Results Both methods HAPMOEA and Nash-EA are implemented on two 2.4 GHz processors. The optimization convergence history obtained by HAPMOEA and Nash-EAs are shown in Figs. 7.3 and 7.4. The HAPMOEA algorithm was allowed to run approximately 1600 function evaluations ($f_{\text{Node }0}=0.0171$) and took 32 CPU hours. The Nash-EAs algorithm ran approximately 315 function evaluations ($f_{\text{NP1-2-3}}=0.0166$) and took eight (8) CPU hours which is 25 % of the computing cost of HAPMOEA.

The optimal surface pressure distributions obtained by HAPMOEA and Nash-EA software are compared to the target pressure distribution in Figs. 7.5 and 7.6.

It can be observed from the results that there is a good agreement between the ONERA M6 wing pressure (baseline) and the optimized pressure distributions obtained by HAPMOEA and Nash-EAs optimizers. Even though Nash-EAs ran only 25 % of HAPMOEA computational cost, the optimal design obtained by Nash-EAs generates lower surface pressure deviation.

7.3 Active Flow Control Bump Design Optimization

Classical challenges in most modern aircraft design technologies include the reduction of the Direct Operation Cost (DOC) by minimizing the drag during cruise and, maximizing the lift for take-off, landing and manoeuvring. Various missions

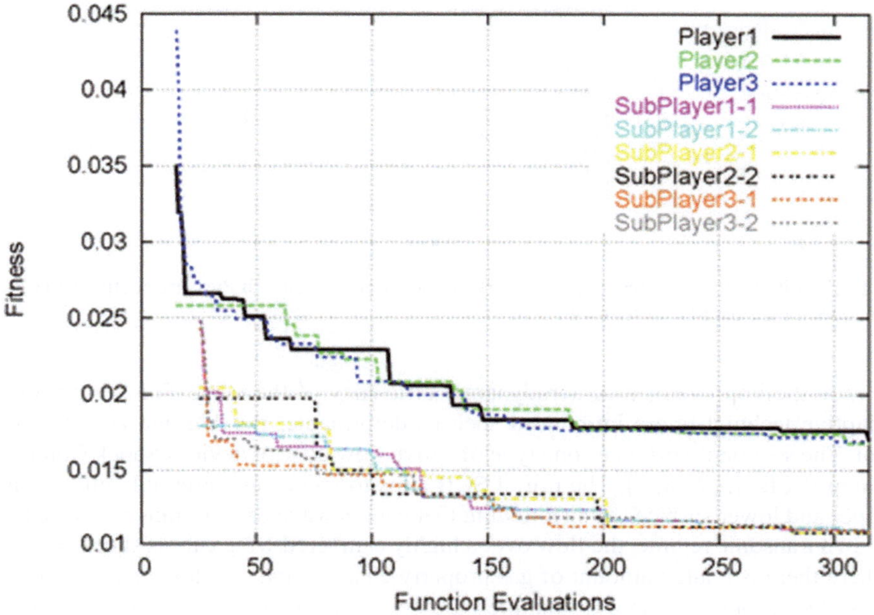

Fig. 7.4 Convergence history obtained by the Nash-EAs software optimizer

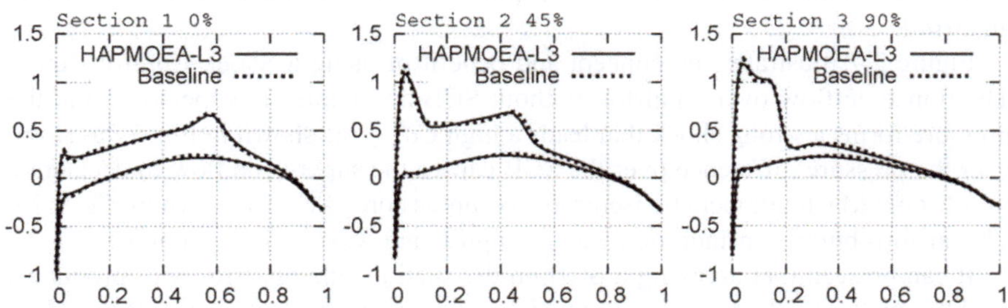

Fig. 7.5 Surface Pressure comparison between ONERA M6 and the optimal wing obtained by HAPMOEA software

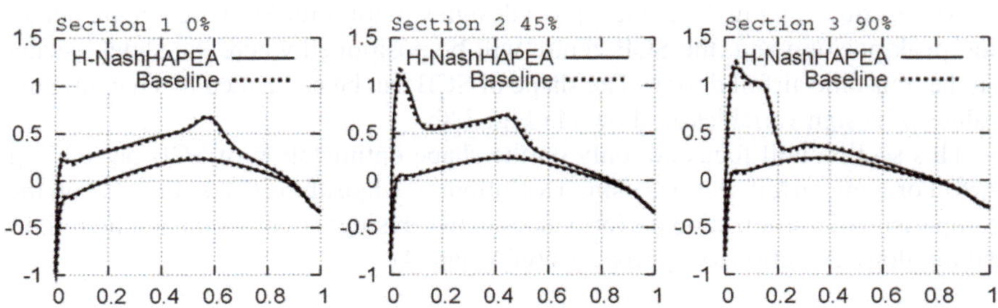

Fig. 7.6 Pressure comparison between ONERA M6 and optimal wing obtained by Nash-EAs software

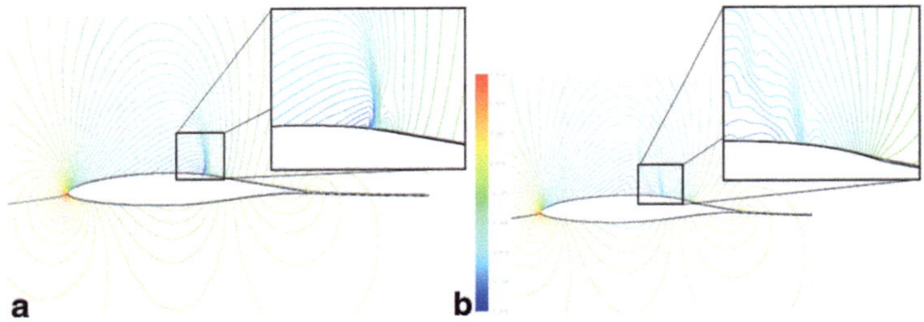

Fig. 7.7 **a** Pressure coefficient (*Cp*) contours for the baseline design. **b** Pressure coefficient (*Cp*) contours for the baseline design with SCB

require the improvement of aerodynamic efficiency of the wing. For this reason, adaptive techniques are introduced such as deformable, flexible and active control. The section investigates one type of active flow control devices Shock Control Bump (SCB) [2, 6, 8, 7]. The use of SCB aims to reduce compressible effects on upper and lower surfaces of airfoil/wing that causes wave drag at transonic speeds.

In a transonic regime, the flow over a highly cambered wing causes shock waves where there is a large amount of gas property changes and the flow becomes irreversible. Through the shock, total pressure decreases and entropy increases which means there is a loss of energy and therefore an increment in wave drag. To resolve this problem, Ashill et al. [1] introduced the concept of a transonic bump named also Shock Control Bump (SCB) by adapting its geometry (shape and position) on an airfoil.

Figure 7.7 illustrates the concept and benefit of using a Shock Control Bump. The transonic flow over an airfoil without SCB accelerates to supersonic and the pressure forms a strong shock that leads a high Cd_{wave} as shown in Fig. 7.7a. However the pressure difference over the SCB causes the supersonic flow to decelerate to subsonic Mach numbers by isentropic compression waves as shown in Fig. 7.7b. This airfoil- bump configuration leads to significant wave drag reduction.

It can be seen that applying a SCB on the suction side of airfoil will produce a slightly thicker thickness ratio (t/c) which causes an increment of the viscous drag ($Cd_{Viscous}$); however it is still beneficial due to a Cd_{Wave} reduction especially when the Mach number is higher than the critical Mach number where the shock starts appearing.

As described on Fig. 7.8a, the typical design variables for SCB are length, height and peak position and, the SCB centre will be positioned where the shock occurs on the transonic airfoil design. The shape of SCB can be controlled by changing the values of design variables as shown in Fig. 7.8b.

This section will focus not only on the shape optimization of SCB on existing airfoil or wing airfoil sections using Evolutionary Algorithms (EAs) but also on the comparison of optimization performances between single population and hierarchical topology evolutionary approaches (cf. Chap. 4).

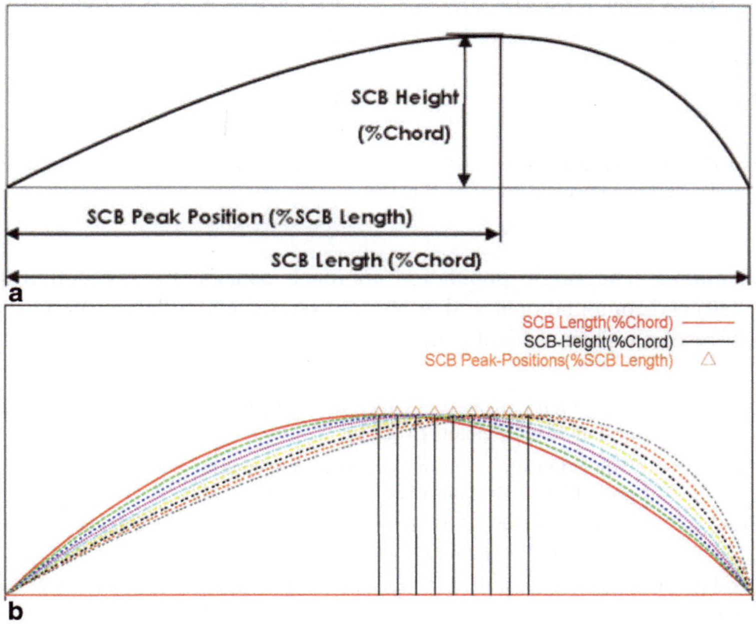

Fig. 7.8 **a** Design components of Shock Control Bump. **b** Geometries of SCB with various peak positions

7.3.1 Suction Side SCB Design Optimization

Problem Definition This test case considers a single objective optimization problem to minimize the total drag at a fixed lift condition by applying a SCB on the upper surface of the RAE 2822 airfoil operating at flow conditions $M = 0.77$, $C_l = 0.524$ and $Re = 17.93 \times 10^6$. Two optimization approaches are considered: the first test uses a single population approach while the second test implements multi-fidelity/resolution based hierarchical populations with three levels. The fitness function is shown in (7.3).

$$f(U\_SCB) = \min(Cd_{Total}) \tag{7.3}$$

where $Cd_{Total} = Cd_{Viscous} + Cd_{Wave}$ and *U_SCB* represents a set of design variables for SCB on the upper surface of RAE 2822. The problem considers only SCB geometry constraints.

Design Variables The design upper-lower bounds for the SCB geometry are given on Table 7.3. The centre of the SCB will be placed at the shock where the flow speed transits from supersonic to subsonic regime.

Table 7.3 Shock Control Bump (SCB) design variables and bounds for SCB geometry

Bounds\Design Variables	SCB_L (%c)	SCB_H (%c)	SCB_P (%SCB_L)
Lower	15.0	0.35	50.0
Upper	30.0	0.65	70.0

SCB_L, SCB_H, SCB_P represent SCB length, height, and peak position respectively. The peak position of SCB is given in percentage of SCB length

Implementation The single population approach considers a population size of twenty (20) and computational grid of 36 × 213 points. The following conditions are for multi-resolution hierarchical populations are:

1st Layer	Population size of 10 and intermediate recombination using a computational grid of 36 × 213 points (Node 0)
2nd Layer	Population size of 20 and intermediate recombination using a computational grid of 24 × 131 points (Node 1, Node 2)
3rd Layer	Population size of 20 and intermediate recombination using a computational grid of 36 × 111 points (Node 3 ~ Node 6)

These grid conditions produce less than 5 % accuracy error compared to precise model at the 1st layer (Node 0)

Numerical Results As illustrated in Fig. 7.9, the algorithm was allowed to run for 24 h and for 2210 function evaluations using single population approach while the hierarchical topology approach was run for 24 h and 1381 of function evaluations. The convergence obtained by single population approach occurred at 815 function evaluations (9.0 h) while the hierarchical topology approach converged at 430 function evaluations after 7.4 h. More precisely, the multi-resolution hierarchical topology approach saves 18 % of the CPU time when compared to the single population approach.

Table 7.4 compares the aerodynamic characteristics obtained by the baseline design RAE 2822 airfoil and the baseline design combined with an upper SCB. Even though the upper SCB increases viscous drag by 9 %, it saves the wave drag by 71 % which leads 26 % of total drag reduction. This optimal SCB improves *L/D* by 35.5 %.

Both single and hierarchical topology approaches capture the same shape of SCB as shown in Table 7.5. Figure 7.10 compares the geometry of the baseline design and baseline with optimal SCB.

Figure 7.11 compares drag distribution obtained by the baseline design and with upper SCB along C_l sweep [0.2:0.7] with fixed $M_\infty = 0.77$, $\text{Re} = 17.93 \times 10^6$. It can be seen that the upper SCB causes a slightly higher total drag when C_l is lower than 0.325. This is due to the fact that applying SCB increases the thickness of airfoil. However, the upper SCB starts to produce a lower total drag compared to the baseline design when C_l is higher than 0.325. In addition, it can be noticed that applying SCB reduces the difference between maximum and minimum drag along lift coefficient which means that the SCB reduces the drag sensitivity with variability of C_l.

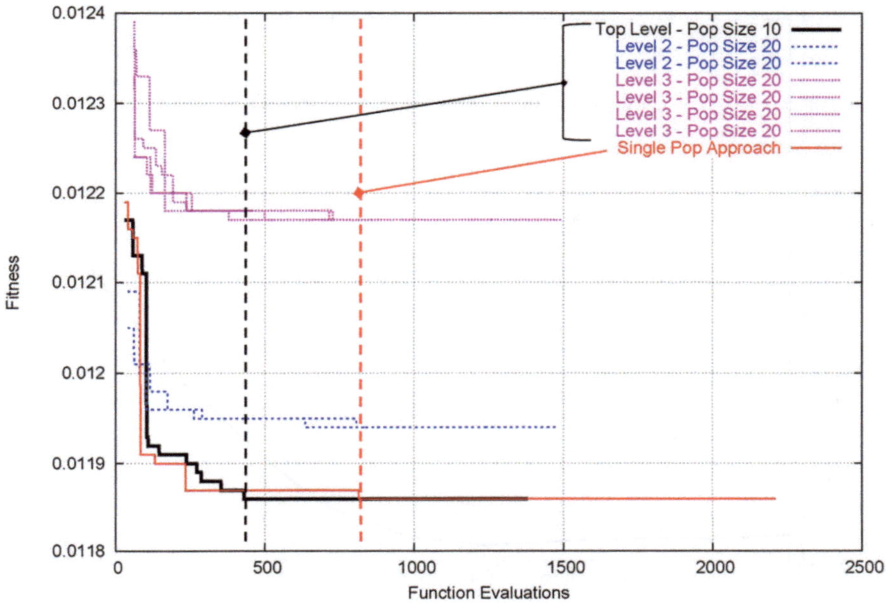

Fig. 7.9 Convergence history for SCB on the upper surface of the RAE 2822 airfoil

Table 7.4 Comparison of aerodynamic characteristics obtained by the clean and optimal aerofoils

Aerofoils	Cd_{Total}	$Cd_{Viscous}$	Cd_{Wave}	L/D
Baseline	0.0160	0.0090	0.007	32.607
with SCB	0.0118 (−26%)	0.0098 (+9.0%)	0.002 (−71%)	44.182 (+35.5%)

Note: Cl_{Fixed} = 0.524. The chosen baseline model is the RAE 2822 airfoil

Table 7.5 Optimal SCB design components

Design components	SCB_L (%c)	SCB_H (%c)	SCB_P (%SCB_L)
Optimal SCB	29.28	0.62	67.35

The optimal SCB design components obtained by single population and hierarchical approaches are the same

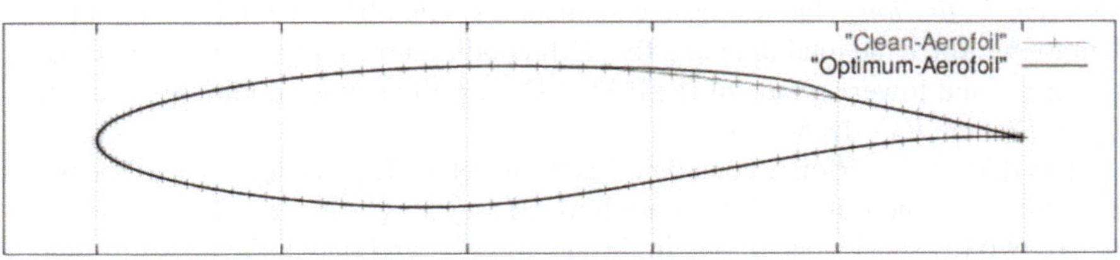

Fig. 7.10 Comparison of clean and optimal airfoil with upper SCB (Clean-airfoil represents RAE 2822 airfoil and RAE 2822 with upper SCB is labelled as Optimal-airfoil)

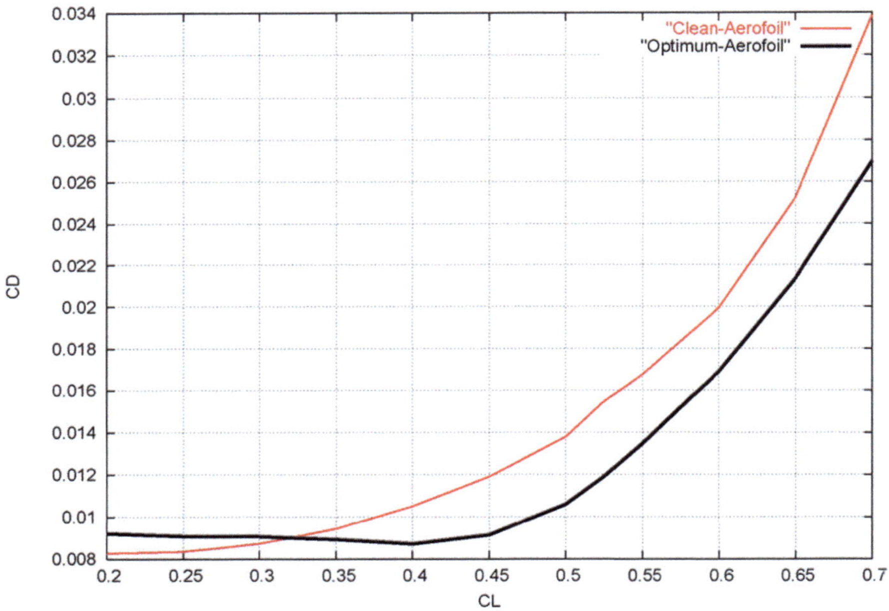

Fig. 7.11 Drag coefficient (*Cd*) distribution obtained by baseline alone and with SCB at increasing *Cl* with fixed $M_\infty = 0.77$, $\text{Re} = 17.93 \times 10^6$

Figure 7.12 compares drag distribution obtained by the baseline design and with upper SCB along Mach sweep [0.55:0.85] with $C_l = 0.524$, $\text{Re} = 17.93 \times 10^6$. It can be seen that the upper SCB starts to produce lower total drag when the Mach number is higher than 0.75 even though it generates slightly higher total drag when the Mach is lower than 0.75. In addition, it can be noticed that applying SCB reduces the interval between maximum and minimum drag along the Mach numbers which means that SCB reduces the drag sensitivity with variation of Mach number.

7.3.2 Suction and Pressure Sides SCB Design Optimization

Problem Definition This test case considers a single objective optimization problem to minimize the total drag at a fixed lift condition by applying two (2) SCBs on the upper and lower surface of the RAE 2822 airfoil at flow conditions $M_\infty = 0.8$, $C_l = 0.175$ and $\text{Re} = 18.63 \times 10^6$.

The RAE 2822 airfoil is considered as the baseline design (Fig. 7.13a). The problem considers the critical flow conditions; $M_\infty = 0.8$, $C_l = 0.175$, Re = 18.63 × 106 where two (2) shocks occur, one in the suction side and one in the pressure side of RAE 2822 airfoil as shown in Fig. 7.13b).

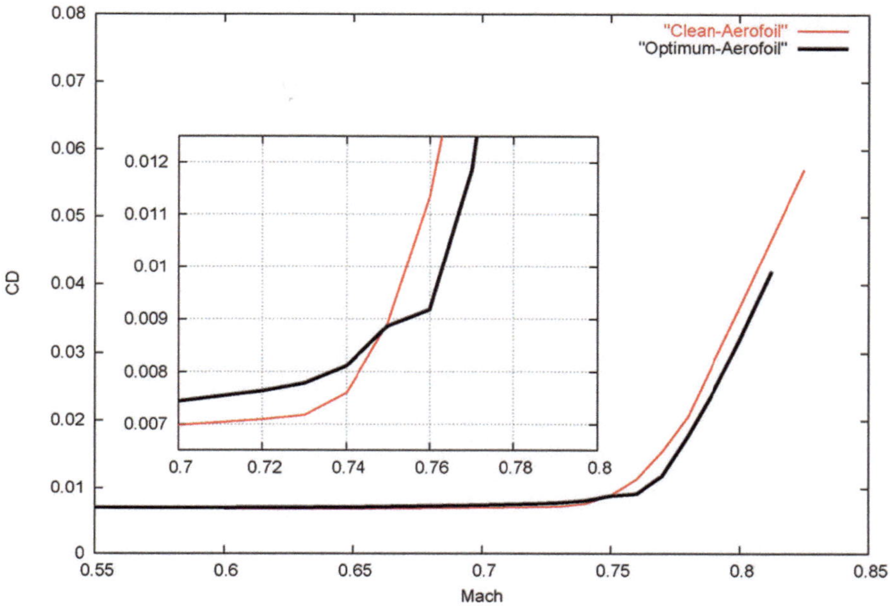

Fig. 7.12 Drag coefficient (Cd) distribution obtained by clean and U-SCB aerofoil at increasing M_∞ with fixed $C_l = 0.524$, $Re = 17.93 \times 10^6$

It can be seen that the upper and lower sonic points occur at 70.6 % and 51.3 % of chord respectively. In the following subsections, double SCB design optimization using HAPMOEA and Hybrid-Game are conducted to minimize the total drag (Cd_{Total}). The aerodynamic analysis tool; MSES will run twice for each function evaluation; the first run will analyse the upper SCB and the second run will analyse both the upper and lower SCB.

The fitness function is the same as the one defined in (7.3). Two test cases are conducted using single population approach and multi-fidelity/resolution based hierarchical populations with three levels to compare optimization performances. The problem considers only SCB geometry constraints.

Design Variables Table 7.3 shows the design variable upper and lower bounds for defining the SCB geometry. Six (6) design variables are considered in total. The centre of both upper and lower SCBs will be placed at the shock position where the flow speed transits from supersonic to subsonic regime.

Implementation The single population approach considers a population size of 20 and computational grid of 36×213 points. The following conditions are used for multi-resolution hierarchical populations.

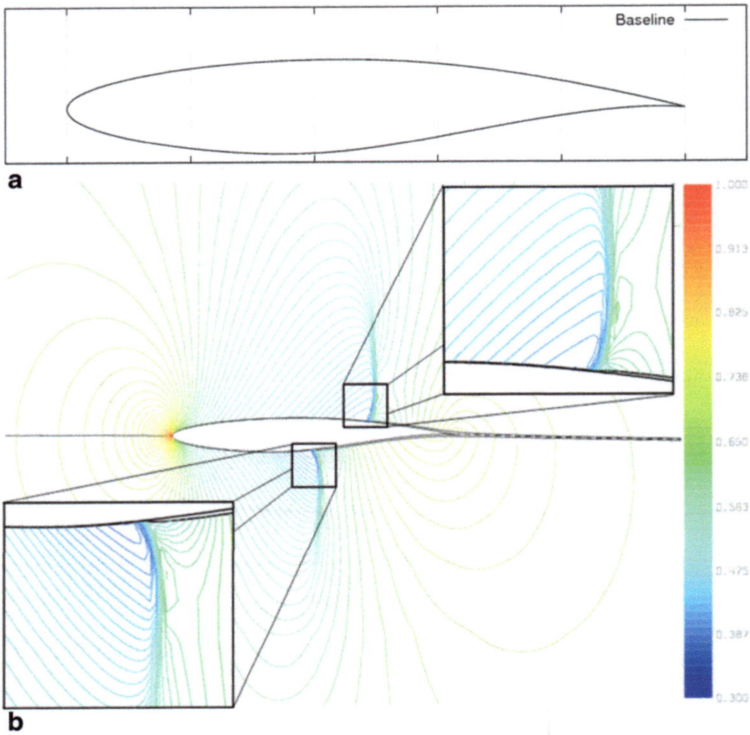

Fig. 7.13 **a** Baseline design with RAE 2822 airfoil. **b** Pressure ratio (P/P_0) contours obtained by the baseline design

1st Layer	Population size of 10 and intermediate recombination using a computational grid of 36 × 213 points (Node 0)
2nd Layer	Population size of 20 and intermediate recombination using a computational grid of 24 × 131 points (Node 1, Node 2)
3rd Layer	Population size of 20 and intermediate recombination using a computational grid of 24 × 111 points (Node 3 ~ Node 6)

These grid conditions produce less than 5 % accuracy error compared to precise model at the 1st layer (Node 0)

Numerical Results The algorithm was allowed to run 3900 function evaluations (72 h) using the single approach while the hierarchical topology approach was run for 527 function evaluations (28 h) as shown in Fig. 7.14. The convergence obtained by single population approach occurs after 1700 function evaluations (31 h) while the hierarchical topology approach converged at 370 function evaluations after 18 h. In other words, the multi-resolution hierarchical topology approach saves 42 % computational time when compared to the single population approach.

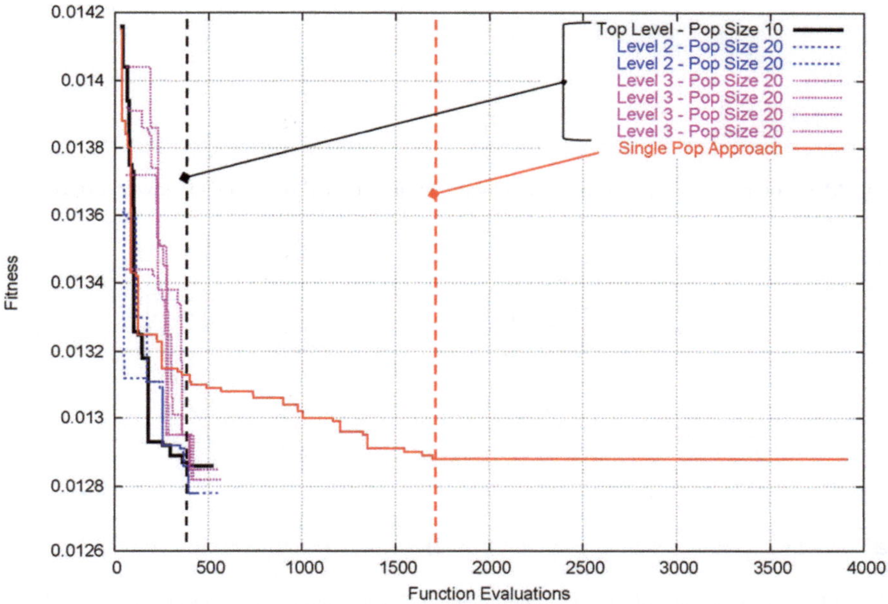

Fig. 7.14 Convergence history for SCB on upper & lower surfaces of the RAE 2822 airfoil

Table 7.6 Comparison of aerodynamic characteristics obtained by clean and the optimal airfoils

Aerofoils	Cd_{Total}	$Cd_{Viscous}$	Cd_{Wave}	L/D
Baseline	0.0192	0.0103	8.87×10^{-3}	9.119
with SCB	0.0129 (−33 %)	0.0106 (+3.0 %)	2.22×10^{-3} (−75 %)	13.587 (+49.0 %)

Cl_{Fixed} = 0.175. Baseline model represents RAE 2822 aerofoil

Table 7.7 Optimal SCB design components

SCBs	Single population			Hierarchical topology approach		
	SCB_L (%c)	SCB_H (%c)	SCB_P (%SCB_L)	SCB_L (%c)	SCB_H (%c)	SCB_P (%SCB_L)
Upper	29.21	0.62	66.95	29.94	0.63	67.50
Lower	29.81	0.64	67.50	29.88	0.64	67.35

Table 7.6 compares the aerodynamic characteristics obtained by the baseline design (RAE 2822 airfoil) and the baseline design with upper and lower SCBs. Even though the upper SCB increases 3 % of viscous drag, it saves the wave drag by 75 % which leads 33 % of total drag reduction. This optimal SCB improves *L/D* by 49 %.

Both single and hierarchical topology approaches capture similar SCB shapes with their design components described on Table 7.7. Figure 7.15 compares the

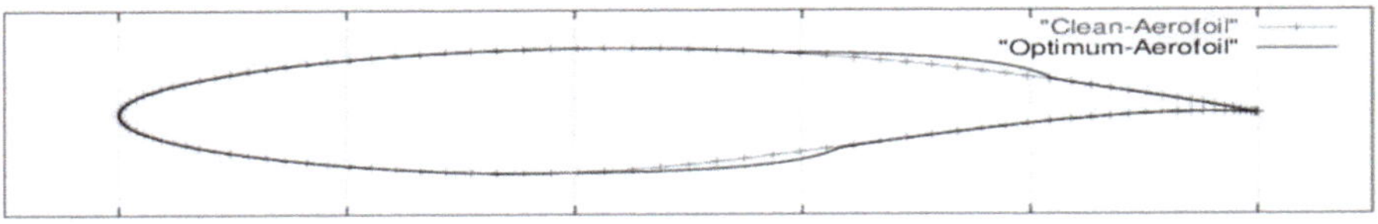

Fig. 7.15 Comparison of clean and aerofoil with optimized SCB on upper and lower surfaces

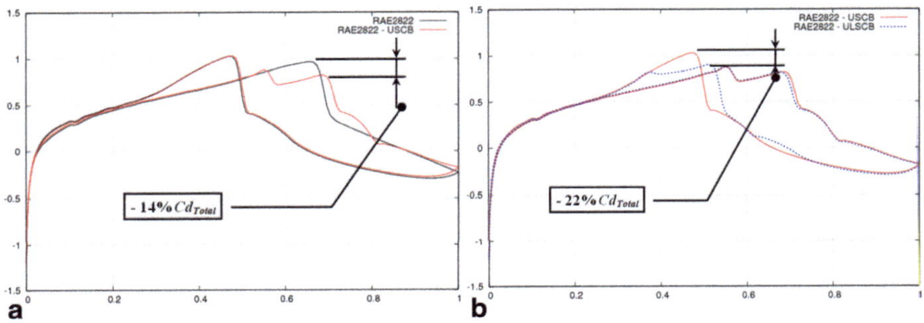

Fig. 7.16 **a** Comparison of pressure coefficients (Cp) distribution obtained by baseline (RAE 2822 airfoil) and with the optimized SCB (RAE 2822-USCB). **b** Comparison of pressure coefficients (Cp) distribution obtained by RAE 2822 airfoil with upper SCB (RAE 2822-USCB) and with upper and lower SCBs (RAE 2822-ULSCB)

geometry of baseline design and with an optimal SCB shape obtained by the multi-resolution hierarchical topology approach.

Figures 7.16a and b compare pressure distributions obtained by the baseline design (RAE 2822 airfoil), with upper SCB and also with upper—lower SCBs. It can be seen that applying SCB on the suction side of baseline design reduces 14 % of total drag (Fig. 7.13a) when compared to the baseline design, and it can be noticed also that both suction and pressure sides SCB reduce 22 % of total drag when compared to the baseline design with only upper SCB.

Figure 7.17 compares drag distribution obtained by the baseline design and with upper & lower SCB along C_l sweep [0.1:0.6] with fixed $M_\infty = 0.8$, Re $= 18.63 \times 10^6$. It can be seen that the upper & lower SCB produces lower drag at the range of C_l [0.1:0.6] even though SCB increases the thickness of aerofoil.

Figure 7.18 compares drag distribution obtained by the baseline design and with upper- lower SCB along Mach sweep [0.55:0.85] with $C_l = 0.175$, Re $= 18.63 \times 10^6$. It can be seen that the upper—lower SCB starts to produce lower total drag when the Mach number is higher than 0.77 even though it causes slightly higher total drag when the Mach is lower than 0.77.

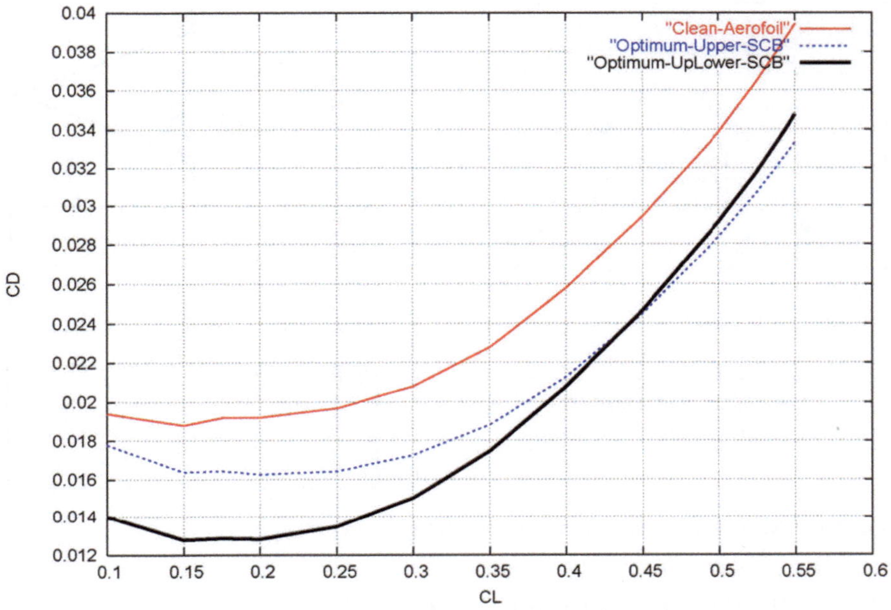

Fig. 7.17 Drag coefficients (*Cd*) distribution obtained by clean and UL-SCB airfoils at increasing C_l with fixed $M\infty = 0.8$, Re $= 18.63 \times 10^6$

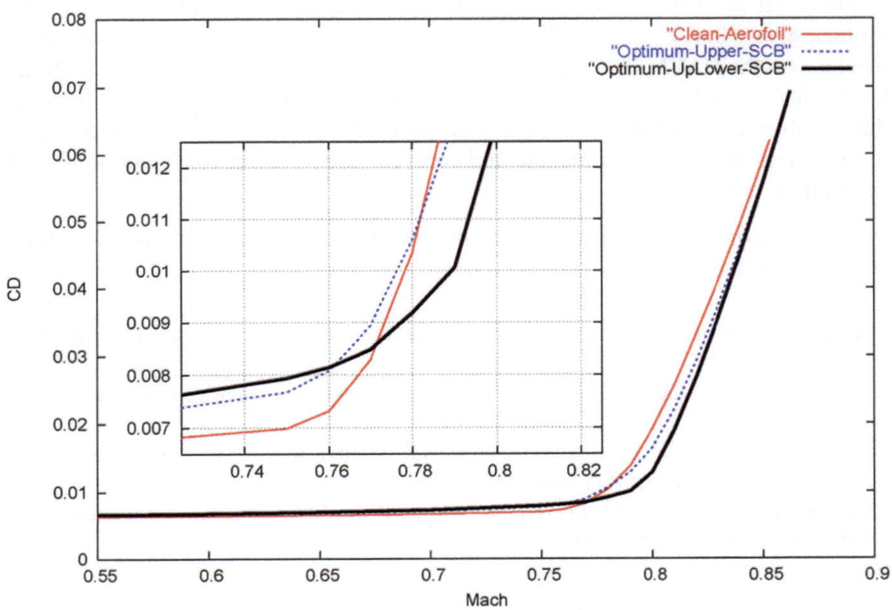

Fig. 7.18 Drag coefficients (*Cd*) distribution obtained by clean and UL-SCB airfoils at increasing $M\infty$ with fixed $C_l = 0.175$, $Re = 18.63 \times 10^6$

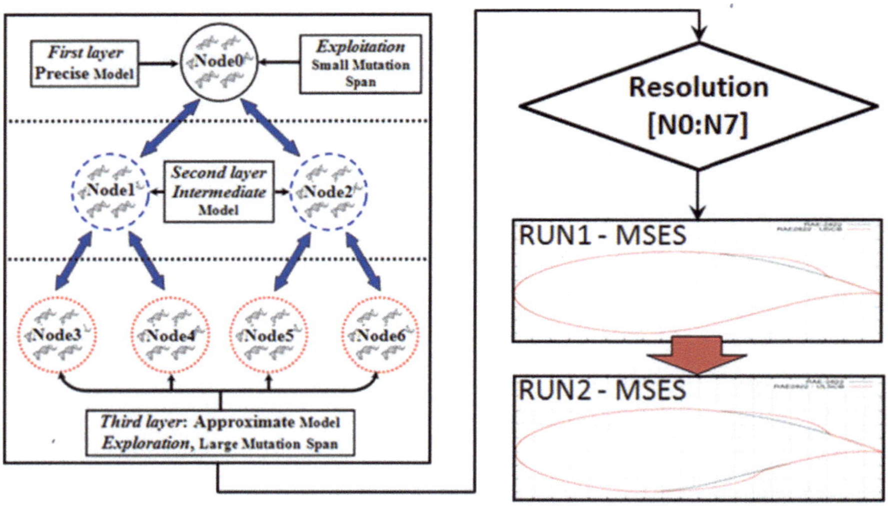

Fig. 7.19 Evaluation mechanism with HAPMOEA software

7.3.3 Double SCB Design using HAPMOEA

Problem Definition The problem considers a single-objective SCB design optimization using HAPMOEA to minimize the total drag defined in (7.6) at flow conditions; $M_\infty = 0.8$, $C_l = 0.175$, $Re = 18.63 \times 10^6$. Figure 7.19 shows the evaluation mechanism for HAPMOEA which consists of hierarchical multi-population (Node 0 ~ Node 6) based on multi-resolution. Each individual will be analysed twice with the aerodynamic analysis tool to evaluate the double SCB design at different resolution layers (precise, intermediate and coarse). The problem considers only SCB geometry constraints.

Implementation The following conditions are for multi-resolution/population hierarchical populations.

1st Layer	Population size of 10 and intermediate recombination using a computational grid of 36×213 points (Node 0)
2nd Layer	Population size of 20 and intermediate recombination using a computational grid of 24×131 points (Node 1, Node 2)
3rd Layer	Population size of 20 and intermediate recombination using a computational grid of 24×111 points (Node 3 ~ Node 6)

These grid conditions (2nd and 3rd layer) produce less than a 5 % accuracy error when compared to the precise model on the 1st layer (Node 0)

Numerical Results As illustrated on Fig. 7.20, the algorithm was allowed to run for 12 h and 1295 function evaluations ($f = 0.01285$) using a single 4×2.8 GHz processor.

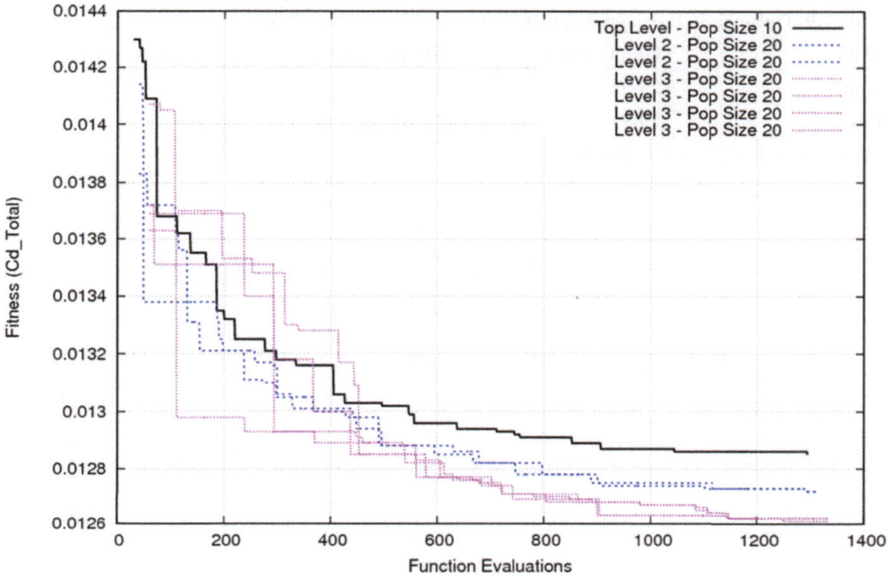

Fig. 7.20 Convergence history obtained by HAPMOEA software

Table 7.8 Aerodynamic characteristics obtained by HAPMOEA

SCB	Cd_{Total}	Cd_{Wave}	*L/D*
Baseline	0.01918	0.00886	9.13
with SCB	0.01285 (−33 %)	0.00225 (−75 %)	13.59 (+49 %)

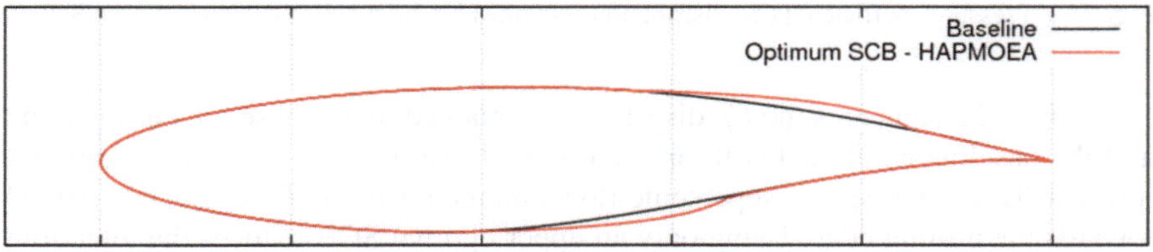

Fig. 7.21 Comparison between the baseline design and baseline with SCB geometries

Table 7.8 compares the aerodynamic characteristics obtained by the baseline design (RAE 2822 airfoil) and baseline with added optimal SCB. It can be seen that the baseline design with optimal SCB can reduce wave drag by 75 % which leads to 33 % of total drag. Such drag reduction improves 49 % of the lift to drag ratio (L/D).

Figure 7.21 compares the shape of baseline design and baseline with SCB geometry. The optimal SCBs obtained by HAPMOEA software are located between (0.5604, 0.0595) and (0.8506, 0.0270) on the suction side and between (0.3633, −0.0591) and (0.6618, −0.0287) on the pressure side. Design parameters for the optimal SCB are shown on Table 7.9.

Table 7.9 Optimal SCB design parameters obtained by HAPMOEA

SCB	Length (%c)	Height (%c)	Peak position
Upper	29.02	0.603	83.8
Lower	29.85	0.643	84.8

Peak Position is in % of SCB length

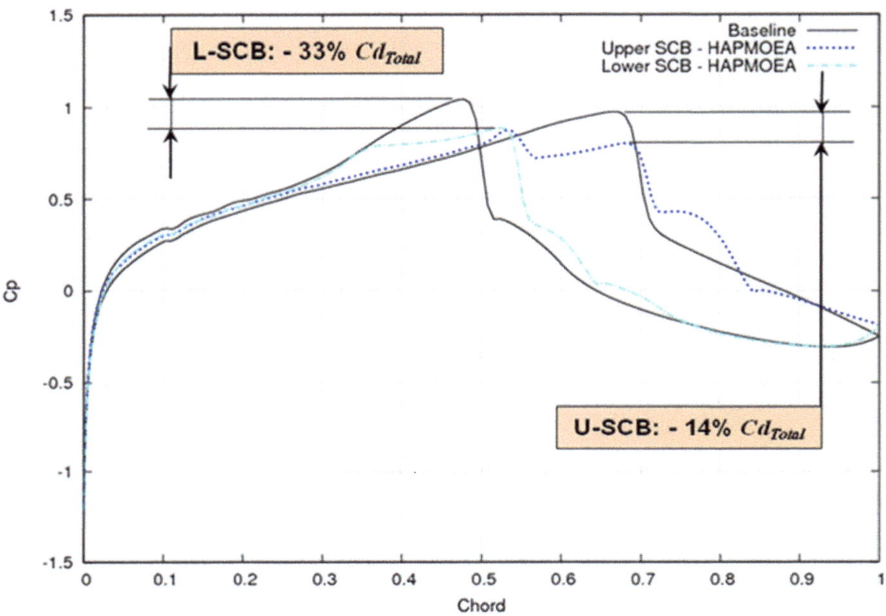

Fig. 7.22 Pressure coefficient (*Cp*) distributions obtained by the baseline design and with SCB

Figure 7.22 compares the *Cp* distributions obtained by the baseline design alone and the baseline combined with optimal SCB. It can be seen that both upper and lower SCB decelerates the supersonic flow and the position of the shock is moved towards the trailing edge. Using only an upper optimal SCB reduces the total drag by 14% while applying both upper and lower optimal SCB produces 33% lower total drag when compared to the baseline design.

7.3.4 Double SCB Design using Hybridized Game

Problem Definition The problem considers a single-objective SCB design optimization using Hybrid-Game (described in Sect. 4.6.3 Chap. 4) employing one Global-Player and two Nash-Players. Therefore the fitness function Eq. (7.3) is decomposed in (7.4, 7.5 and 7.6) expressions due to Nash- Players:

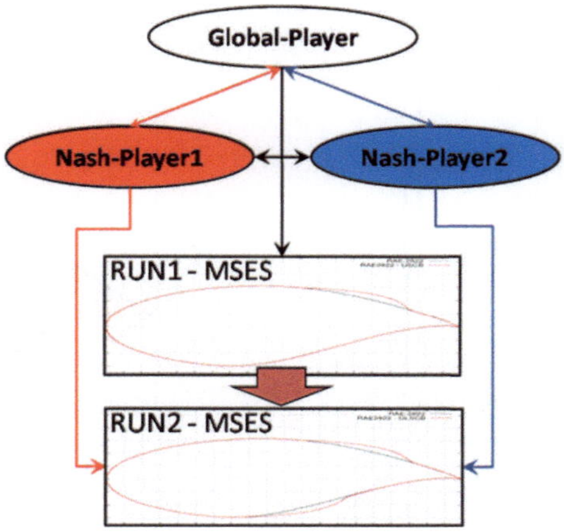

Fig. 7.23 Evaluation mechanism using Hybridized Games

$$f_{GP}(U\_SCB, L\_SCB) = \min(Cd_{Total}) \tag{7.4}$$

$$f_{NP1}(U\_SCB, L\_SCB^{*}) = \min(Cd_{Total}) \tag{7.5}$$

$$f_{NP2}(U\_SCB^{*}, L\_SCB) = \min(Cd_{Total}) \tag{7.6}$$

where U_SCB* and L_SCB* represent elite SCB designs obtained by Nash-Player 1 and Nash-Player 2 respectively.

Figure 7.23 shows the evaluation mechanism for Hybridized Games with employs three players: one Global-Player and two Nash-Player 1 and Nash Player 2. The Global-Player runs the aerodynamic analysis tool twice since its optimization domain includes both upper and lower SCB. However, the analysis tool is run once for each Nash-Player 1 and 2 due to the Nash-Game decomposition characteristics. Nash-Player 1 (7.5) will only optimize a SCB on the suction side with an elite lower SCB obtained by Nash-Player 2 while Nash-Player 2 will optimize a SCB on the pressure side using an elite upper SCB design from Nash-Player 1 shown in (7.6). The elite designs obtained by Nash-Players are seeded in the Global-Player population. This optimization mechanism allows a significant speed up of the optimization procedure used by the Global-Player.

Implementation The conditions for Hybridized Games are:

Global Layer	Population size of ten (10) and intermediate recombination with a computational grid of 36×213 points
Nash-Player 1	Population size of ten (10) and intermediate recombination with a computational grid of 36×213 points
Nash-Player 2	Population size of ten(10) and intermediate recombination with a computational grid of 36×213 points

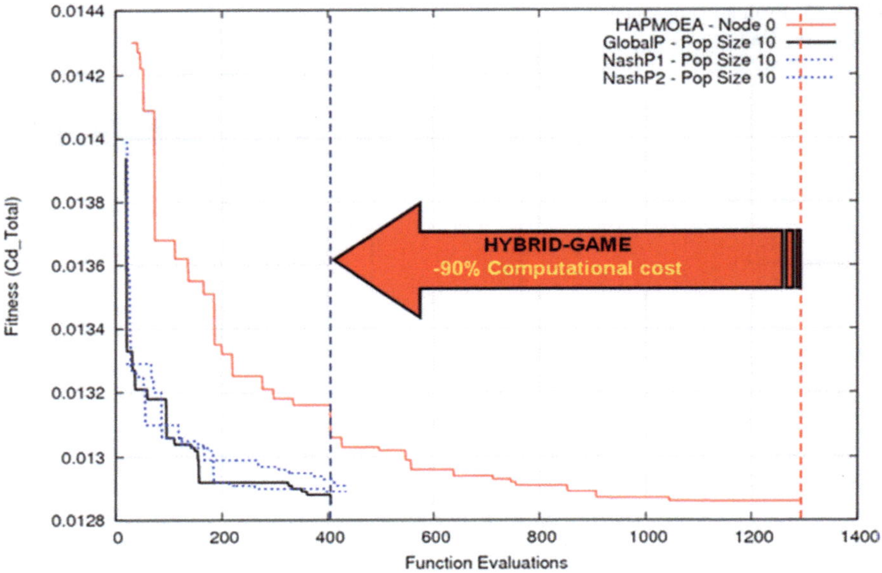

Fig. 7.24 Convergence history obtained by Hybridized -Games

Table 7.10 Aerodynamic characteristics obtained by Hybrid-Game

SCB	Cd_{Total}	Cd_{Wave}	*L/D*
Baseline	0.01918	0.00886	9.13
with SCB	0.01282 (−33 %)	0.00214 (−76 %)	13.65 (+50.0 %)

Numerical Results The algorithm is run for 1.15 h and 406 function evaluations to reach the same fitness function value ($f = 0.01282$) with the first optimization method (Sect. 5.1) using a single 4×2.8 GHz processor. Figure 7.24 shows the convergence history obtained by Hybridized Games and also compares the computational cost obtained by HAPMOEA and Hybridized Games. The computational cost for Hybridized Game is only 9.5 % of HAPMOEA computation cost due to Nash-Game characteristics (decomposition of design problem) and evaluation mechanism. In other words, the use in parallel of Nash- Games linked to hierarchical topology improves the MOEA efficiency by 90 %. In addition, the Hybridized Games produces slightly better solution even though it has lower computational cost. Table 7.10 compares the aerodynamic characteristics obtained by the baseline design (RAE 2822) and baseline with optimal SCB. It can be seen that the baseline design with optimal SCB can reduce wave drag by 76 % which leads to 33 % of total drag. Such a drag reduction provides a 50 % increase of the lift to drag ratio (L/D).

Figure 7.25 compares the shape of the baseline design and the baseline coupled with a SCB geometry obtained by Hybridized Games. The optimal double SCBs are located between (0.5566, 0.0597) and (0.8543, 0.0263) on the *suction* side, and be-

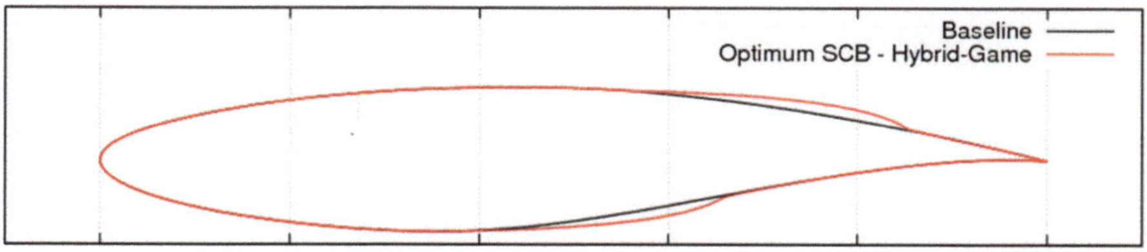

Fig. 7.25 Comparison between the baseline design and baseline with SCB geometries

Table 7.11 Optimal SCB design parameters obtained by Hybrid-Game

SCB	Length (%c)	Height (%c)	Peak position
Upper	29.77	0.644	84.8
Lower	29.94	0.643	84.8

Peak Position is in % of SCB length

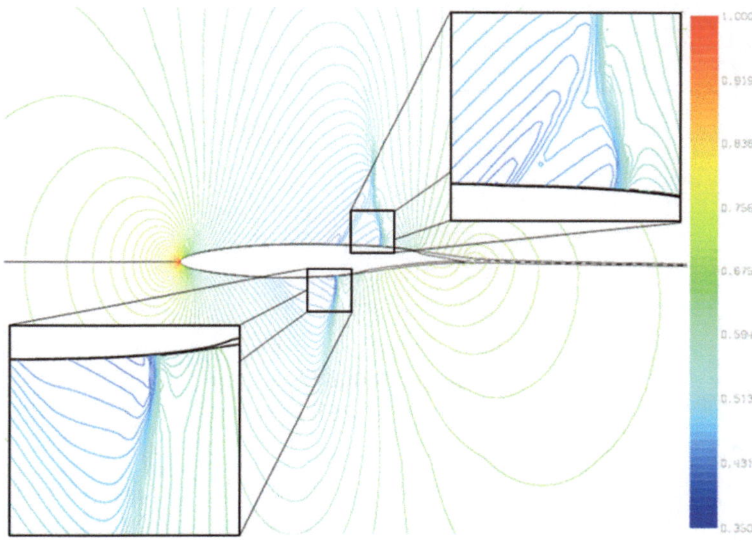

Fig. 7.26 Pressure ratio (P/P_0) contour obtained by Hybridized Games

tween (0.3624, −0.0591) and (0.6619, −0.0287) on the *pressure* side of RAE 2822 airfoil. The design parameters for optimal SCB are given on Table 7.11.

Figure 7.26 compares *Cp* contours obtained by the baseline design and baseline with optimal SCB. It can be seen that both upper and lower SCB decelerates the supersonic flow and the position of shock is moved towards the trailing edge when compared to the baseline design shown in Fig. 7.13b. Using only upper optimal SCB reduces the total drag by 14 % while applying both upper and lower optimal SCB produces 33 % lower total drag when compared to the baseline design.

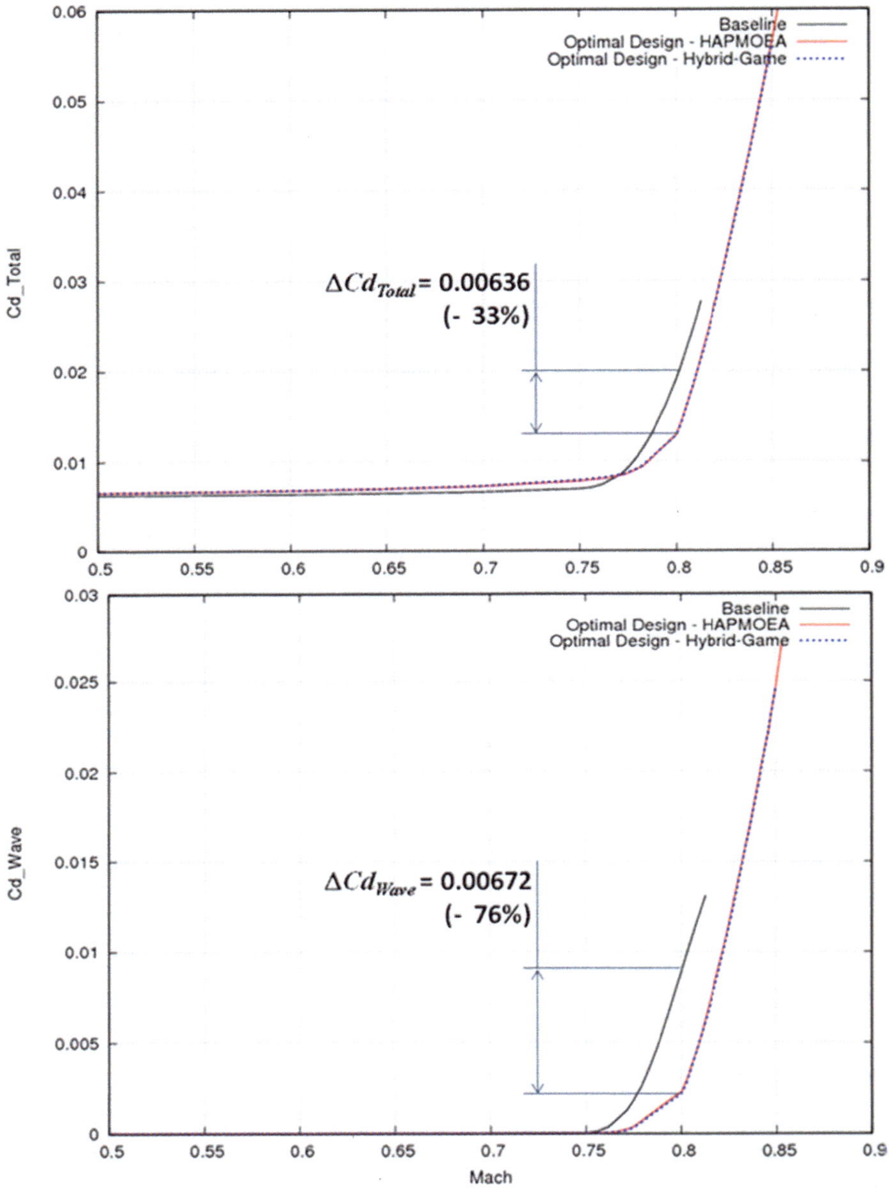

Fig. 7.27 Total drag coefficients (Cd_{Total}—upper), wave drag coefficients (Cd_{Wave}—lower) vs. Mach numbers

Figures 7.27a and b compare the total drag (Cd_{Total}) and wave drag (Cd_{Wave}) distributions obtained by the baseline design and baseline with the optimal SCB found by HAPMOEA software and Hybridized Game software. Flow conditions are: $M\infty$

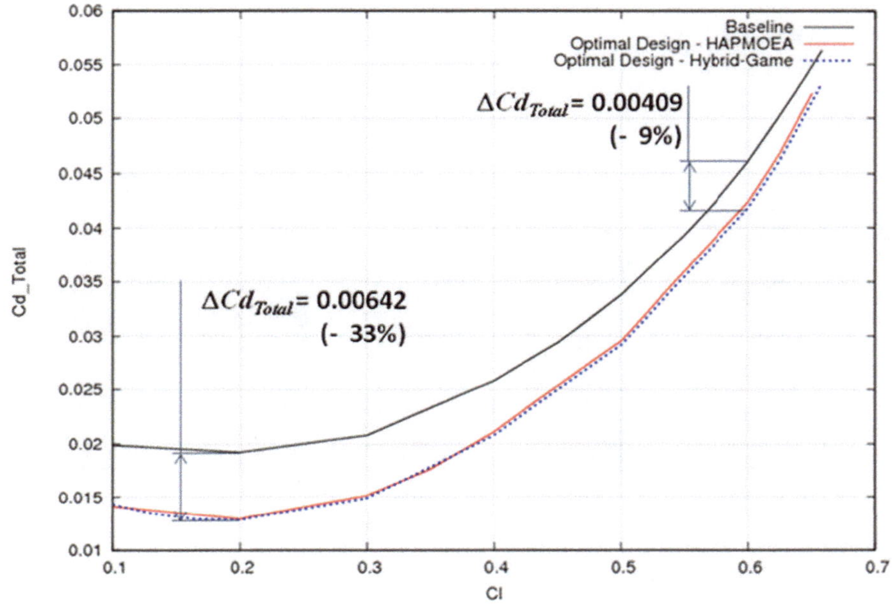

Fig. 7.28 Total drag coefficient (Cd_{Total}) vs. lift coefficient (C_l)

$\in$ [0.5:0.85] with $Cl_{Fixed}=0.175$, Re $=18.63\times10^6$. The optimal SCB provided by Hybridized Games perform solutions similar to HAPMOEA software and starts to produce lower total drag when Mach number is higher than 0.77. It should be noticed from Fig. 7.27b is that the critical Mach number ($M_C=0.75$) for the baseline design is extended to Mach number 0.77.

Figure 7.28 compares the total drag distribution obtained by the baseline design and baseline with the optimal SCB found by HAPMOEA software and Hybrid-Games software for $C_l \in$ [0.1:0.85] with $M_{Fixed}=0.8$, Re$=18.63\times10^6$. Both optimal SCB from HAPMOEA and Hybridized Game optimizers produce lower drag for a range of C_l values when compared to the baseline design. In addition, the optimal SCB obtained with Hybridized Games software has a slightly lower total drag when compared to the optimal SCB obtained with HAPMOEA software despite the Hybridized Games used only 9.5 % of HAPMOEA computational cost.

This optimal double SCB configuration is also tested at five different flight conditions to show the benefit of this configuration. The histogram shown in Fig. 7.29 shows the comparison of total drag and the ratio of lift to drag (L/D) obtained by the baseline design and optimal double SCB from HAPMOEA and Hybrid-Game. It can be seen that the optimal double SCB at critical flight conditions reduces more total drag by 15–33 % and improves lift to drag ratio (L/D) by 18–49 % at normal flight conditions when compared to the baseline design.

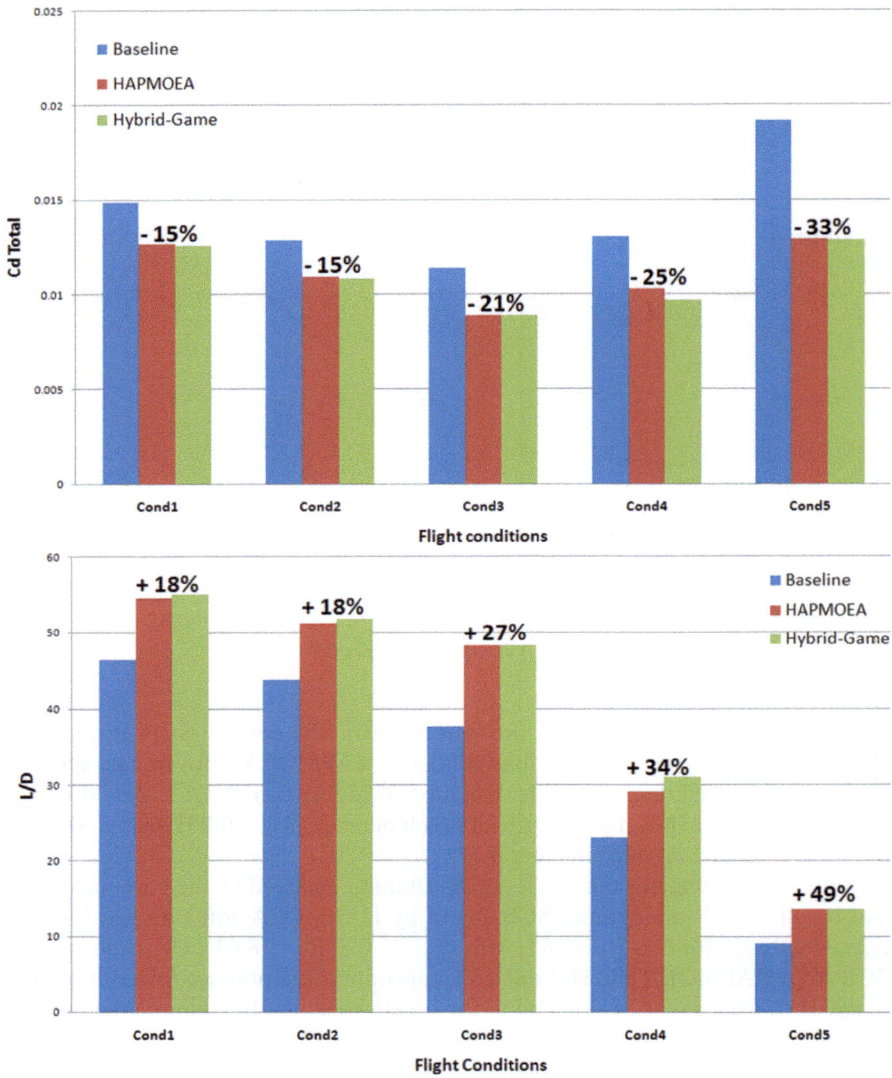

Fig. 7.29 Drag reduction (*upper*) and lift to drag ratio (*lower*) improvement at normal flow conditions. Note: Condi represents *i*-th flow conditions: $Cond_1$: $M\infty=0.750$, $Cl=0.690$, $Re=18.63\times10^6$, $Cond_2$: $M\infty=0.760$, $Cl=0.560$, $Re=18.63\times10^6$ $Cond_3$: $M\infty=0.770$, $Cl=0.430$, $Re=18.63\times10^6$, $Cond_4$: $M\infty=0.785$, $Cl=0.300$, $Re=18.63\times10^6$ $Cond_5$: $M\infty=0.800$, $Cl=0.175$, $Re=18.63\times10^6$

To summarize the results of the optimization test case, the use of optimal SCB obtained by HAPMOEA software and Hybridized Games software is beneficial at both normal and critical flow conditions due to significant transonic drag reduction. In addition, Hybridized Games improve MOEA efficiency while generating high quality optimal solutions when compared to the HAPMOEA approach.

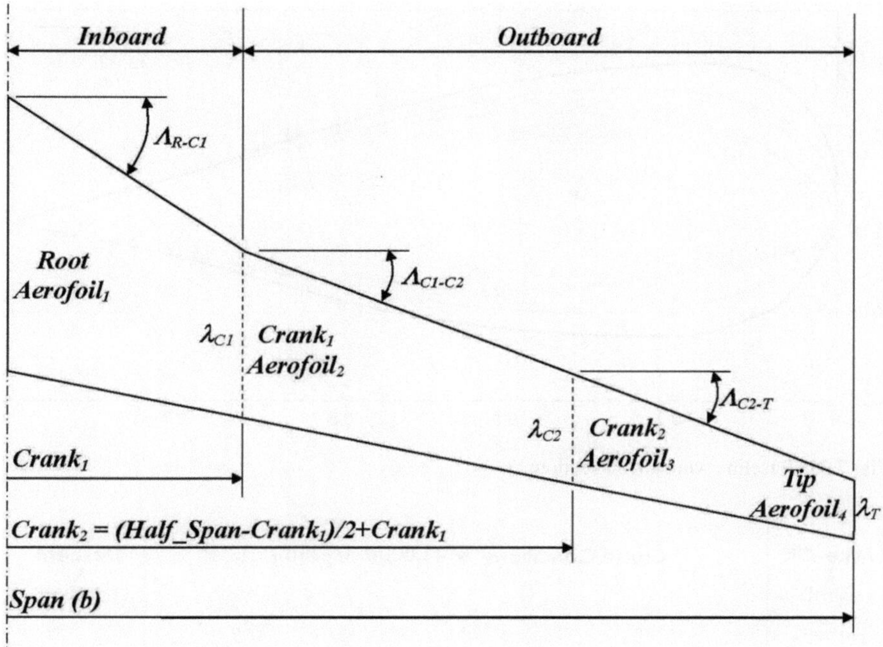

Fig. 7.30 Baseline wing geometry

Table 7.12 Baseline wing configurations

AR	*b (m)*	*Crank$_1$*	*Crank$_2$*	Λ_{R-C1}	Λ_{C1-C2}	Λ_{C2-T}	λ_{C1}	λ_{C2}	λ_T	$\Gamma_{Overall}$
11.57	34.32	28.12%	64.06%	34.03°	21.38°	21.38°	0.60	0.41	0.22	6

7.4 Generic Aircraft Wing Aerofoil Section Design Optimization

Problem Definition This test case considers the design and optimization of generic aircraft wing airfoil section geometries at one operating condition. The optimization initially considers a baseline wing design of generic commercial aircraft presented on Fig. 7.30. The baseline design is that of a generic wing similar to the Boeing 737 [3, 4]. The wing specifications are obtained from the reference [3, 4]. The wing geometry parameters are indicated in Table 7.12. The aspect ratio L/D is 11.57, the inboard and outboard sweep angles are 34.04° and 21.38° respectively. The crank position 1 is where nacelle is located and crank 2 is assumed the middle of outboard. The coordinates of the baseline airfoil sections at four stations (root, crank1, crank2 and tip) are obtained from reference [9]. These airfoils are used on a Boeing 737 which has similar size and characteristics. Their airfoil sections are indicated in Fig. 7.31.

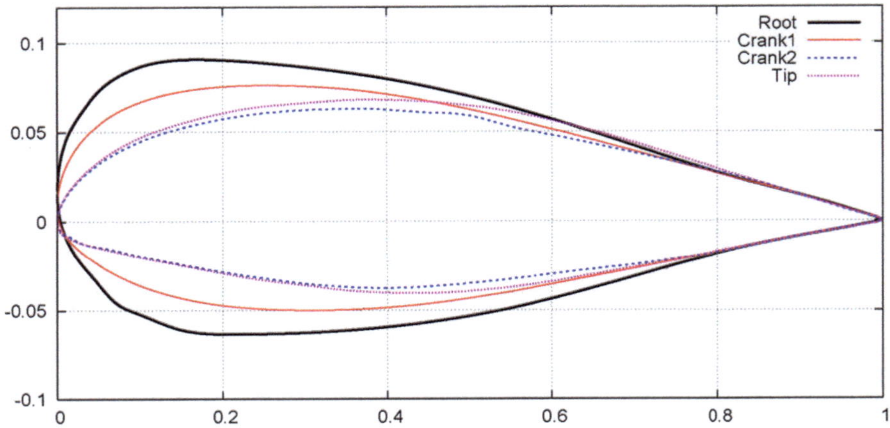

Fig. 7.31 Baseline wing airfoil sections

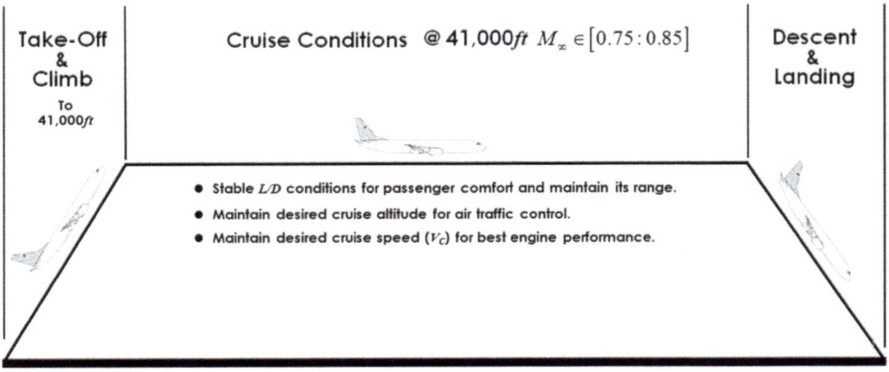

Fig. 7.32 Mission profile for a generic aircraft

A generic aircraft is a logistic long range aircraft. A mission profile for a generic aircraft is given on Fig. 7.32. The main objectives of the mission are: Initially, the generic aircraft climbs up to a desired altitude 41,000 ft then cruises at the flight conditions transit to $M_\infty \in 0.7:0.82$. Its objective can maintain cruise altitude and cruise speed to have safe air traffic control and reduction of operating cost. Once the cruise is completed, it descends and lands at a pre-determined location.

The aircraft maximum take-off weight is approximately 79,002 *kg* with max payload of 20,240 *kg*. The minimum lift coefficient requires 0.691 $(C_{L_{\min}} = 0.691)$ (baseline design). The Breguet range Eq. 7.7 is considered to calculate the minimum lift to drag ratio;

$$R = \frac{V_\infty}{gSFC}\left(\frac{L}{D}\right) ln\left(\frac{W_{Initial}}{W_{Final}}\right) \rightarrow R = \frac{V_\infty}{TSFC}\left(\frac{L}{D}\right) ln\left(\frac{W_{Initial}}{W_{Final}}\right) \tag{7.7}$$

where:

R is the distance flown and
$V\infty$ is the velocity
SFC stands for Specific Fuel Consumption
TSFC stands for Thrust Specific Fuel Consumption ($TSFC = 1.07 \times 10^{-4}$/s)

$$TSFC = \frac{FF \times g}{Thrust} = \frac{1.28 \times 9.81}{117.3 \times 10^3} = 1.07 \times 10^{-4}/\text{s}$$

g is the acceleration of gravity ($g = 9.81$ m/s^2)
L/D is the lift to drag ratio which is obtained using aerodynamic analysis tool
$W_{Initial}$ is the gross weight of aircraft at the start of cruise:

$$W_{Initial} = W_{MaxTO} - 10\% \text{ of } W_{Fuel} = 76902 \text{ kg where } W_{Fuel} = 2100 \text{ kg}$$

W_{Final} the gross weight of aircraft at the end of cruise

$$W_{Final} = W_{MaxTO} - 85\% \text{ of } W_{Fuel} = 61152 \text{ kg}$$

$$\therefore\ R = \frac{0.82 \times 340}{1.07 \times 10^{-4}}\left(\frac{L}{D}\right)\ln\left(\frac{76902}{61152}\right) = 597\left(\frac{L}{D}\right)km$$

The range of the baseline aircraft is 2728 nm which requires a minimum lift to drag ratio of 8.46. This minimum *L/D* ratio is applied in the optimization as one of inequality constraints;

$$\frac{L}{D}_{\min} \geq 8.46$$

This is a single objective problem where the fitness function is to maximise lift-to-drag ratio (L/D). Results of this application are compared to the baseline design. The optimization for this test case is based on Algorithm 3 described in Sect. 6.5.4 of Chap. 6. The fitness function and flight conditions are:

$$f_1 = \min\left(1/(L/D_1)\right) + Penalty \tag{7.8}$$

$$\text{at } M_\infty = 0.82,\ C_{L_{\min}} = 0.691$$

where a Penalty (2.4) and (2.5) in Chap. 2 is computed and added to the fitness function shown in (7.8) if the thickness (*t/c*) is higher than 20% and less than 10.0%

of the chord and also if the ratio of lift to drag at flight conditions is lower than 8.46. As a geometry constraint, the thickness ratio should be chosen in the interval $10\% \leq t/c \leq 22\%$ and as a performance constraint, the lift to drag ratio (L/D) should be $L/D_{\min} \geq 8.46$.

Design Variables Four airfoils at root, crank1, crank2 and tip section are shown on Figs. 7.33a, b, c, d. For the thickness design bounds, 25 % of chord is assigned for the upper thickness bounds (blue triangles) and 10 % for the lower bounds (red inverse triangles). For the mean line, the design bounds use 5 % of chord for upper (blue circles) and lower (red circles) bounds. The airfoils between wing sections are interpolated by the CFD FLO22 flow analyser.

The external wing geometry is fixed as shown in Fig. 7.31 and Table 7.12. Four airfoil sections including root, crank 1, crank 2 and tip are considered as design variables and there are eighty eight (88) design variables (eleven (11) mean line + eleven (11) thickness at each section) in total. The control points of mean and thickness distribution are represented on Figs. 7.33a, b, c, d.

Implementation The FLO22 and FRICTION solvers (cf./1/) are utilised and the following specific parameters are considered for HAPMOEA software.

1st Layer	Population size of 15 and intermediate recombination Computational grid of $96 \times 12 \times 16$ cells used by the FLO22 solver
2nd Layer	Population size of 40 and intermediate recombination Computational grid of $82 \times 12 \times 16$ cells used by the FLO22 solver
3rd Layer	Population size of 60 and intermediate recombination Computational grid of $68 \times 12 \times 16$ cells used by the FLO22 solver

Numerical Results The algorithm was allowed to run approximately 1100 function evaluations and took 150 h on a single 1.8 GHz processor. The evaluation convergence history is shown in Fig. 7.34. It can be observed that the problem converged after 450 function evolutions.

The lift to drag ratio (L/D) of the optimal solution is compared to the baseline design in Table 7.13 where the optimal wing produces 23 % of the lift to drag ratio (L/D) improvement while reducing 19 % of the total drag.

The optimal airfoils are compared to the baseline design on Fig. 7.35. Airfoil sections: root, crank1, tip are thinner approximately by 10 % of the chord than the baseline aerofoils and negative cambers are observed from the root to crank2 sections while the airfoil at the tip section has a positive camber.

The pressure coefficients distribution is compared at 0, 20, 40, 60, 80 and 90 % of the span in Figs. 7.36a, b, c. It can be seen that the optimal wing has weaker shock from 20 ~ 90 % of the span when compared to the baseline design marked as blue circles.

Figures 7.37a and b show the comparisons of Mach distribution vs. the drag coefficient and lift to drag ratio obtained by the optimal wing and the baseline design respectively. The optimal wing produces lower drag coefficient and higher lift to drag ratio at all Mach range 0.75–0.85 when compared to the baseline. The base-

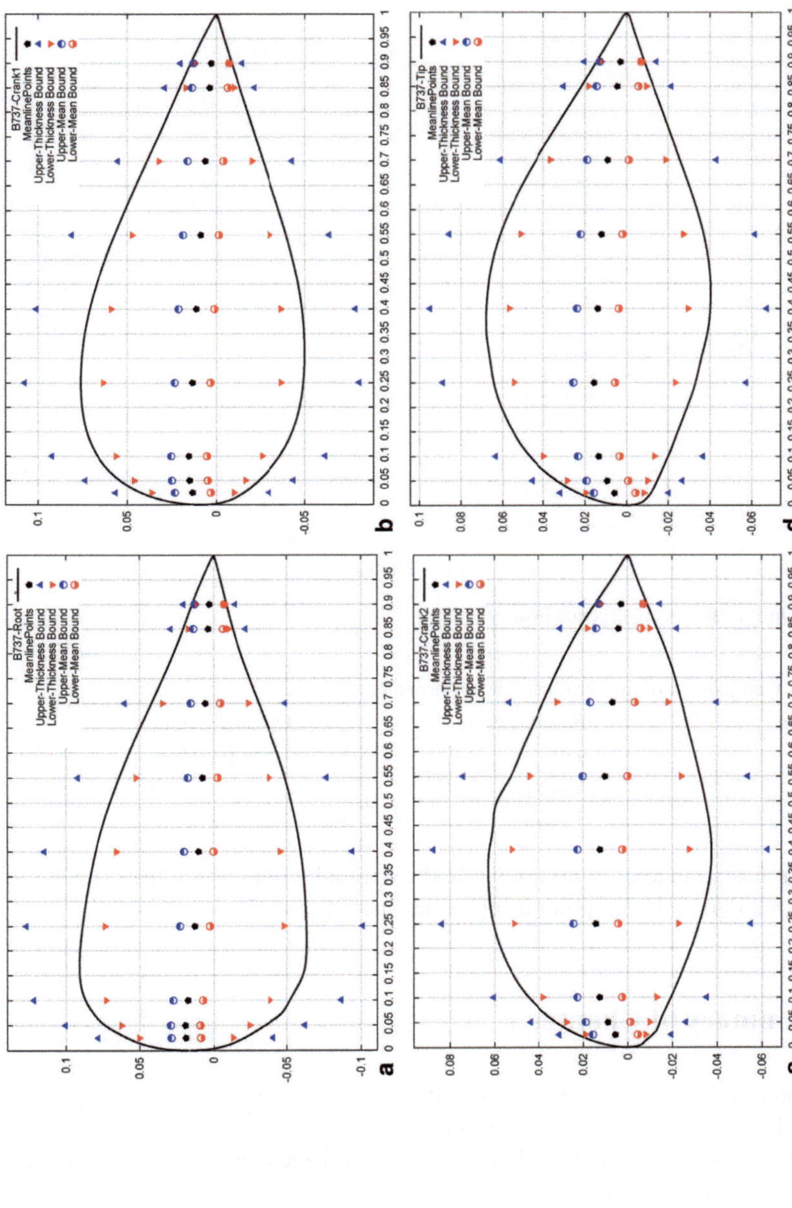

Fig. 7.33 **a** Root section mean line and thickness control points design envelope. **b** Crank1 section mean line and thickness control points design envelope. **c** Crank2 section mean line and thickness control points design envelope. **d** Tip section mean line and thickness control points design envelope

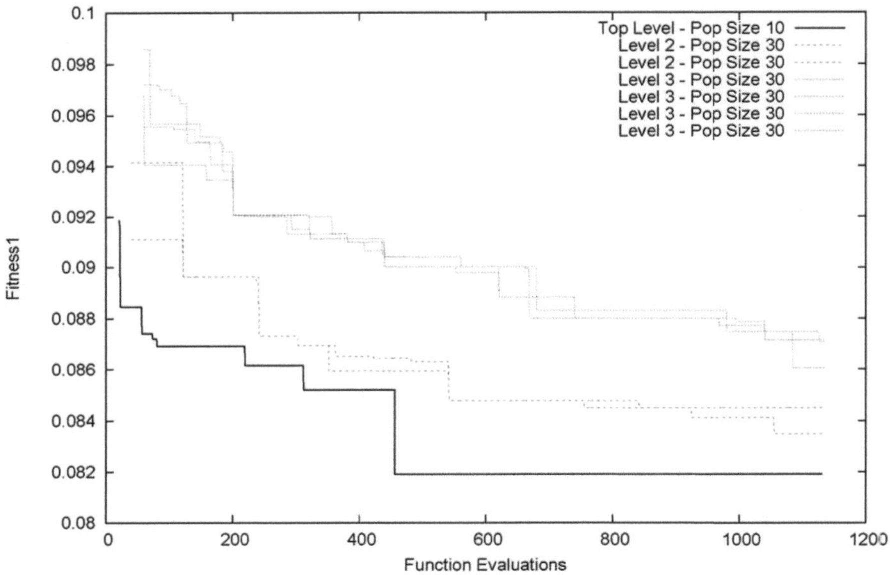

Fig. 7.34 Optimization convergence history

Table 7.13 Comparison of the aerodynamic performances

Description	Baseline design	Optimal Solution
L/D	9.95	12.21 (+23 %)
CD_{Total}	0.0694	0.0565 (−19 %)

line design has a range of 5940 km while the optimal wing can fly the distance of 7290 km (+23 %) at design point using same amount of fuel.

7.5 Summary of Chapter 7

This chapter is dedicated to applications of Evolutionary Algorithms to single-objective and inverse design problems in aeronautics.

Numerical results obtained by HAPEA and Hybridized Games-EAs software are compared in terms of computational efficiency and solution quality. These two optimization techniques offer alternative choices to the designer when solving airfoil space reconstruction/inverse design problems. It is seen that splitting the design space into Nash Players design territory can be high beneficial by reducing significantly the computational expense of the optimization procedure.

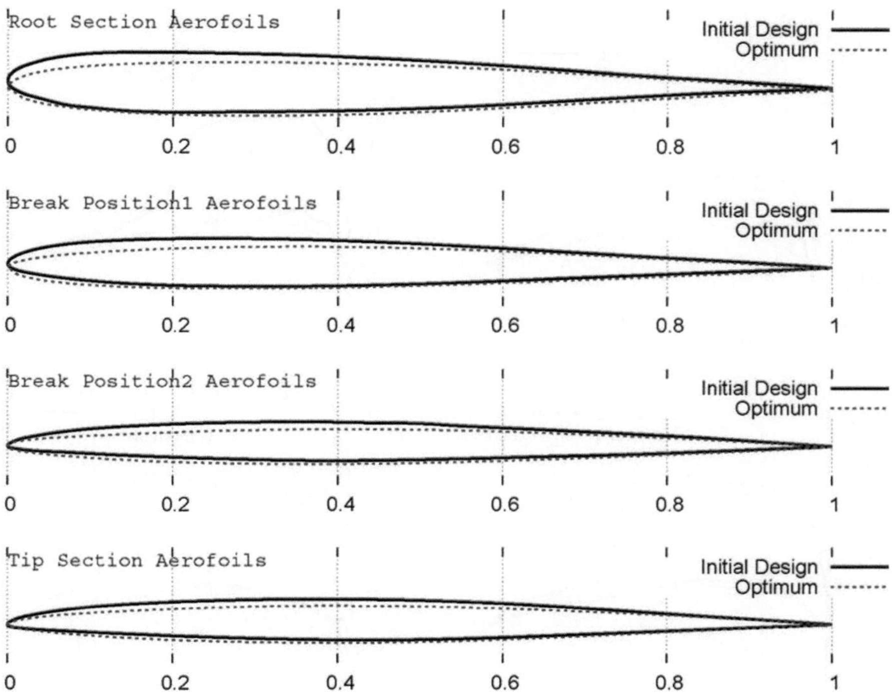

Fig. 7.35 Airfoils sections of compromised solutions

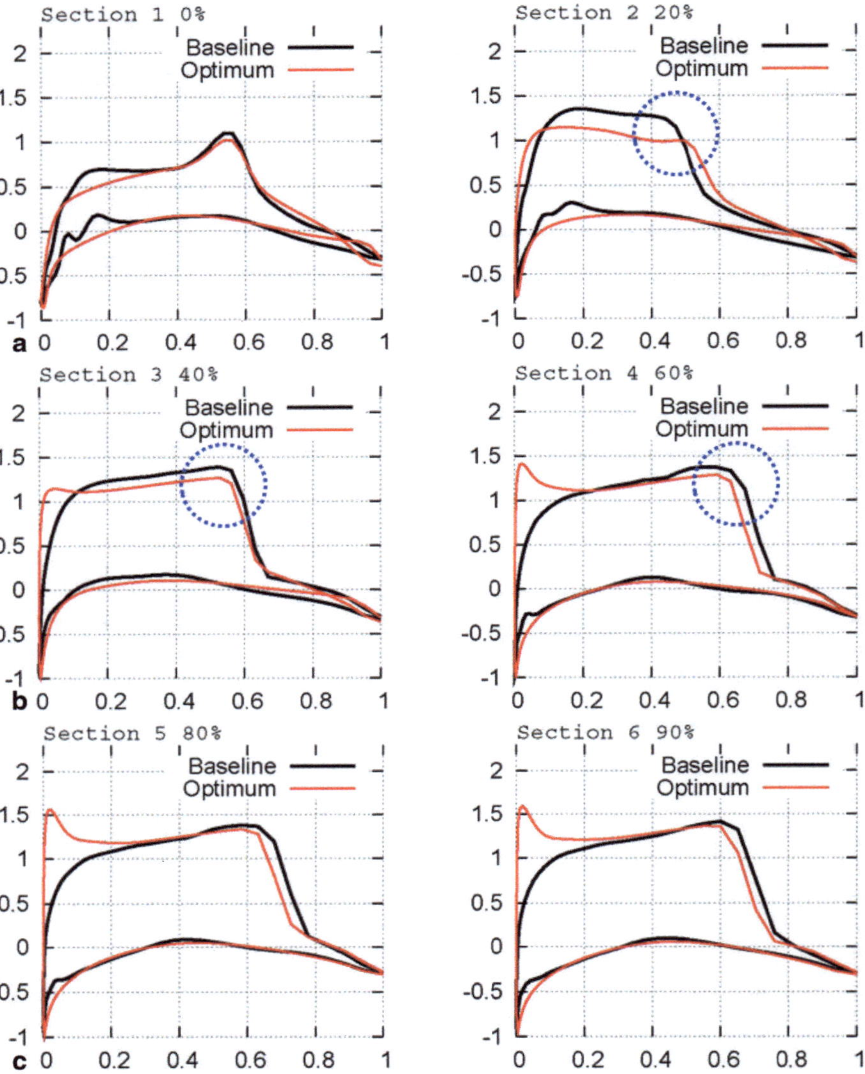

Fig. 7.36 **a** Comparison of pressure coefficient (*Cp*) distribution at 0 and 20 % of span. **b** Comparison of pressure coefficient (*Cp*) distribution at 40 and 60 % of span. **c** Comparison of pressure coefficient (*Cp*) distribution at 40 and 60 % of span

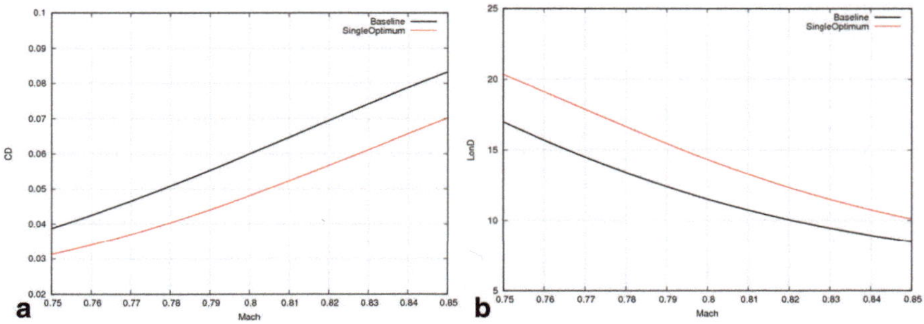

Fig. 7.37 **a** Comparison of total drag coefficients $(C_{D_{Total}})$ vs. Mach number. **b** Comparison of *L/D* vs. Mach number

References

1. Ashill PR, Fulker J, Shires A (1992) A novel technique for controlling shock strength of laminar-flow aerofoil sections. Proceedings 1st European Forum on Laminar Flow Technology, pp 175–183, Hamburg, Germany, DGLR, AAAF, RAeS, March 16–18
2. Bauer M, Peltzer I, Nitsche W, Gölling B (2010) Active flow control on an industry-relevant civil aircraft half model. In: Active flow control ii notes on numerical fluid mechanics and multidisciplinary design, vol 108. Springer, Berlin, pp 95–107
3. Boeing. Defence, Space & Security: P-8 A Poseidon. http://www.boeing.com/defense-space/military/p8a/index.html. Accessed 17 Dec 2010
4. Boeing 737 Technical Specifications. http://www.b737.org.uk/techspecsdetailed.htm. Accessed 17 Dec 2010
5. Jameson A, Caughey D, Newman P, Davis RA (1976) Brief Description of the Jameson Caughey NYU Transonic Swept-Wing Computer Program FLO22, NASA Technical Memorandum, NASA TM X-73996
6. Lee DS, Periaux J, Srinivas K, Gonzalez LF, Onate E, Qin N (2010) Shock control bump design optimization on natural laminar aerofoil. In Kuzmin A (ed) computational fluid dynamics: Proceedings of the Sixth International Conference on computational fluid dynamics, ICCFD6, Springer, St Petersburg
7. Nübler K, Colliss SP, Lutz T, Babinsky H, Krämer E (2013) Numerical and experimental examination of shock control bump flow physics. High performance computing in science and engineering '12, pp 333–349
8. Patzold M, Lutz T, Kramer E, Wagner S (2006) Numerical optimization of finite shock control bumps, 44th AIAA Aerospace Sciences Meeting and Exhibit, Reno, Nevada, USA, 9-12 January
9. UIUC Airfoil Coordinates Database Version 2.0. http://www.ae.uiuc.edu/m-selig/ads/coord_database.html. Accessed 17 Dec 2010

Chapter 8
Multi-Objective Optimization Model Test Case Problems

8.1 Overview

In this chapter, applications solved by EAs software and advanced EAs software introduced in previous chapters and dealing with six (6) multi-objective design optimization problems are described. Different MOEAs are used and the efficiency of their optimizers are compared.

8.2 Pareto Reconstruction: Two Airfoils at Two Different Design Points

Problem Definition A classic example in inverse airfoil design is to reconstruct shapes corresponding to prescribed surface pressure distribution. In this example, the problem is extended to reconstruct the pressure distribution at two design points for reconstructing two airfoil geometries.

Practical design engineers in aerodynamics often guess the appropriate pressure distribution to their problems, but do not always have tools for finding a feasible solution. This problem has been studied extensively by Jameson [1], Kim and Rho [2], and Obayashi [3, 4], Obayashi and Takanashi [5] and Zetina et al [6], Oyama et al. [7] among many others.

This case considers minimization of the difference of the surface pressure coefficient distribution on a candidate airfoil with the prescribed pressure distribution over two airfoils (NACA0012 and RAE2822) operating at different design points. Flow conditions for the two points analyzed are:

$$\text{Flow Condition 1}: \mathrm{M}_\infty = 0.2, \mathrm{Re} = 2.7\times10^6,\ \alpha = 1.25°(\text{NACA0012}) \quad (8.1)$$

$$\text{Flow Condition 2}: \mathrm{M}_\infty = 0.75, Re = 9.0\times10^6,\ a = 1.25°(\text{RAE2822}) \quad (8.2)$$

J. Periaux et al., *Evolutionary Optimization and Game Strategies for Advanced Multi-Disciplinary Design,* Intelligent Systems, Control and Automation: Science and Engineering 75, DOI 10.1007/978-94-017-9520-3_8

This reconstruction problem can be solved faster using a classical deterministic optimizer if the two objectives are sought independently. In this case, the goal is to find the Pareto in one single run and without having to choose weights in advance. This optimization procedure uses Algorithm 4 described in Sect. 6.5.5 of Chap. 6.

Design Variables The airfoil geometry is represented by two (2) Bezier curves for the parameterization of the mean line and one (1) for the thickness distribution. The mean line–thickness distribution is defined by a standard spline method for representing airfoils [8], as it closely couples the representation with results; the mean line has a powerful effect on cruise lift coefficient and pitching moment, while the thickness distribution has a powerful effect on the cruise drag. Defined simply, the airfoil is obtained by perpendicular offset of the thickness distribution about the mean line.

For a given mean line point (x_m, y_m) and matching thickness distribution height y_t, upper and lower surface points can be obtained by (8.3) and (8.4):

$$x_{u,l} = x_m \pm y_t \sin(\theta) \tag{8.3}$$

$$y_{u,l} = y_m \pm y_t \cos(\theta) \tag{8.4}$$

where θ is the mean line angle at (x_m, y_m). This is shown in Fig. 8.1. The x-positions of the Bezier control points are predefined in advance; y- positions remain unknowns. The only restrictions are that the first and last points are fixed to (0,0) and (1,0) to provide leading and trailing edges respectively, and that the first control point on the thickness distribution must be directly above the leading edge (i.e. (0,yc,1)) to provide a rounded geometry (Bezier curves are by definition always tangent to the extreme edges of their defining envelopes).

The vertical heights to range $y_c \in [0.01,0.1]$ offering a very wide range of possible geometries (theoretically spanning airfoils from 2 to 20 % thick) are bounded.

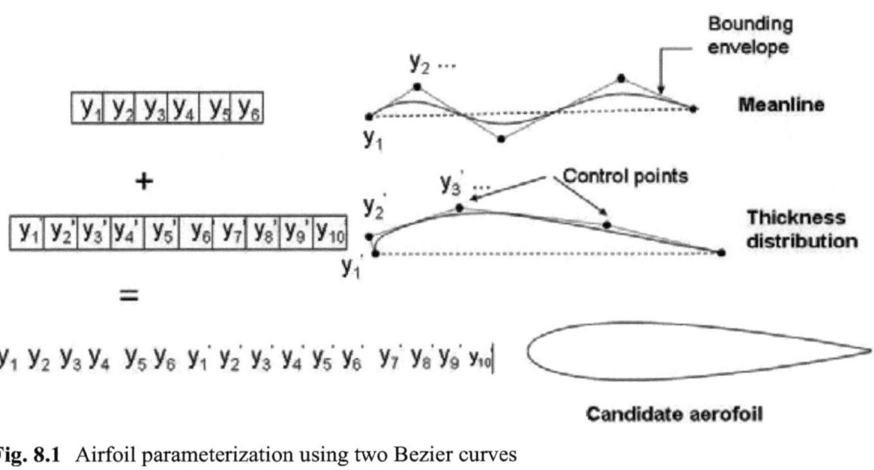

Fig. 8.1 Airfoil parameterization using two Bezier curves

The advantage of using single high-order Bezier curves for the representation rather than piecewise splines or others is their geometrical stability. A Bezier curve must by definition always be contained within the bounding envelope of control points. Furthermore, if the bounding envelope is not re-entrant, then the curve will also have this property. Also, Bezier curves do not 'kink' like a piecewise spline, and the defining equations are not stiff (ill-conditioned). Therefore, a small change in control point location will always result in a small change in surface representation. This mathematical property provides a favourable interface between the optimizer and the flow analyzer. For this case, four (4) evenly spaced interior (free) control points were considered for the mean line, and five (5) for the thickness distribution, giving a problem with nine (9) unknowns.

The fitness functions are the Root Mean Square Error (RMSE) of the surface pressure coefficients distribution between candidate shape (current) and target shape (objective 1) and surface pressure coefficients between candidate (current) and target –two (objective 2). The problem is solved when the positive value of the fitness decreases below a prescribed value.

$$\min(f_1): f_1 = \frac{1}{N}\sqrt{\sum_{i=1}^{N}(Cp_{current} - Cp_{Target-one})^2}, \tag{8.5}$$

$$\min(f_2): f_2 = \frac{1}{N}\sqrt{\sum_{i=1}^{N}(Cp_{current} - Cp_{Target-two})^2}. \tag{8.6}$$

In this case a flow solver needs to compute the aerodynamic drag. The MSES solver [9] evaluate candidate solutions, is based on a structured quadrilateral streamline mesh and coupled to an integral boundary layer with a multi-layer velocity profile representation.

Implementation The problem was implemented using Alg. 4 in Sect. 6.5.5 of Chap. 6.

Details on the multi-fidelity–hierarchical tree are: (EA with CMA/Pareto tournament selection, asynchronous evaluation) on each node of the hierarchical tree with the following parameter settings for the EAs and CFD solver are:

1st Layer	Population size of 20 and intermediate recombination using a computational grid of 215×36 cells
2nd Layer	Population size of 20 and intermediate recombination using a computational grid of 165×27 cells
3rd Layer	Population size of 20 and intermediate recombination using a computational grid of 151×24 cells

Numerical Results This problem was run with 3000 function evaluations of the head node, and took approximately ten (10) hours on the cluster with twelve PC cores. Figure 8.2 shows the Pareto front captured for this test case. Figures 8.3 and 8.4 show a comparison of the target geometries and Fig. 8.5 surface pressure dis-

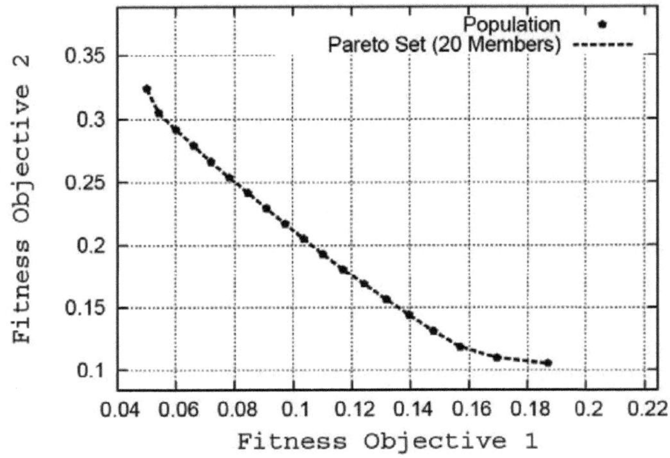

Fig. 8.2 Pareto front for the multi-point airfoil design

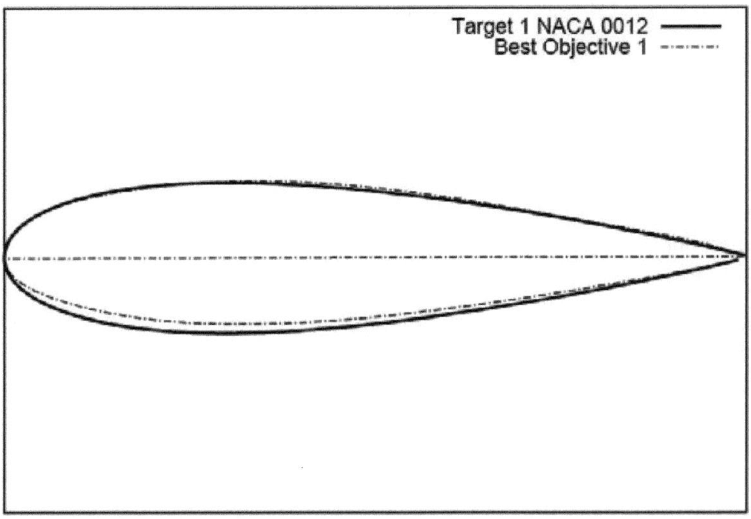

Fig. 8.3 Target and computed geometries, multi-point airfoil design, Objective 1

tributions respectively. A good match of the computed and target surface pressure distribution was observed.

Comparisons of performances of the algorithm with an increasing number of computers are represented. The parallel environment used is the BORGS cluster of PCs. The message-passing model used is the Parallel Virtual Machine (PVM) [9]. Figure 8.6 shows the speed-up of the computation as the number of computers increases and also the comparison with a linear speed-up for reference purposes.

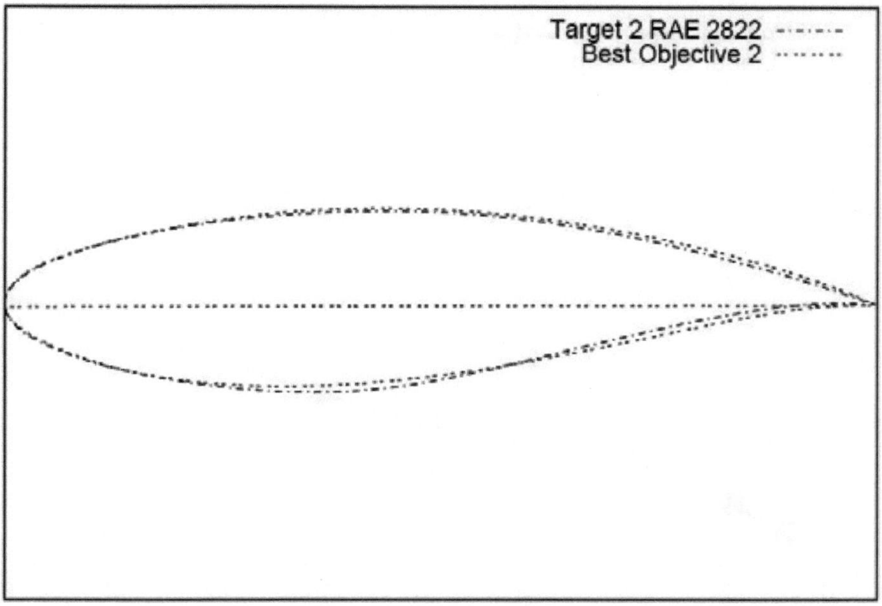

Fig. 8.4 Target and computed geometries, multi-point airfoil design, Objective 2

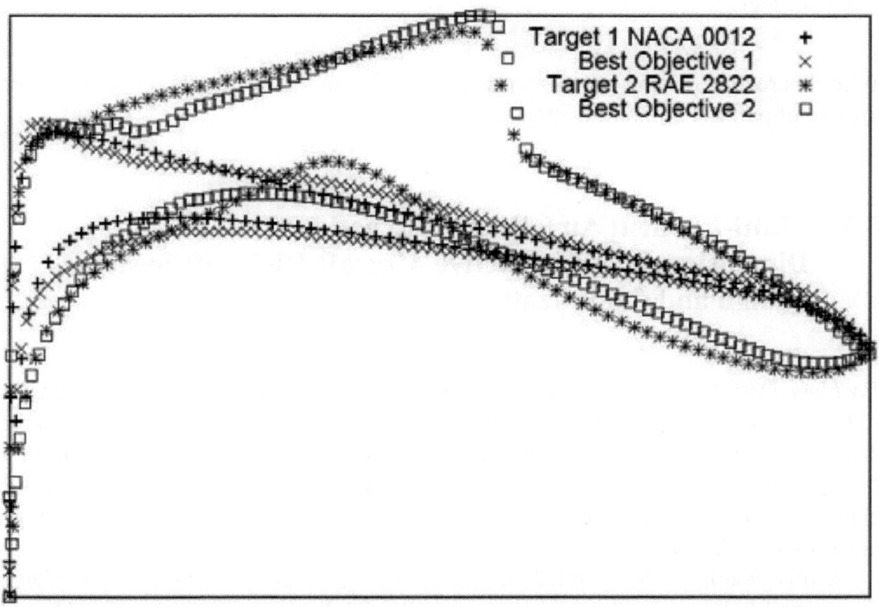

Fig. 8.5 Target and computed pressure distribution for multi-point airfoil design

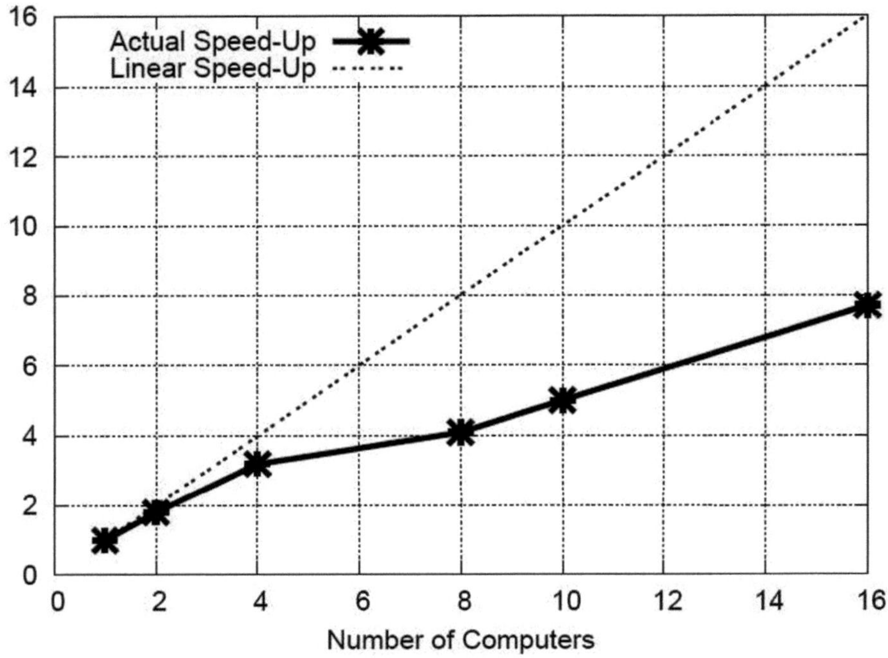

Fig. 8.6 Computation speed-up of with an increasing number of computers and comparison with linear speed-up

With this reconstruction test case, it is clearly shown that the method benefits from the use of parallel computing strategies.

8.3 Multi-Element Airfoil Reconstruction: Two-Dimensional Two Objective Aircraft High Lift System Design and Optimization

Problem Definition The use and development of Unmanned Aerial Vehicles for military and civil applications are rapidly increasing but there are difficulties in the design of these vehicles due to the varied and non-intuitive nature of the configurations and missions that can be performed. Similar to their manned counterparts, the challenge is to develop trade-off studies of optimal configurations to produce a high performance aircraft that satisfy mission requirements.

In this example, the detailed design of a single element airfoil is considered for a small UAV application similar to the RQ-7A Shadow 200 Tactical UAV and uses the airfoil design module for this task. The aircraft maximum gross weight is approximately 320 lbs, it has a wingspan of approximately 12.8 ft, a mean chord

Table 8.1 UAV data and operating conditions

Aerofoil section	NACA4415
b, ft	12.8
cr (approx.), *ft*	2.0
Length, *ft*	11.2
Cruising altitude, *m*	3000

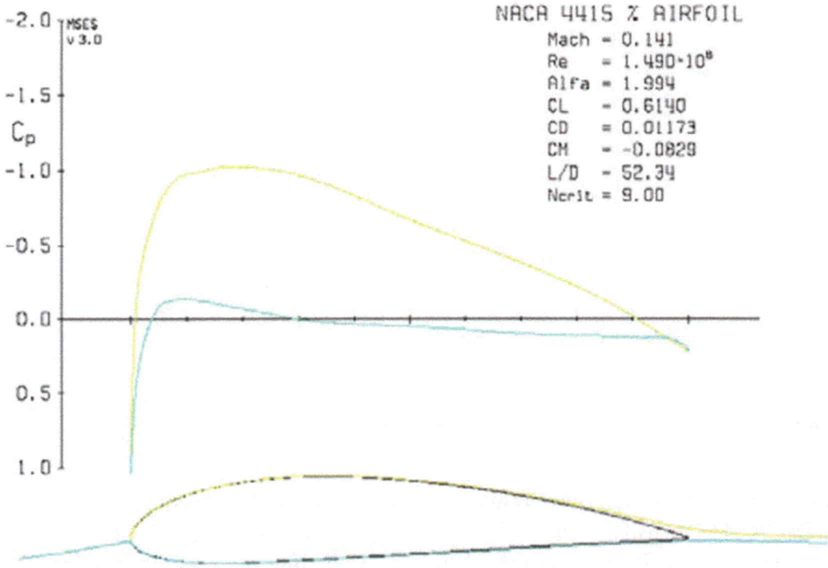

Fig. 8.7 NACA 4415—Flight condition 2—Fast Cruise

of approximately 2 ft, length of 11 ft, and a plan form shape without sweep. It is assumed that the aircraft is operating between a slow cruise 33.3 m/s and fast cruise 46.6 m/s approximately. This results in airframe, flight parameters and operating conditions indicated on Table 8.1. These conditions assume an aircraft at mid weight-cruise during and extended cruise phase at intermediate altitude.

Description	Flight condition one—slow cruise	Flight condition two—fast cruise
$M\infty$	0.1025	0.141
Re	1.085×10^6	1.490×10^6
c_l	1.18	0.6140

For the optimization, it is initially assumed that an existing air foil geometry is a operating at two suggested design points, and then a design and air foil that preserves the original thickness while reducing the drag coefficient. The assumed baseline air -foil geometry is the NACA4415. This airfoil thickness is 15 % of the chord. Figures 8.7 and 8.8 show the pressure coefficient (Cp) distribution and some

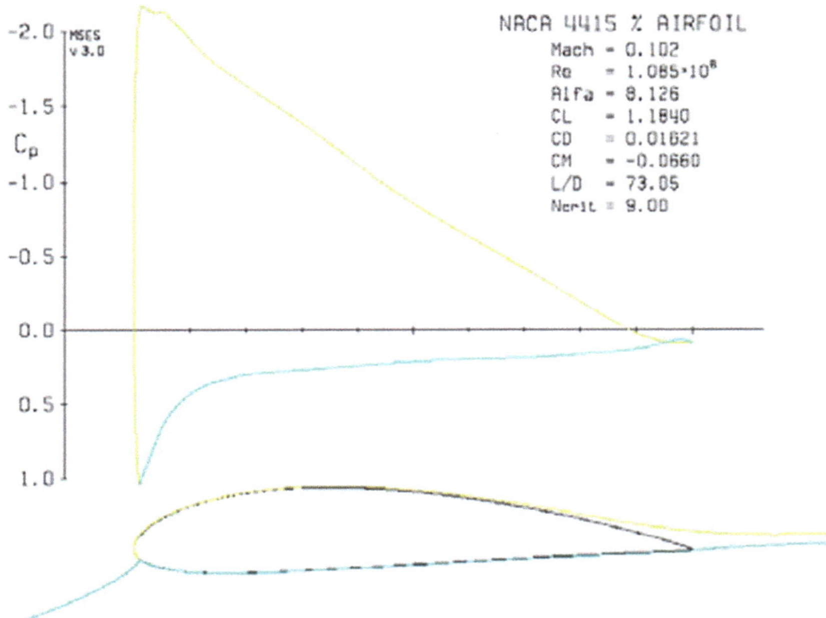

Fig. 8.8 NACA 4415—Flight condition 1—Slow Cruise

aerodynamic data for the two flight conditions considered. The combined polars for the NACA4415 airfoil are shown in Fig. 8.9. It is noted that both cruise points operate inside the invariant drag region of the airfoil; the low speed cruise condition giving approximately $Cd=0.016$ and the high speed giving approximately $Cd=0.012$.

The complexity, the non-linearity and multi-objective characteristics of this problem make it suitable to be solved with an EA optimizer. The computational cost for this problem is an important consideration due to the wide upper and lower bounds in the search space and dependency of the computing facilities used, in particular in industrial design environments. Therefore it is also desirable to use parallel computing and a multi-fidelity approach. In this case it is desired to use a multi-objective parallel EA (MOPEA) and to select the HAPEA approach which has all these capabilities; however, other EA approaches would perform similarly, such as NSGAII or VEGA [10, 11] for instance. This optimization uses Algorithm 4 in Sect. 6.5.5 of Chap. 6.

Design Variables The airfoil geometry is represented by two Bezier curves, one for the mean line and another for the thickness distribution. The mean line-thickness distribution is a standard method for representing airfoils [1], as it closely couples the representation with results; the mean line has a powerful effect on cruise lift coefficient and pitching moment, while the thickness distribution has a powerful effect on the cruise drag. In this case six (6) free control points are considered on the mean line and ten (10) free control points on the thickness distribution.

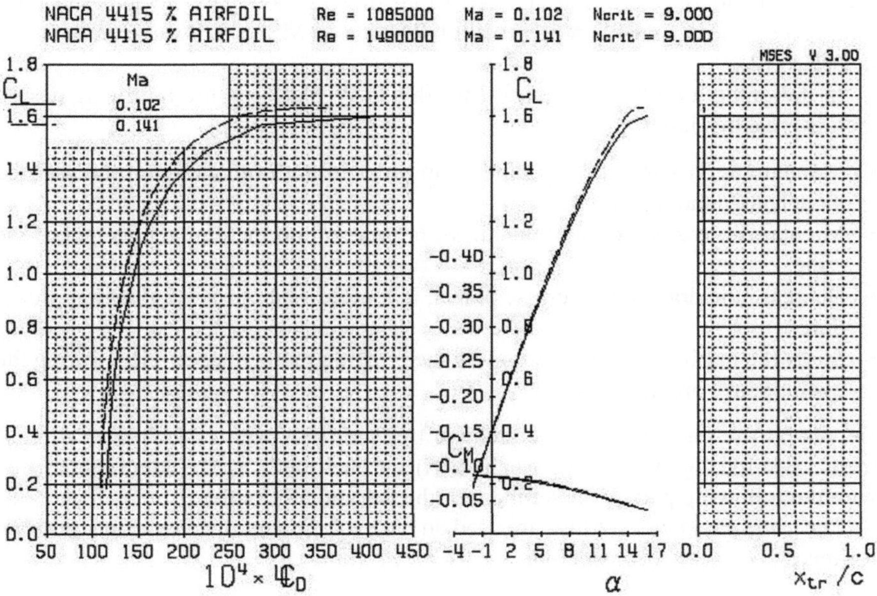

Fig. 8.9 NACA 4415—Polar

An individual as an object containing a vector of the sixteen (16) design variables is defined and a vector of strategy variances (σ) and rotation angles (α) for the evolution strategy:

$$\begin{aligned} I &= (\mathbf{x}, \sigma, \alpha) \\ &= ((y_1, y_2, \ldots, y_6, y_1', y_2' \ldots, y_{10}'), (\sigma_1, \sigma_2, \ldots, \sigma_N), (\alpha_1, \alpha_2, \ldots, \alpha_{N(N-1)/2})). \end{aligned}$$

The two fitness functions to be optimised are defined as minimization of drag (cd) at the two flight conditions:

$$\mathrm{Min}(f_1) = Cd \text{ subject to } \mathrm{Re} = 1.085 \times 10^6, C_l = 1.184, \tag{8.7}$$

$$\mathrm{Min}(f_2) = Cd \text{ subject to } \mathrm{Re} = 1.490 \times 10^6, C_l = 0.614. \tag{8.8}$$

There are three types of constraints: maximum thickness, maximum thickness location and pitching moment (cm). The thickness and maximum thickness location of each airfoil must exceed 15 % (t/c ≥ 0.15) and be between 20 and 40 % of the chord, respectively. If a constraint on pitching moment is applied this must not be more severe than −0.0660 (cm ≥ −0.0660).

Implementation The aerodynamic characteristics of the candidate airfoils are evaluated using the MSES [12] software. This solver is based on a structured quadrilateral streamline mesh and coupled to an integral boundary layer based on a multi-layer velocity profile representation. Details on MSES software can be found in Drela [12].

In designing a replacement airfoil for this UAV platform, the following design factors are considered:

- Maintain approximately the same Cl in order to not impinge upon the assisted launch and landing length.
- Maintain at least the current thickness, in order not to increase the weight of the wing.
- Decrease the drag at both cruise points, in a multi-objective optimization procedure.
- Study the implication of constraining the pitching moment coefficient during the evolutionary optimization.

Details on the multi-fidelity–hierarchical tree are: (EA with CMA/Pareto tournament selection, Asynchronous Evaluation) on each node of the hierarchical tree with the following parameter settings for the EA and CFD solver:

The same parameter settings are used on the evolutionary optimisation algorithm for the two test cases considered:

1st Layer	Population size of 20 and intermediate recombination with a computational grid of 215×36 cells
2nd Layer	Population size of 20 and intermediate recombination with a computational grid of 165×27 cells

Numerical Results This test case was run with 2000 function evaluations on the top level and took approximately 8 h in a cluster of 4×2.8 GHz processors. Figure 8.10 and 8.11 shows the set of airfoils on the Pareto front. From this front, three airfoils are selected; objective one optimal, objective two optimal and compromise airfoils from the middle of the front. These geometries are shown and compared to the NACA 4415 airfoil in Fig. 8.12.

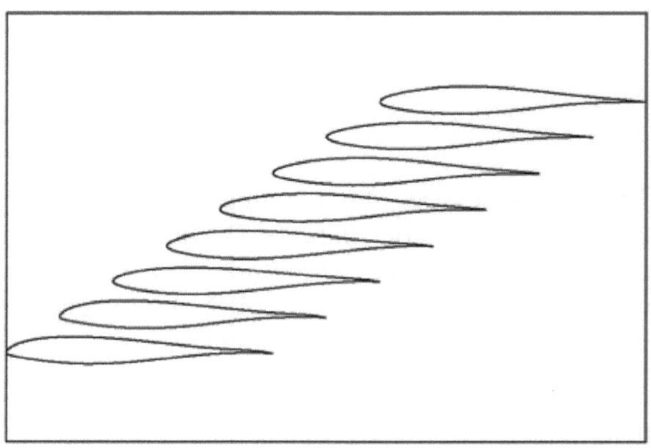

Fig. 8.10 The Ensemble of Pareto airfoils—constrained pitching moment coefficient (C_m)

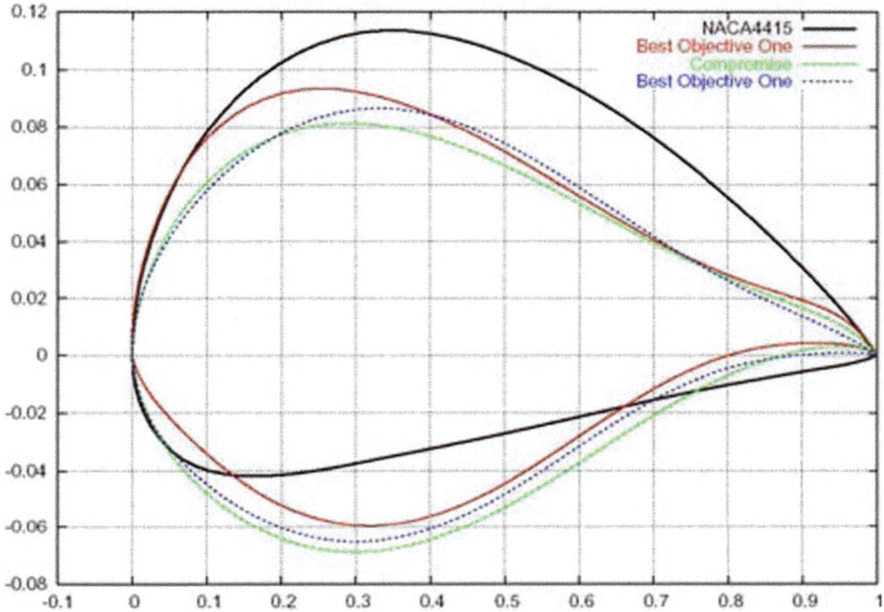

Fig. 8.11 Comparison of selected geometries—constrained pitching moment coefficient (C_m)

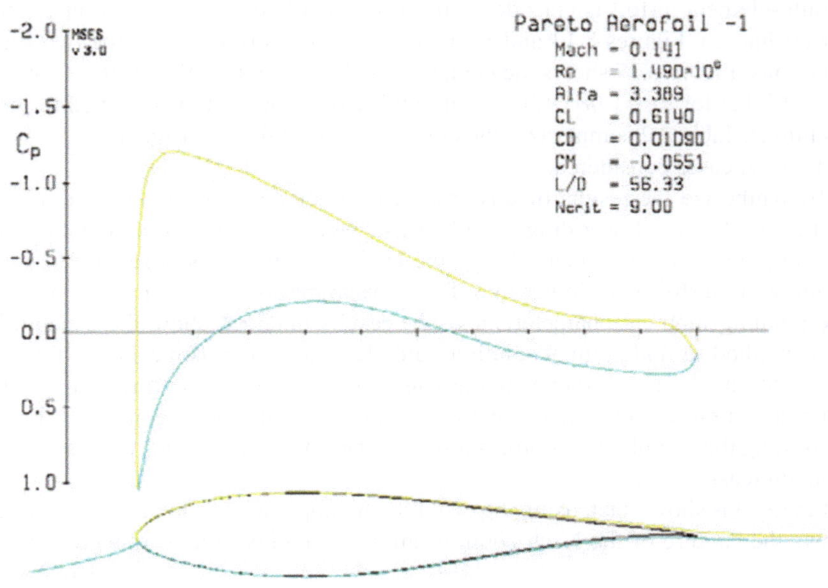

Fig. 8.12 Pareto 01 Flight condition 1—constrained pitching moment coefficient (C_m)

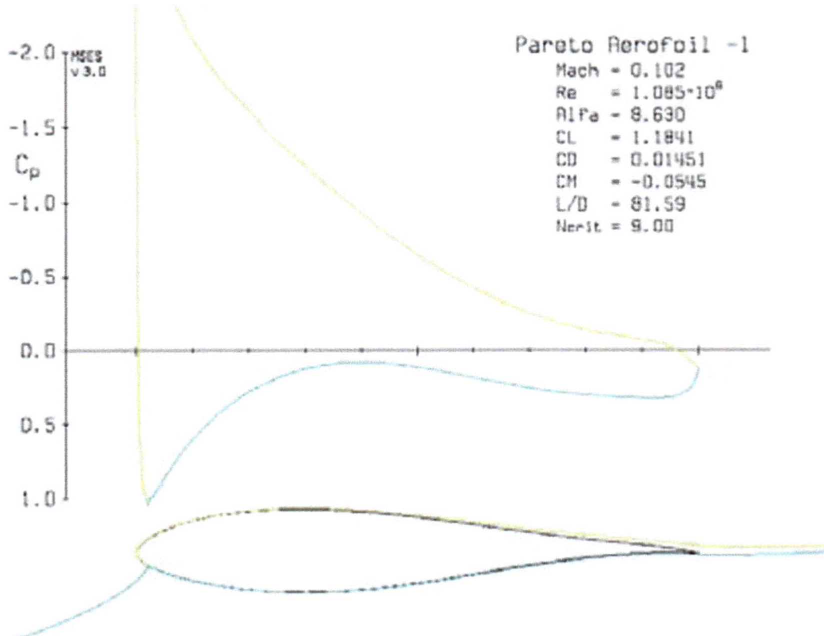

Fig. 8.13 Pareto 01 Flight Condition 2—constrained pitching moment coefficient (C_m)

A multi-element airfoil is considered from the middle of the Pareto front for further evaluation. Figures 8.13 and 8.14 show the *Cp* distribution for the two flight conditions. Figure 8.17 shows the comparative drag polars for Re = 1.480 × 106 and Fig. 8.15 that for Re = 1.085 × 10 6. But in this case a lower cm is obtained for both conditions. Table 8.2 summarizes the drag reduction at the two flight conditions for the two test cases considered.

To synthesize the results of this case, it is concluded that the evolved airfoils offer a significantly lower drag at both cruise conditions. The requirement of constraining the pitching moment during the evolution process is necessary to avoid obtaining an airfoil with lower drag for some flight conditions but with undesirable pitching moment characteristics. The results obtained show the capabilities of the method to find optimal solutions and classical aerodynamic shapes for flow drag. The importance of sound engineering judgement before, during and after the optimization must not be under-estimated: a proper definition of constraints before performing the evolutionary optimisation and the final results needs to be evaluated to obtain feasible designs.

Finally it is shown that, using the evolutionary approach and methods developed, allows the capture of the Nash equilibrium and complex Pareto fronts describing

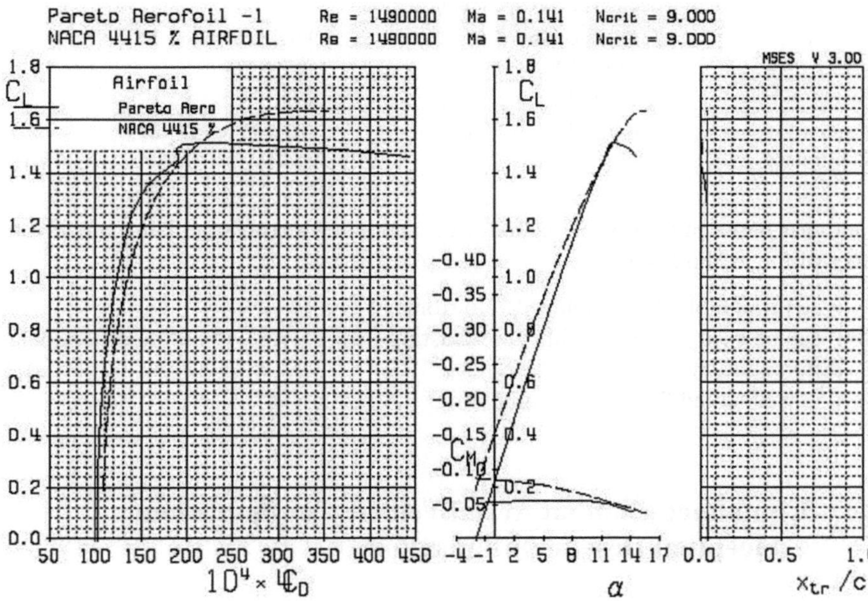

Fig. 8.14 Comparative Polars—Compromised airfoil and NACA 4415 Re = 1.490×10^6—constrained pitching moment coefficient (C_m)

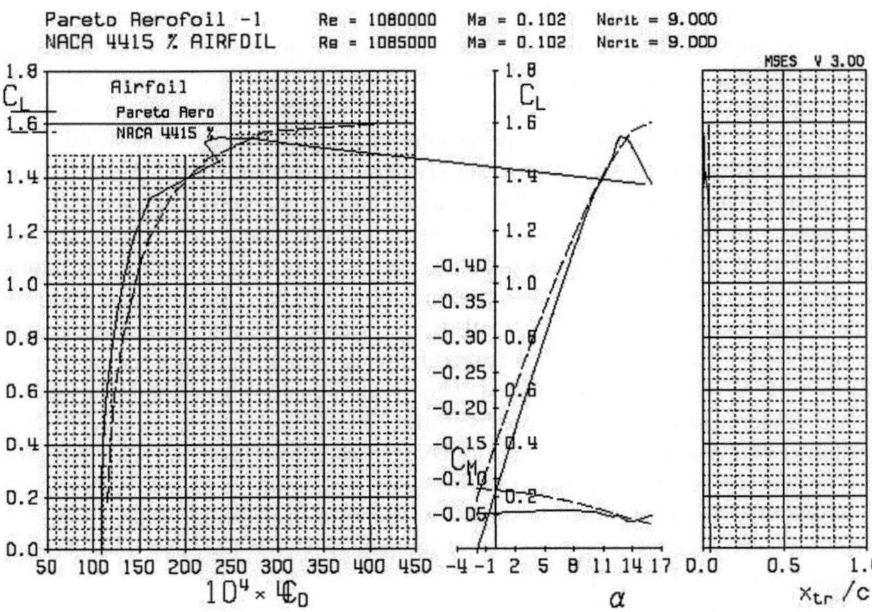

Fig. 8.15 Comparative Polars—Compromised airfoil and NACA 4415 Re = 1.085×10^6—constrained pitching moment coefficient (C_m)

Table 8.2 UAV Drag reduction at two operating conditions

Description	Flight condition 1 slow cruise	Flight condition 2 fast cruise
NACA 4415	0.01621	0.01173
Pareto 01 Constrained	0.01451 [−10.48 %]	0.01090 [−7.07 %]

trade-offs between the objectives for direct and inverse problems. The main advantages of the method is that:

I. it can find globally optimum Pareto fronts,
II. it can be easily coupled with a pre-existing flow-solver without modification or differentiation,
III. it is well suited for parallel computing.

8.4 Unmanned Combat Aerial Vehicle Configuration: Conceptual Design Optimisation

Problem Definition This test case considers the design and optimization of an Unmanned Combat Aerial Vehicles (UCAV) wing airfoil sections and plan form geometry operating at cruise regime and ingress conditions. UCAVs have the potential to reduce the risk of human life in combat operation, satisfy redundancy requirements and improve the effectiveness and efficiency of the mission. The missions for an UAV/UCAV can be categorized in terms of air-air, air-ground or both. Successful completion of these missions requires a UAV/UCAV exhibit high survivability which can be achieved by the improvement of manoeuvrability and stealth. The manoeuvrability is directly related to aerodynamic performance of UAV/UCAV and can be improved by optimising the shapes for the wing aerofoil sections [13, 14] and wing plan form [15–18]. In terms of stealth, Radar Cross Section (RCS) must be considered and the minimisation of the visual signature, infrared heat signature, acoustic signature, radio transmission signature and radar echo signature can provide a low observable aircraft [19, 20]. In this section, the multi-objective design optimisation of the wing aerofoil sections and wing plan form shapes for the survivability of UAV/UCAV is investigated. The visual signature from the front view is only considered to achieve a low observability. The method couples the HAPMOEA optimizer with aerodynamic analysis tools for transonic aerofoil and wing design.

The type of vehicle considered in this section is Joint Unmanned Combat Air Vehicle (J-UCAV) [21] which is similar in shape to the Boeing X-45C, Dassault Aviation—PETIT DUC and Northrop Grumman X-47A/B [22, 23]. The baseline design is shown in Fig. 8.16.

The wing plan form shape is assumed to be an aero-diamond shape with jagged trailing edge. This is convenient for stealth purposes and deflecting radar echoes away from the source [24, 25]. The aircraft maximum gross weight is approxi-

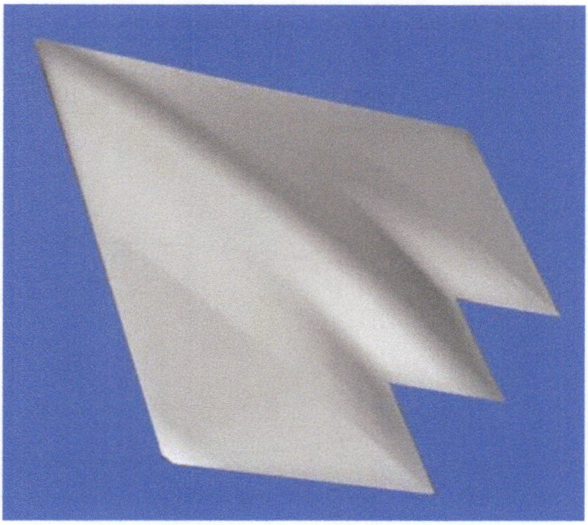

Fig. 8.16 Baseline UCAV

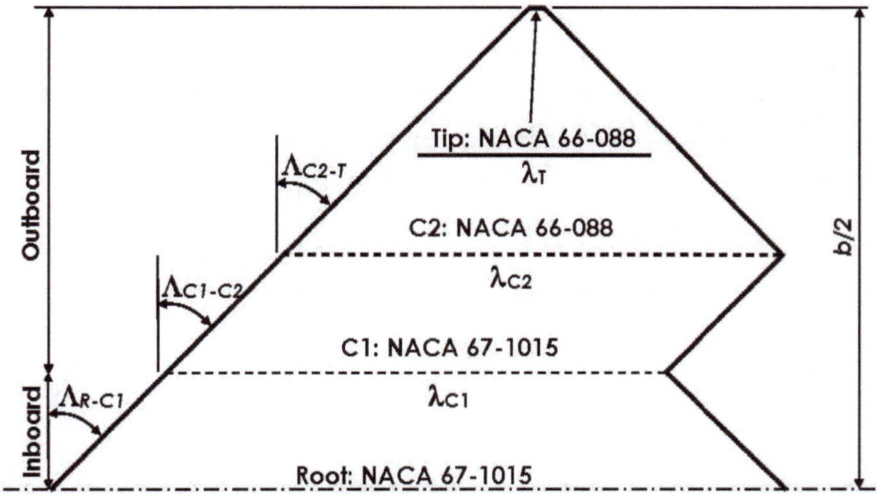

Fig. 8.17 Baseline UCAV wing configurations

Table 8.3 Baseline UCAV wing configurations

AR	b	$\Lambda_{R\text{-}C1}$	$\Lambda_{C1\text{-}C2}$	$\Lambda_{C2\text{-}T}$	λ_{C1}	λ_{C2}	λ_T	$\Gamma_{Overall}$
1.65	18.2 m	55°	55°	55°	0.68	0.68	0.02	0°

mately 5190 lb (2,354 kg) and empty weight is 3249 lb (1473 kg). The wing design parameters for the baseline wing configuration are given in Fig. 8.17 and Table 8.3. The inboard and outboard sweep angles are 55°. Inboard and outboard taper ratios are 0.68. These identical sweep angles and taper ratios provide a plan form alignment such that the leading edge is parallel with trailing edge.

Table 8.4 Baseline UCAV wing airfoil sections

Sections	Root	Crank1	Crank2	Tip
Aerofoils	NACA 67-1015	NACA 67-1015	NACA 67-088	NACA 67-088

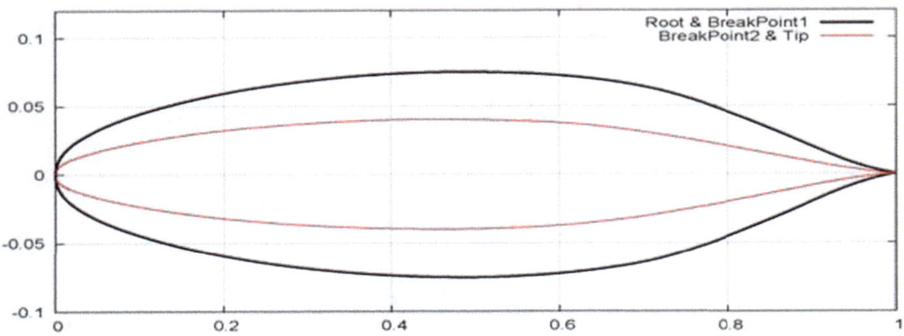

Fig. 8.18 Airfoil sections for baseline design

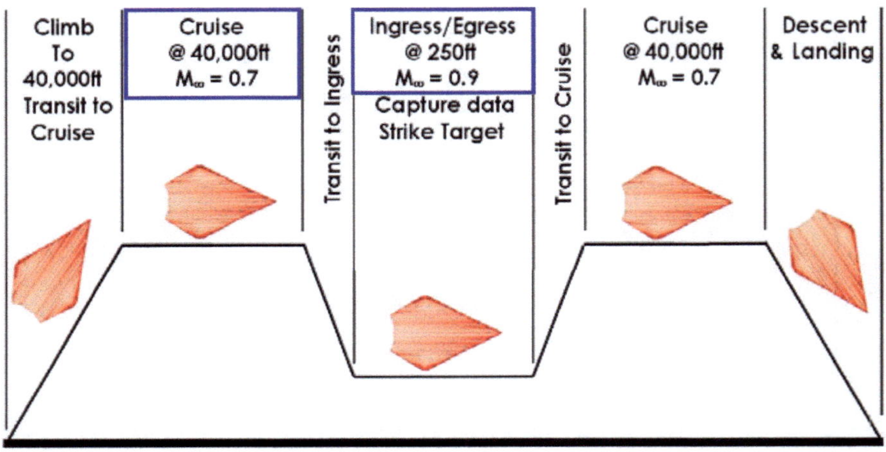

Fig. 8.19 Mission profile for baseline UCAV

It is assumed that the baseline design contains two types of airfoils as illustrated in Table 8.4 and Fig. 8.18; NACA 67–1015 and NACA 67–008 at the inboard and outboard sections.

The mission profile of the UCAV considers Reconnaissance, Intelligence, Surveillance and Target Acquisition (RISTA) as illustrated in Fig. 8.19. Initially the UCAV climbs up to 40,000 ft (12,192 m) then cruises close to the target range at $M\infty = 0.7$ and then the flight conditions transit at $M\infty = 0.8$. It will perform a stealth ingress to the target point at $M\infty = 0.9$ and altitude = 250 ft (76 m) and then destroy the target or establish surveillance of the target range. The UCAV climbs back to 40,000 ft and returns to a predetermined location.

Table 8.5 Flight conditions

Mission flight section	*M∞*	*Altitude*	*L/D Baseline*
Flight condition 1—Cruise	0.7	40,000 *ft*	16.565
Flight condition 2—Ingress	0.9	250 *ft*	4.729

Two flight conditions including the cruise and ingress are investigated during optimization. These baseline flight conditions are given on Table 8.5. For the cruise/ingress flight condition, the minimum lift coefficients are calculated to be 0.1914 and 0.0215. These C_L values are applied as inequality constraints. To achieve these lift coefficients for the baseline, it was found that the angle of attack should be at 5.408° for flight condition 1 (cruise) and 0.5733° for flight condition 2 (ingress).

In addition, one of the factors for stealth and RCS geometry area is also considered. Radar cross section (σ) is most easily visualized as the product of three factors; {AU: "GEOMETRIC" BELOW??}

$$\sigma = \begin{pmatrix} \text{Geometirc} \\ \text{Cross Sections} \end{pmatrix} \times (\text{Reflectivity}) \times (\text{Directivity}) \tag{8.9}$$

where the area of the target presented to the radar can be referred to geometric cross sections. Reflectivity refers to the fraction of the intercepted power that is reflected by the target, regardless of direction. Directivity is related to reflectivity but refers to the power scattered back in the direction of the transmitting radar.

In this test, the only factor considered is the frontal area of the model as the geometric cross section. The objectives are maximization of the lift to drag ratio (L/D) at cruise and ingress conditions and minimization of frontal area. The fitness function and flight conditions are:

$$fitness(f_1) = \min\left(\frac{1}{L/D_1}\right) + Penalty \tag{8.10}$$

$$\text{at } M_\infty = 0.7 \text{ and } \alpha = 5.408°$$

$$fitness(f_2) = \min\left(\frac{1}{L/D_2}\right) + Penalty \tag{8.11}$$

$$\text{at } M_\infty = 0.9 \text{ and } \alpha = 0.5733°$$

$$fitness(f_3) = \min\left(Area_{Frontal} / 100\right) + Penalty \tag{8.12}$$

$$Penalty_{Visual} = Area_{Frontal} + Area_{Frontal} \times Error_{Area} \tag{8.13}$$

where $Error_{Area}$ is the ratio of the frontal area error;

where *Penalties* (2.4) and (2.5) introduced in Chap. 2 are computed and added to fitness functions.

- if the thickness ratio (*t/c*) at root and crank 1 sections are higher than 20 % and less than 10.0 % of the chord $\left(10\% \geq t/c_{Root\&C1} \geq 20\%\right)$;
- if the ratios of lift to drag at cruise and ingress conditions are lower than 16.565and 4.729 $(L/D_{Cruise} \leq 16.565$ and $L/D_{Ingress} \leq 4.729)$ then the multiple penalty (8.10 and 8.11) is applied.
- if the frontal area exceeds 23.28 $(Area_{Frontal} \geq 23.28)$. Then the multiple penalty (Eq. 8.13) is applied.

This optimization uses Algorithm 4 in sect. 6.5.5 of Chap. 6.

Design Variables The airfoil geometry is represented using Bézier curves [8] with a combination of a mean line and thickness distribution as mentioned in Sect. 8.2. Four (4) airfoil sections are considered at root, crank 1, crank 2 and tip. Upper and lower bounds for mean and thickness control points at root, crank 1, crank 2 and tip sections are given in Figs. 8.20a, b, c.

The wing plan form shape is parameterized with variables described in Fig. 8.21 where three sweep angles, three sectional wing wetted areas and three taper ratios are considered. These lead to the different span length and Aspect Ratio (*AR*). The taper ratio at crank 2 will not be higher than the taper ratio at crank 1 i.e. $(\lambda_{C2} \leq \lambda_{C1})$.

The upper and lower bounds for the aerofoil design variables are illustrated in Figs. 8.20a, b, c. Seven control points are taken on the mean line and ten control points are taken for the thickness distribution respectively. For the wing planform design, six (6) design variables are considered three (3) trapper ratios and three sweep angles. To maintain the same wing wetted area as baseline design, crank positions are recalculated without washout and dihedral effects. Upper and lower wing design bounds are illustrated in Table 8.6 and Fig. 8.22. Fifty seven (57) variables are considered in total.

Implementation The FLO22 + FRICTION solvers are utilised and the following specific parameters are considered for HAPMOEA.

1st Layer	Population size of 200 and intermediate recombination Computational grid of 96 × 12 × 16 cells used by FLO22 solver
2nd Layer	Population size of 40 and intermediate recombination Computational grid of 82 × 12 × 16 cells used by the FLO22 solver
3rd Layer	Population size of 60 and intermediate recombination Computational grid of 68 × 12 × 16 cells used by the FLO22 solver

Numerical Results The algorithm was allowed to run approximately 1600 function evaluations and took 200 h on a single 1.8 *GHz*. The resulting Pareto set is shown in Fig. 8.23 where it can be seen that the shape of Pareto fronts formed is a linear function for objectives 1 and 2.

Fig. 8.20 **a** Root mean line and thickness control points. **b** Crank 1 mean and thickness control points. **c** Crank 2 and Tip mean and thickness control points

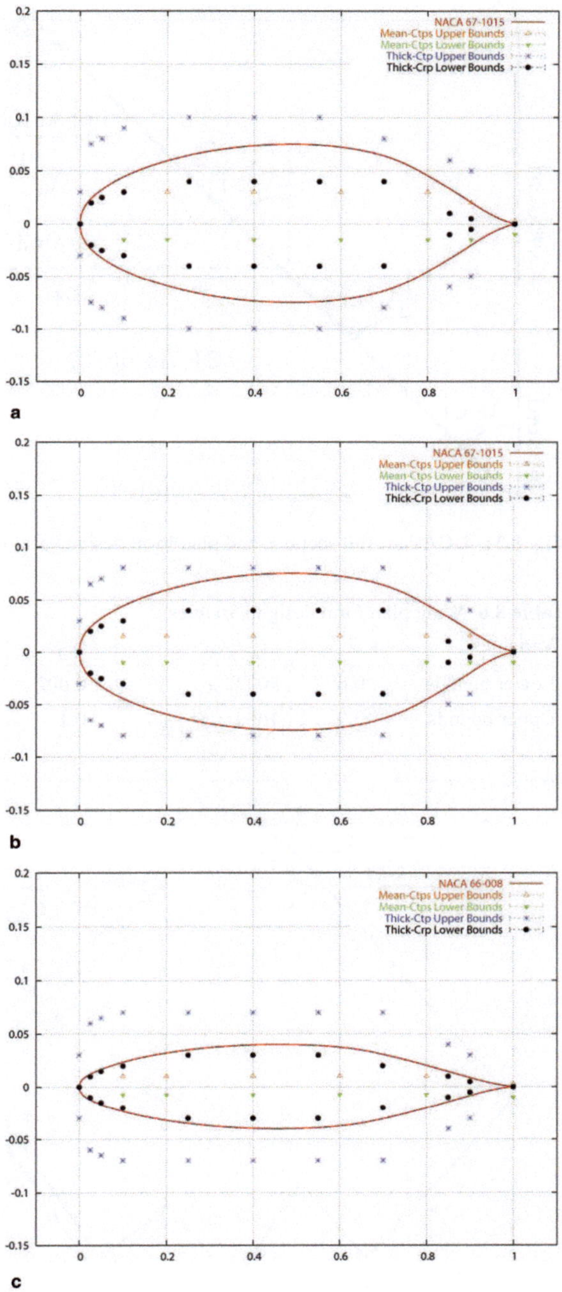

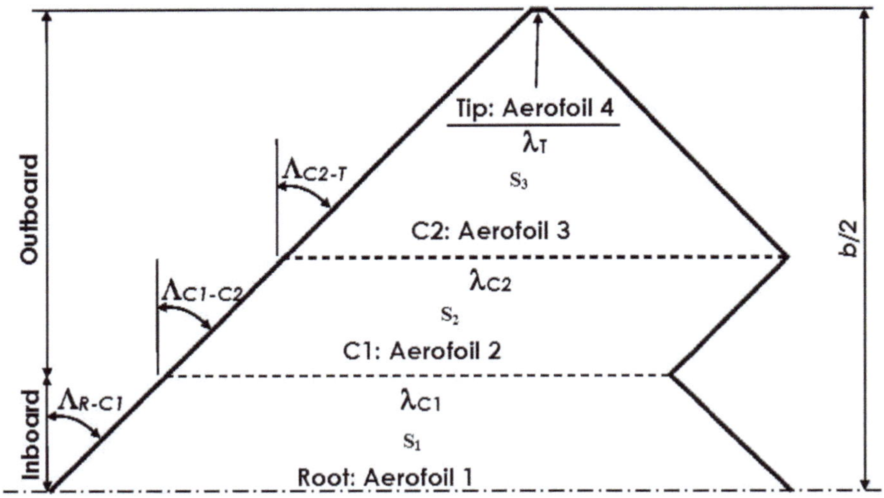

Fig. 8.21 UCAV air foil sections and plan form design variables

Table 8.6 Wing plan form design variables

Variables	λ_{R-B1}	λ_{B1-B2}	λ_{B2-T}	Λ_{R-B1}	Λ_{B1-B2}	Λ_{B2-T}
Lower bounds	0.6	80 (% λ_{R-B1})	0.002	45°	45°	45°
Upper bounds	0.9	100 (% λ_{R-B1})	0.1	65°	65°	65°

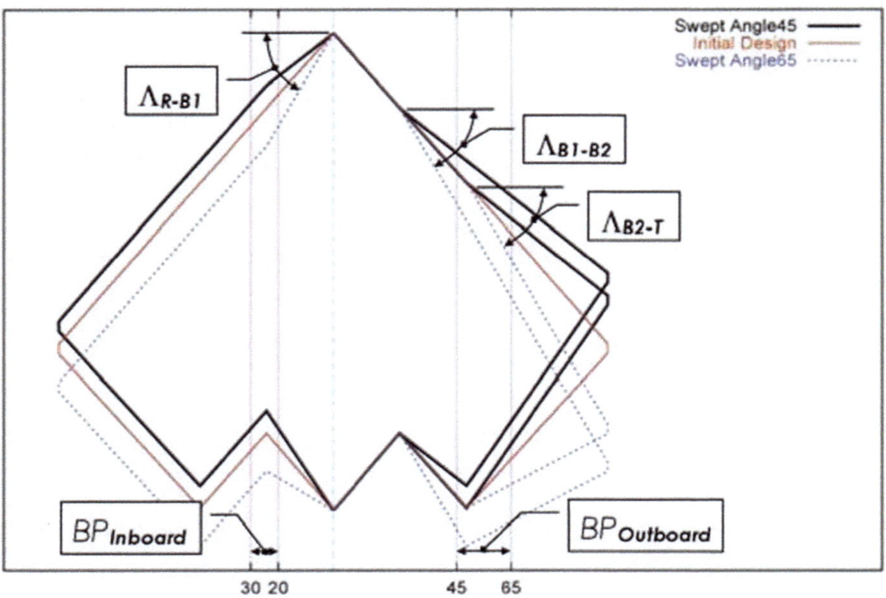

Fig. 8.22 Wing plan form shape corresponding to design variables

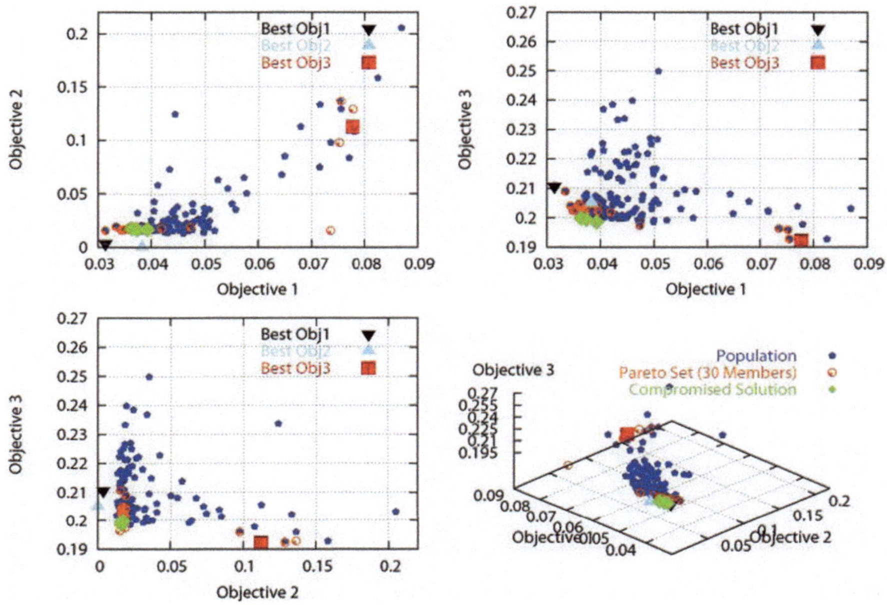

Fig. 8.23 Pareto optimal fronts for UCAV wing airfoil sections/plan form

Table 8.7 Comparison of the aerodynamic performances

Description	Baseline	Pareto M1	Pareto M10	Pareto M19	Pareto M22	Pareto M30
L/D_{Cruise}	16.565	31.849	*27.477*	*26.775*	*25.462*	12.871
L/D_{Ingress}	4.729	63.879	*57.250*	*61.032*	*60.104*	7.764
$Area_{\text{Front}}$	23.28	21.08	*19.98*	*20.11*	*20.15*	19.28

The lift to drag ratios of five Pareto members (1, 10, 19, 22 and 30) were selected and compared to the baseline design in Table 8.7 where it can be seen that Pareto members provide the improvement of aerodynamic performance. Pareto members 10, 19 and 22 are selected for further evaluation since they produce higher ratios of lift to drag for both cruise and ingress conditions and contain lower frontal area by 14 % when compared to the baseline design; in other words, a reduction in RCS. The lift coefficient distribution of Pareto sets for cruise and ingress conditions are illustrated in Figs. 8.24a and b.

The wing plan form shapes corresponding to Pareto fronts are shown in Fig. 8.25 where it can be seen that all Pareto members have a longer span compared to the baseline. The outcome is derived from an ignorance of the bending moment in the fitness function. In addition, their wing configurations are compared to the baseline design in Table 8.8. It can be seen that the aspect ratio of Pareto members (10, 19 and 22) are 1.5 or 2 times the baseline design. The span length is increased by 27 % when compared to the baseline design. Hence these modifications explain the

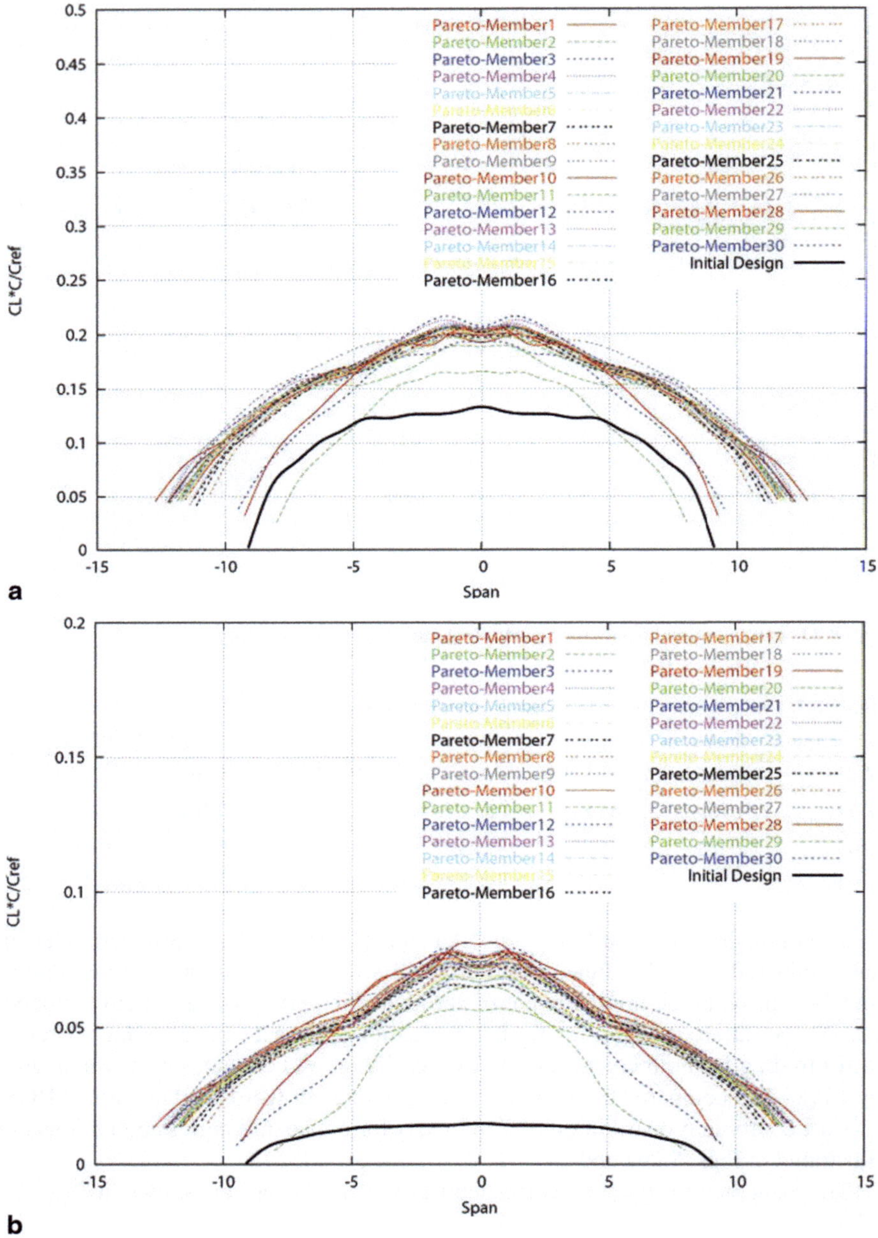

Fig. 8.24 **a** Lift coefficient distribution-Cruise. **b** Lift coefficient distribution-Ingress

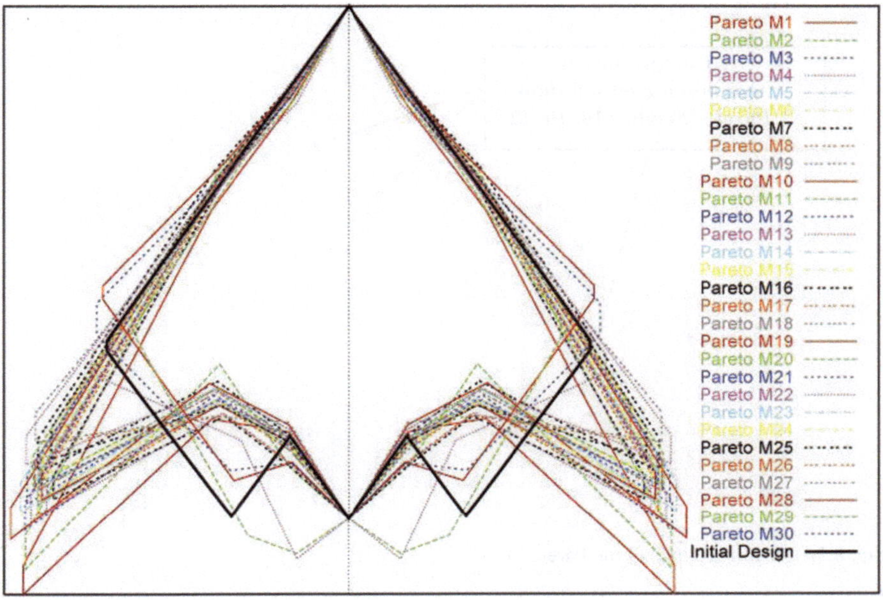

Fig. 8.25 Optimal wing plan form shapes

Table 8.8 Comparison of wing configurations

Description	Baseline	Pareto M1	Pareto M10	Pareto M19	Pareto M22	Pareto M30
AR	1.66	3.02	*3.25*	*2.68*	*2.96*	1.80
b	18.22	24.55	*25.49*	*23.15*	*24.32*	18.99

gap discrepancies in the lift distribution of cruise (Fig. 8.24a) and during ingress (Fig. 8.24b).

Figure 8.26 shows the shape of Pareto airfoil sections. It can be seen that classical aerodynamic shapes for transonic speeds have evolved, even considering that the optimization procedure is randomly started and the evolution algorithm had no problem specific knowledge of appropriate solution. The airfoil sections are compared on Table 8.9 and Fig. 8.27 where it can be seen that the airfoils for the root and crank1 have less thickness ratio when compared to the baseline design, however for the crank 2 and tip, the thickness of the airfoil is higher and positive cambers are observed.

The span wise *Cp* distribution between the compromised solutions (Pareto members 10, 19 and 22) and the baseline design for cruise and ingress conditions are compared in Figs. 8.28 and 8.29. *Cp* distributions are plotted at 0, 20, 40, 60, 80, 90 % of the span where it can be seen that the compromised solutions are the supercritical wings at ingress conditions. Top, front, side and three-dimensional views for optimal wing geometry (Pareto Member 22) are presented in Fig. 8.30. In addition,

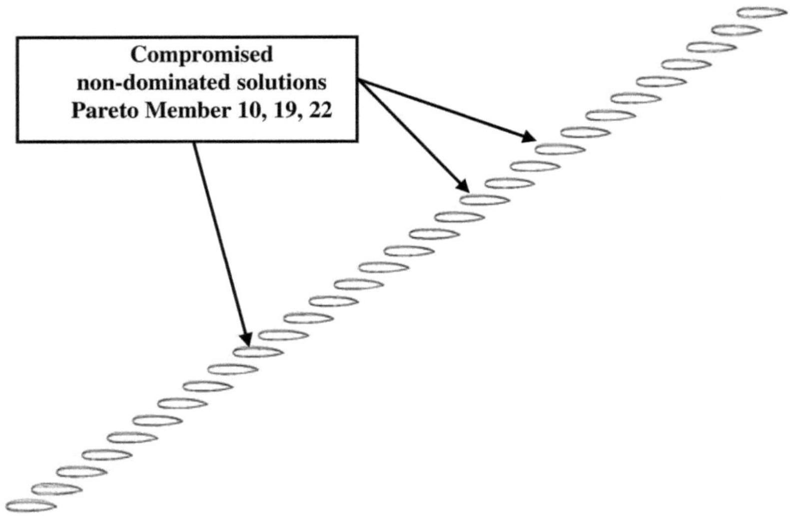

Fig. 8.26 Airfoil sections of the Pareto Front

Table 8.9 Comparison of airfoil sections configurations

Description	*Thick* Root	*Camber* Root	*Thick* C1	*Camber* C1	*Thick* C2-Tip	*Camber* C2-Tip
Baseline	0.1499	0.0	0.1499	0.0	0.0799	0.0
Pareto M1	0.1402	0.0139	0.1085	0.0090	0.0855	0.0036
Pareto M10	*0.1217*	*0.0121*	*0.1025*	*0.0111*	*0.0867*	*0.0038*
Pareto M19	*0.1201*	*0.0089*	*0.1032*	*0.0101*	*0.0890*	*0.0044*
Pareto M22	*0.1249*	*0.0118*	*0.1054*	*0.0075*	*0.0892*	*0.0033*
Pareto M30	0.1246	0.0102	0.0979	0.0106	0.0823	0.0055

their wing geometries are quite similar to the stealth fighters Boeing X-45C and Northrop Grumman X-47B, and F-117 Nighthawk as shown in References [26].

8.5 Unmanned Aerial Vehicle Mission Path Planning System (Hybridized Game/NSGA-II)

UAS mission path planning and EAs for mission path planning is an active field of research [26–33]. In this section, EAs are implemented to explore a path planning system of a generic national security stealth type of Unmanned Aerial System (UAS), or a civilian UAS that needs to fly through mountainous terrain during for example a gas sensing, environmental or search and rescue mission. Two optimisation methods NSGA-II and Hybrid-Game (explained in Sect. 4.6) applied

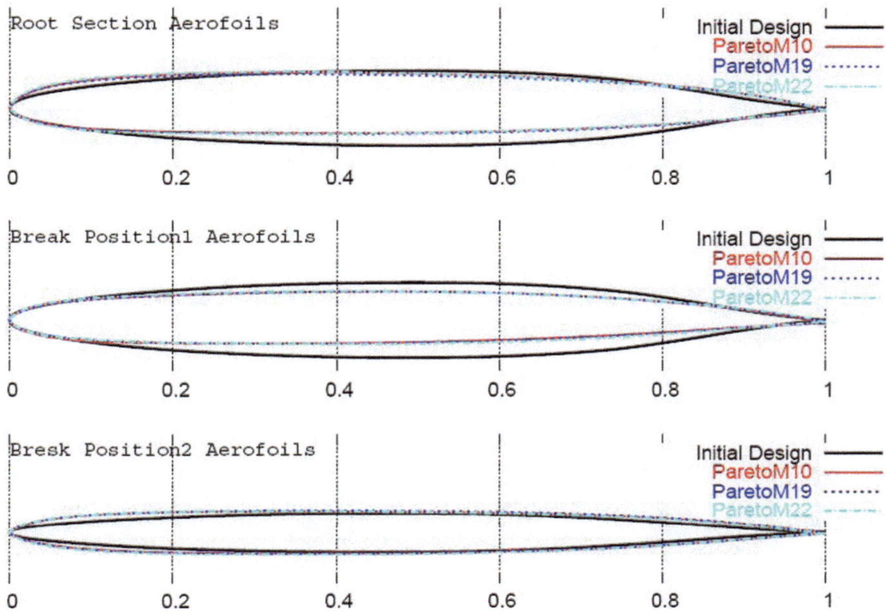

Fig. 8.27 Air foils sections for the compromised solutions

to NSGA-II are implemented for a UAS Mission Path Planning System (AMPPS). Results obtained by NSGA-II and Hybridized Games are compared in terms of solution quality and computational expense.

The terrain is represented by meshing three-dimensional surfaces with obstacles and altitude constraints. The example terrain is shown in Fig. 8.31 where there are obstacles in 90 % of the region where the UAS will be flying. The x and y distances are scaled down by 100 in x-axis and y-axis. The red square is the starting position and the blue square is the target position. This artificial terrain is randomly generated however it can be based on real geographical data.

For this application, a constraint of flying below 60 % of maximum altitude (surface (Fig. 8.32): 180 m) is applied due to the possible existence of enemy radar or to the limitations in the sensing capabilities of the UAS or because the mission requires air sampling below a given altitude.

The generated z coordinates of the trajectories should be below this altitude constraint and should not be through obstacles in order to be considered as a collision-free trajectory. Two terrains are considered in the applications: Test 1 (Fig. 8.33) and Test 2 (Fig. 8.33). The difference between them is on the return trajectory: Test 1 is a Target-Start while Test 2 is a Target-End.

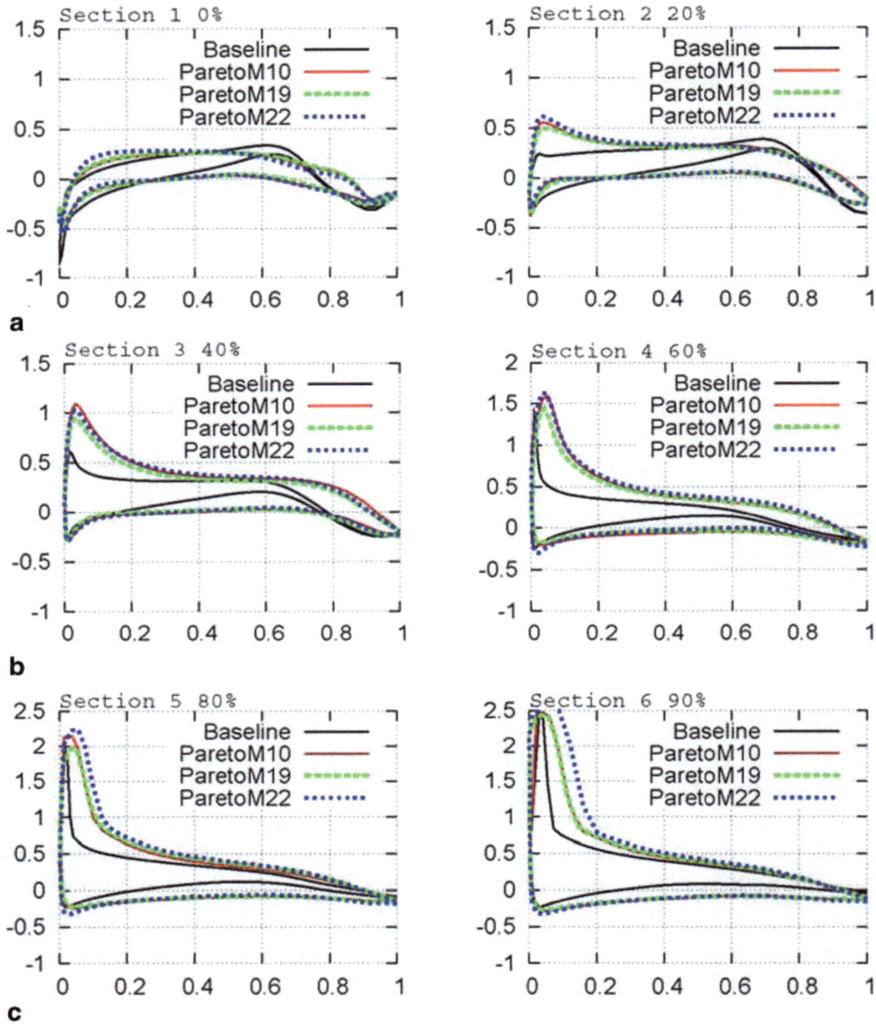

Fig. 8.28 **a** Comparison of *Cp* distributions at 0 and 20 % of span—Cruise. **b** Comparison of *Cp* distributions at 40 and 60 % of span—Cruise. **c** Comparison of *Cp* distributions at 80 and 90 % of span—Cruise

Two applications are considered:

- Test case 1 considers minimization of two trajectories one from the start position (100, 100, 100) to the target position (1800, 1800, 100) and another one from the target to the start position (Fig. 8.32).
- Test case 2 considers minimization of two trajectories one from the start position (900, 100, 150) to the target position (1800, 1800, 100) and another one from the target to the end position (100, 900,150) (Fig. 8.33).

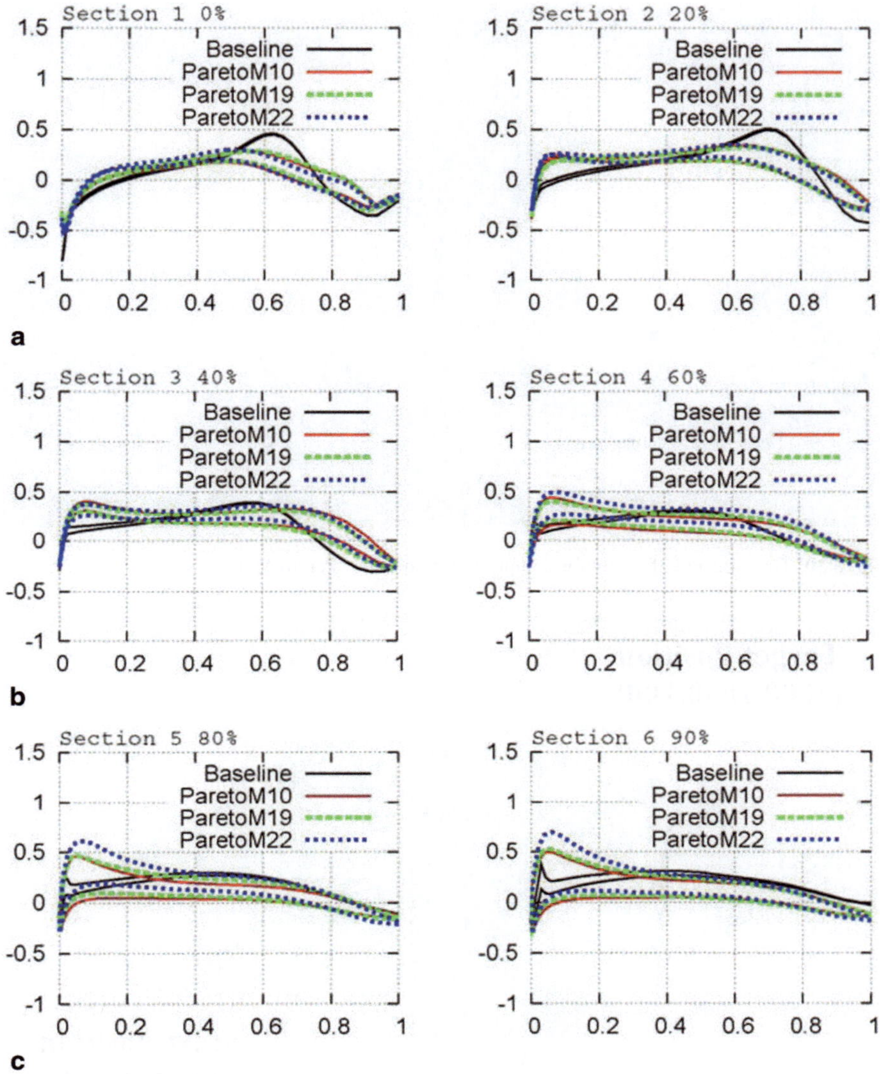

Fig. 8.29 **a** Comparison of *Cp* distributions at 0 and 20 % of span—Ingress. **b** Comparison of *Cp* distributions at 40 and 60 % of span—Ingress. **c** Comparison of *Cp* distributions at 80 and 90 % of span—Ingress

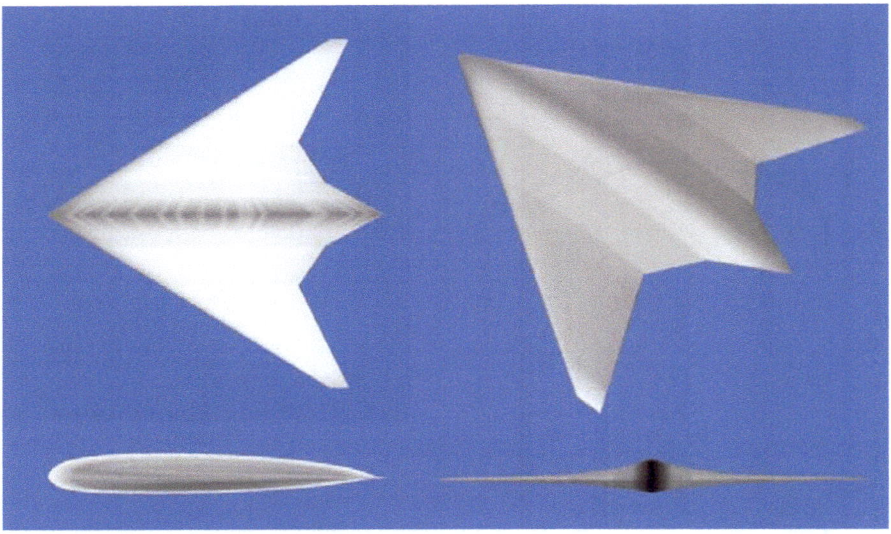

Fig. 8.30 Top/side and 3D view for compromized solution (Pareto Member 22)

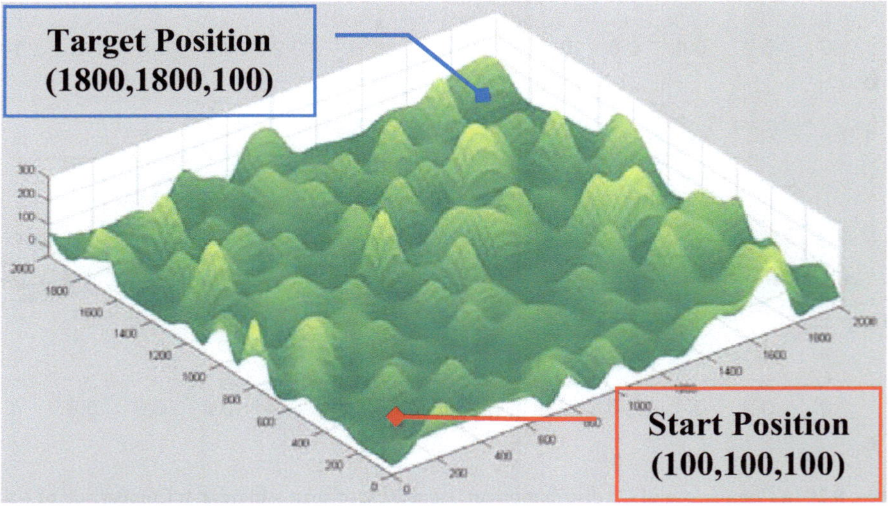

Fig. 8.31 Baseline terrain

8.5.1 Test 1: Start to Target to Start Position Trajectory Optimization

Problem Definition The test consists of minimizing the total length of the trajectory from the start position (100, 100, 100) to the target position (1800, 1800, 100) and

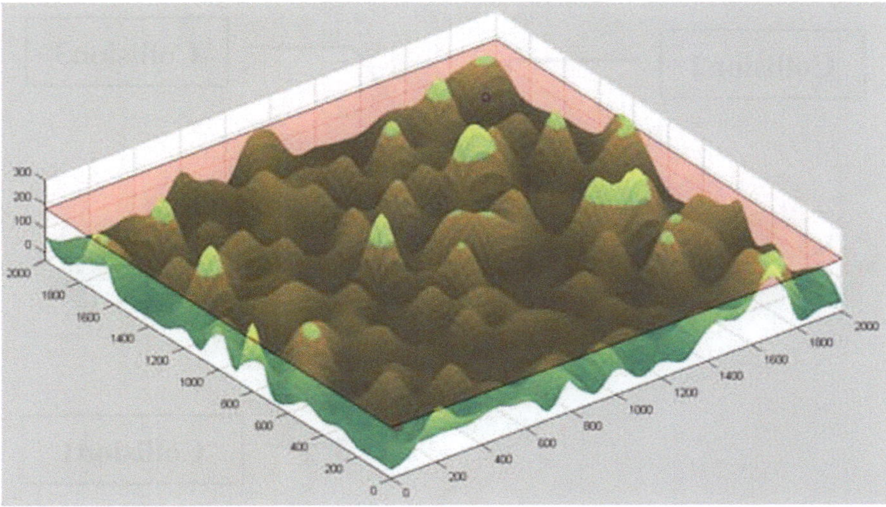

Fig. 8.32 Baseline terrain with altitude constraints (Test 1)

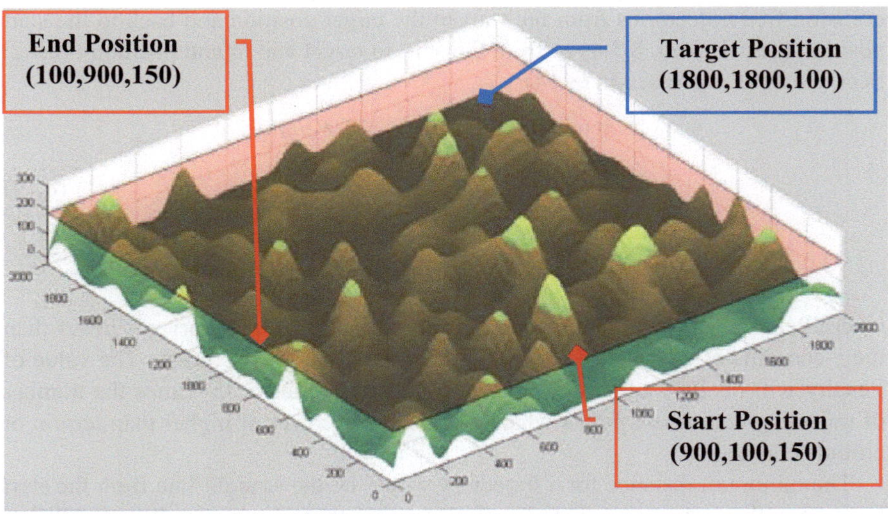

Fig. 8.33 Baseline terrain with altitude constraints (Test 2)

back to the start position. The aim of the optimization is to find a set of collision-free trajectories whose length is less than the straight line + 10 % of the length of this straight line. The minimum straight line distance for a trajectory between the start and target and back in this case is 4809 m. Therefore in this case, optimal trajectories should be less than 10 % of this value ($Path_{min+10\%}$ = 5289 m).

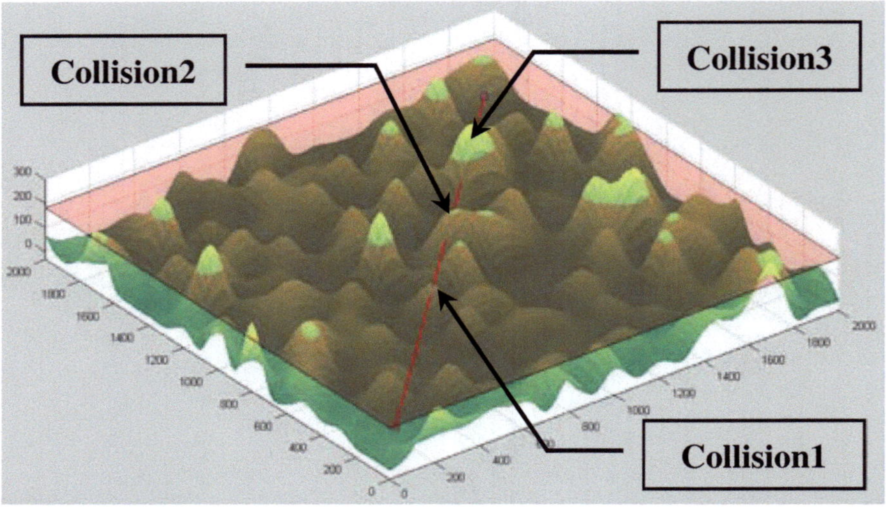

Fig. 8.34 Trajectory for minimum distance (Test 1)

The multi-objective optimisation consists of finding the length of Pareto optimal collision-free trajectories from the start to the target position and back to the start position (Test 1: Sect. 8.5.4) or from the start to target and to end position (Test 2: Sect. 8.5.5) as expressed in (8.14).

$$\begin{aligned} f_1 &= \min\left(length(Path_{S-T})\right) + Penalty \\ f_2 &= \min\left(length(Path_{T-S})\right) + Penalty \\ &\quad or \\ f_2 &= \min\left(length(Path_{T-E})\right) + Penalty \end{aligned} \tag{8.14}$$

where Penalty will be applied when the z-coordinates of trajectories is lower than the *z*-coordinates of obstacles or higher than the altitude constraints. The value of Penalty will be 10 % of straight distance (Figs. 8.38 and 8.35) times the number of trajectory coordinate points which are located at lower or higher than terrain or altitude constraints.

The minimum distance for a trajectory would be the straight line from the start position to the target position and back. However, there is a possibility of collision with obstacles when the UAV is flying as shown in Figs. 8.34 and 8.38.

The fitness functions are;

$$f_1 = \min(length(Path_{S-T})) + Penalty \tag{8.15}$$

$$f_2 = \min(length(Path_{T-S})) + Penalty \tag{8.16}$$

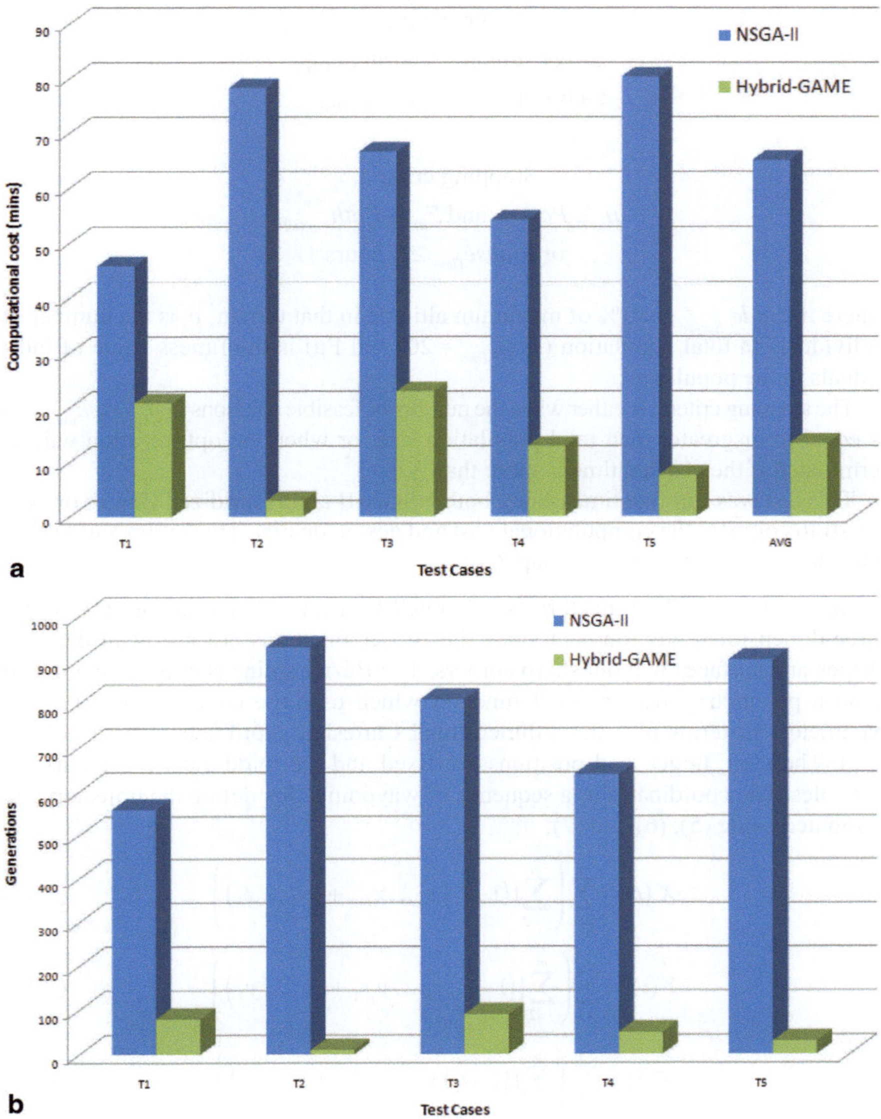

Fig. 8.35 **a** Comparison of performance (time in minutes) between NSGA-II and Hybridized Games applied to NSGA-II. Note: AVG represents the average computational cost of NSGA-II and Hybridized Games for five tests. **b** Comparison of performance (generations) between NSGA-II and Hybridized Games applied to NSGA-II software

Subject to;

$$z\text{- } coordinates < Altitude_{Radar}$$

$$\text{z-coordinates} > Altitude_{Obstacles}$$

Stopping criteria;

$$\mu_i \geq Pop_{Total} \text{ and } F_{\mu i} \leq Path_{min+10\%}$$

$$\text{or } Elapse_{Time} \geq 3 \text{ hours}$$

where $Altitude_{Radar}$ is 60 % of maximum altitude in that terrain, μ_i is the number of individuals in total population (Pop_{Total}=20) and Fμi is the fitness value of individuals in the population.

The stopping criteria is either when the number of feasible solutions ($F\mu_i \leq Path_{min+10\%}$) is equal to or greater than total population size, or when the optimization will be terminated if the elapsed time is more than 3 h

This test was run five times using both NSGA-II and Hybridized Games on NSGA-II to compare the computational cost and design quality. This optimization uses Algorithm 4 in Sect. 6.5.5 of Chap. 6.

Design Variables The trajectory is generated by using Bézier spline curves in a three-dimensional environment since the Bézier functions are useful in defining shapes and surfaces without sharp corners. The Bézier spline curves are computed from a parametric mathematical function which uses the control points (P_n) as parameters in terms of a three-dimensional Cartesian coordinate system i.e., (x, y, z). The start, target, end positions are fixed and the middle control points are variables. The coordinates of a sequence of waypoints that define the trajectory are computed using (5), (6) and (7);

$$X(t) = \sum_{i=1}^{n}\left(\sum_{j=2}^{m}\left((1-t_i)x_0 \ldots x_{j-1} + t_i x_1 \ldots x_j\right)\right) \tag{8.17}$$

$$Y(t) = \sum_{i=1}^{n}\left(\sum_{j=2}^{m}\left((1-t_i)y_0 \ldots y_{j-1} + t_i y_1 \ldots y_j\right)\right) \tag{8.18}$$

$$Z(t) = \sum_{i=1}^{n}\left(\sum_{j=2}^{m}\left((1-t_i)z_0 \ldots z_{j-1} + t_i z_1 \ldots z_j\right)\right) \tag{8.19}$$

where n is the number of coordinate points in the trajectory and m represents the number of control points for the Bezier spline curve. The value of parameter t is between 0 and 1 i.e., $t \in [0{:}1]$.

Results from simulations have shown that ten control points are sufficient to produce a detailed trajectory. The five coordinates of x-direction are fixed while five y and z coordinates are optimised for each trajectory. The range of y and z coordinates is from 0 to 2000 and from 0 to 300 respectively. The trajectory from the start position to the target position will be marked as red lines while the blue line will represent the trajectory from the target to the start position.

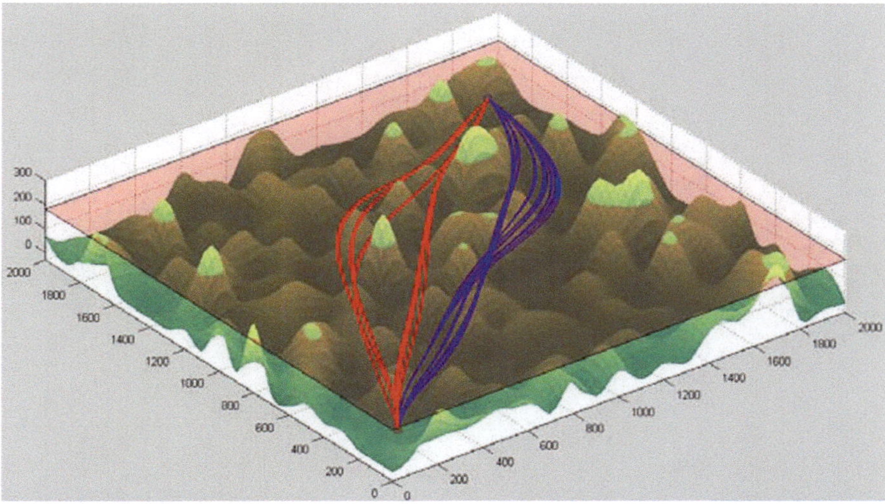

Fig. 8.36 Trajectories obtained by NSGA-II (T1)

The trajectory is generated using ten (10) control points for Start-Target and Target-Start. The y and z-coordinates are design variables while the x-coordinates are fixed. Twenty (20) design variables are considered in total.

Numerical Results The computational cost obtained by NSGA-II and Hybridized Games are compared in Figs. 8.35a (CPU time) and b (number of generations). It can be seen in the last column (AVG: average computational cost of five tests) in Fig. 8.35a that the Hybridized Games take 13.5 min while the computational cost of NSGA-II is 65.2 min for five tests (T1 ~ T5). In other words, Hybridized Games improve NSGA-II software efficiency by 80 % in this test case.

The trajectories (T1) obtained by NSGA-II are shown in Fig. 8.36 where the red lines represent the trajectory from the start position to the target position while the blue lines are for the trajectory from the target to the start position. The average distance of collision-free trajectories is 4989 m which is only 3.7 % higher than the minimum distance ($Path_{min}$).

Figure 8.37 shows that the trajectories (T4) obtained by Hybridized Games and the average distance of collision-free trajectories is 4980 m which is only 3.5 % higher than the minimum distance ($Path_{min}$).

However, trajectories (Fig. 8.37) obtained by Hybridized Games are not as diverse as the trajectories (Fig. 8.36) obtained from NSGA-II software, the computational cost of Hybrid Game is much lower and have shorter collision-free trajectories. There is only 9 m difference between NSGA-II software and Hybridized Games however this difference could be as much as 900 m if the terrain is scaled down in x and y-axis by 100 %.

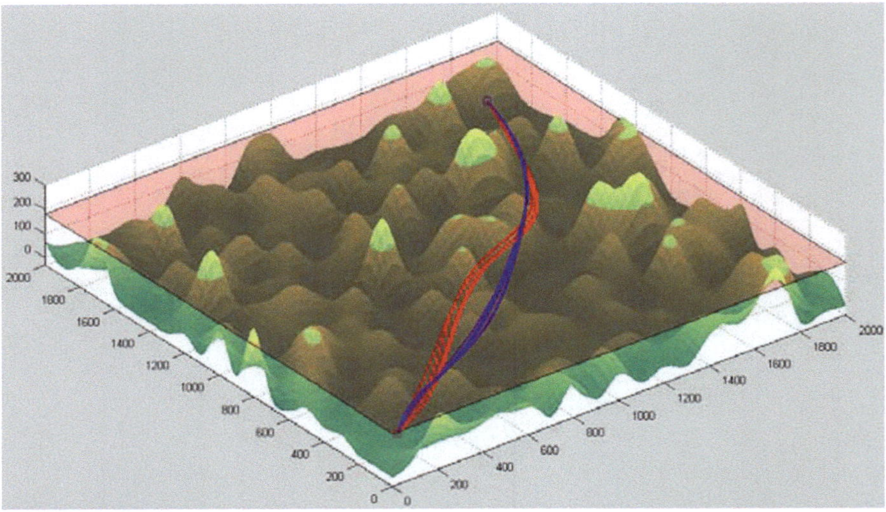

Fig. 8.37 Trajectories obtained by Hybridized Games (T4)

8.5.2 Test 2: Start to Target to End Position Trajectory Optimization

Problem Definition

- The test is to minimize the total trajectory length from the start position (900, 100, 150) to a target position (1800, 1800, 100) and then to an end position (100, 900,150). Test 2: The minimum straight line distance for a trajectory between the start to target and target to end point in this case is 3849 m; therefore in this case, optimal trajectories should be less than 10% longer than this value ($Path_{min+10\%}$=4233 m).

The fitness functions are;

$$f_1 = \min\left(length(Path_{S-T})\right) + Penalty \tag{8.20}$$

$$f_2 = \min(length(Path_{T-E})) + Penalty \tag{8.21}$$

Subject to;

$$z - coordinates < Altitude_{Radar}$$
$$z - coordinates > Altitude_{Obstacles}$$

Stopping criteria;

$$\mu_i \geq 0.5 \times Pop_{Total} \text{ and } F_{\mu i} \leq Path_{min+10\%}$$
$$\text{or } Elapse_{Time} \geq 3\text{hours}$$

where $Path_{min}$ is 3849 m and $Path_{min+10\%}$ is 4233 m.

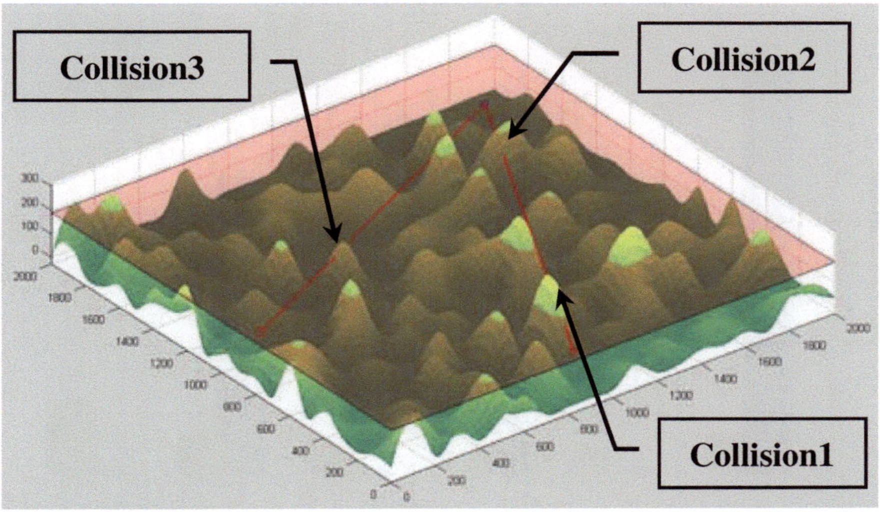

Fig. 8.38 Trajectory for finding Minimum distance (Test 2)

The same stopping criteria are applied as the previous test case in Sect. 4.1, one difference condition is that the optimisation will stop if the population of the Pareto Player (μ_i) contains 10 of the collision-free members out of total population (Pop_{Total}=20). This test was run five times using both NSGA-II and Hybridized Games on NSGA-II to compare the computational cost and solution design quality. This optimisation uses Algorithm 4 in Sect. 6.5.5 of Chap. 6.

Design Variables The trajectory is generated using ten (10) control points for Start-Target and Target-End as shown in Fig. 8.33. The *y* and *z*-coordinates for each control points are considered as design variables, while the *x*-coordinates are fixed. Twenty design variables are considered in total. See sect. 8.5.4 for details.

Numerical Results The computational cost obtained by NSGA-II software and Hybridized Games are compared in Figs. 8.39a (CPU time) and b (generations). It can be seen that the NSGA-II software failed to find ten collision-free path ways in Test 2 (T2) and Test 4 (T4) within a 3 h run in contrast to the Hybridized Games which successfully found the collision-free trajectories for five test cases (T1 ~ T5).

The shortest average distances of collision-free trajectories obtained by NSGA-II (T3: 41 min, 521 generations) and Hybridized Games (T2: 2 min, 7 generations) are 3967 m 3924 m respectively. In addition, the distance of Nash-Game inside Hybridized Games (T2) is 3922 m. The population of NSGA-II (T3: blue diamonds) and Hybrid-Game (T2: red squares) including Nash-Equilibrium (green triangles) are compared in Fig. 8.40. It can be seen that Hybridized Games have a better convergence than NSGA-II. Three members of NSGA-II failed the condition ($F\mu i \leq Path_{min} + 10\,\%$) from the start to target position while the entire population of

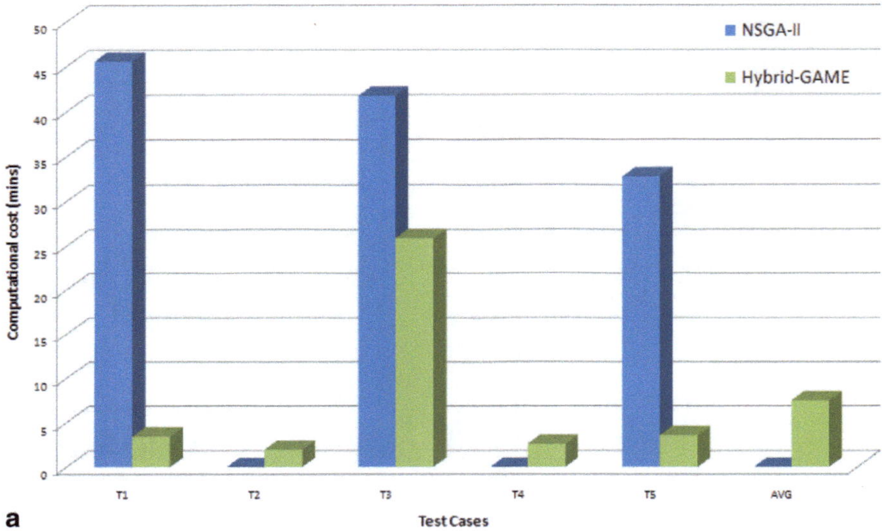

a

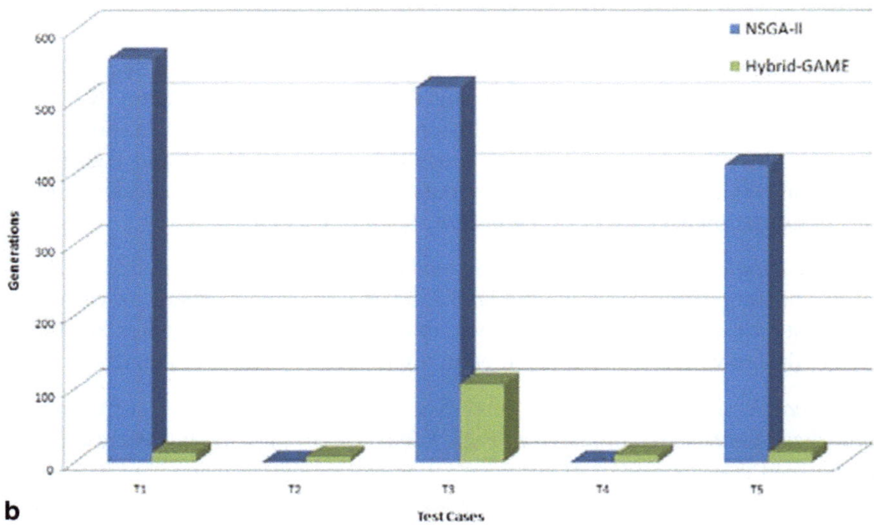

b

Fig. 8.39 a Comparison of performance (CPU time: mins) between NSGA-II software and Hybridized Games coupled to NSGA-II. **b** Comparison of performance (generations) between NSGA-II and Hybrid-Game coupled to NSGA-II

Hybridized Games satisfied this condition. The computational cost of Hybridized-Games (T2) is only 5 % of the minimum computational cost of NSGA-II (T5).

Figures 8.41 and 8.42 show the collision-free trajectories obtained by NSGA-II (T3: 17 members) and Hybrid-Game (T2: 20 members) respectively. Figure 8.41b and 8.42b show the trajectories on the top view where the return trajectories (T-E) of Hybridized Games are shorter than NSGA-II ones.

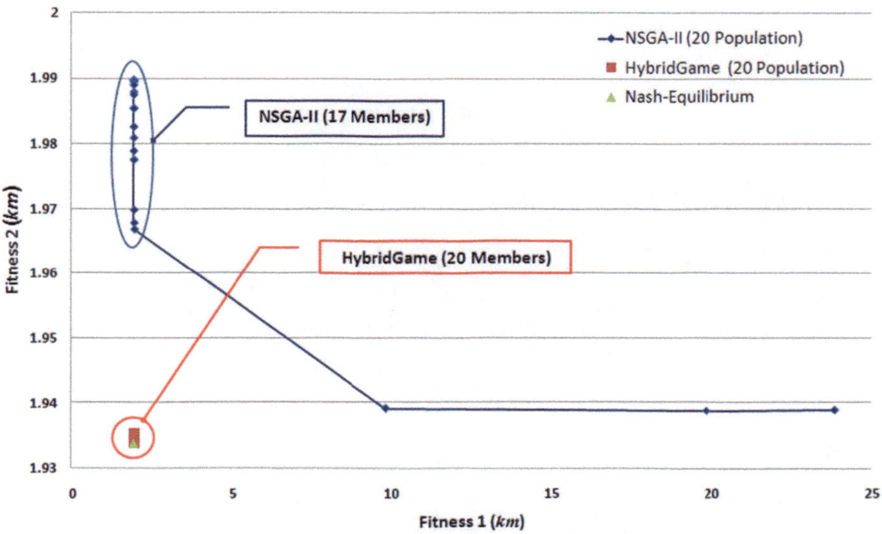

Fig. 8.40 Comparison of Pareto front between NSGA-II, Hybridized Games coupled to NSGA-II and Nash-Games

The reason why Hybridized Games have faster convergence than NSGA-II can be found in Test 1. For test case 1, NSGA-II (45.5 min) and Hybridized Games (3.4 min) are terminated after 560 and 13 generations respectively. The elite design obtained by Nash-Game and the population of Hybridized Games are compared in Fig. 8.43a where the elite design of Nash-Game dominates all individuals of Hybridized Games. The initial population distribution of Hybridized Games is similar to NSGA-II (Fig. 8.43b). However there is big difference between Hybridized Games (Fig. 8.44a) and NSGA-II (Fig. 8.44b) after the 10th generation: six members of Hybridized Games already have less than the distance of $Path_{min} + 10\,\%$ while none of NSGA-II members satisfy the distance conditions ($F\mu i \leq Path_{min} + 10\,\%$). This can be explained by the fact that the use of dynamic Nash-Game accelerates the searching speed of the Hybridized Games to capture the global solutions.

8.6 Unmanned Aerial Vehicle (Uav) Configuration: Detailed Design Optimisation

The type of vehicle addressed in this section is a generic Joint Unmanned Combat Air Vehicle (J-UCAV) which is similar in shape to Northrop Grumman X-47B [34,35] as shown in Fig. 8.45.

The wing plan form shape is assumed as an arrow shape with jagged trailing edge. The aircraft maximum gross weight is approximately 46,396 *lb* (21,045 *kg*) and empty weight is 37,379 *lb* (16,955 *kg*). The wing design parameters for the

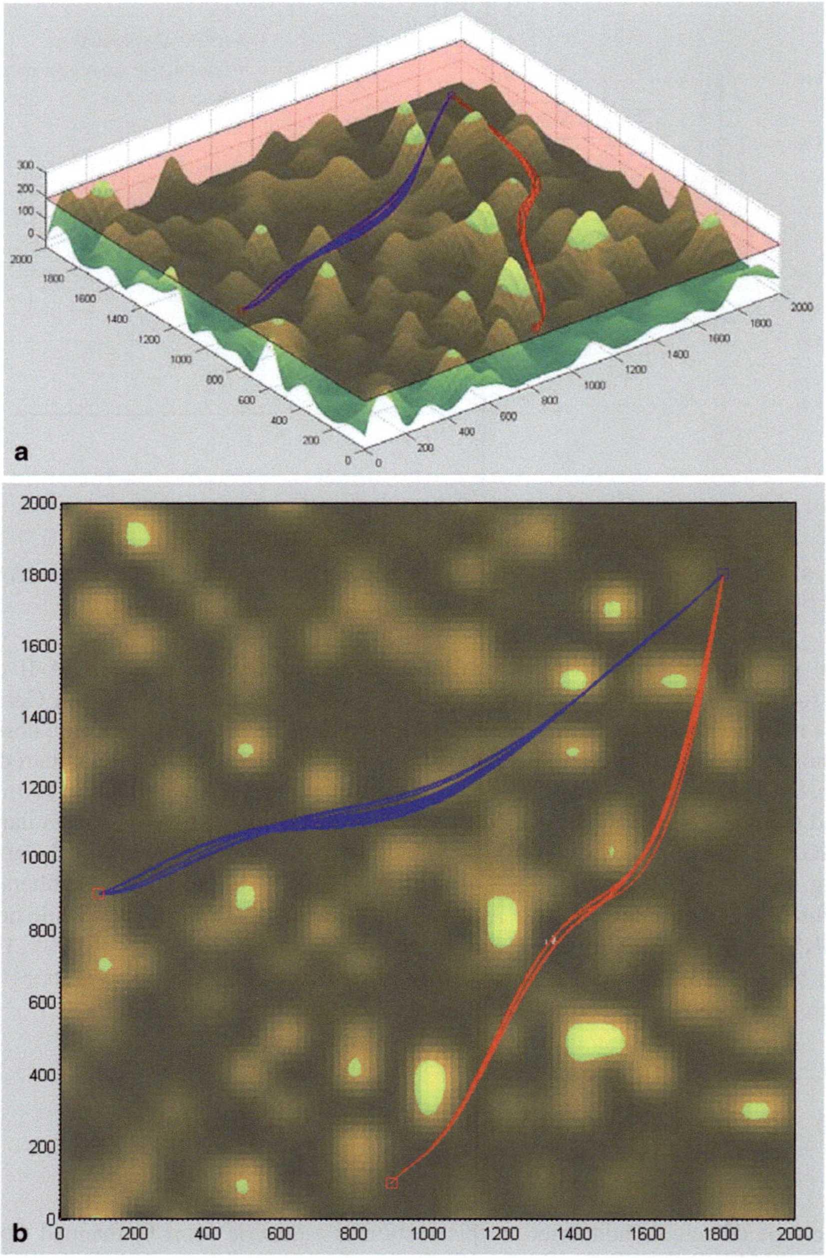

Fig. 8.41 **a** Trajectories obtained by NSGA-II (T3: 17 members). **b** Trajectories obtained by NSGA-II (T3)

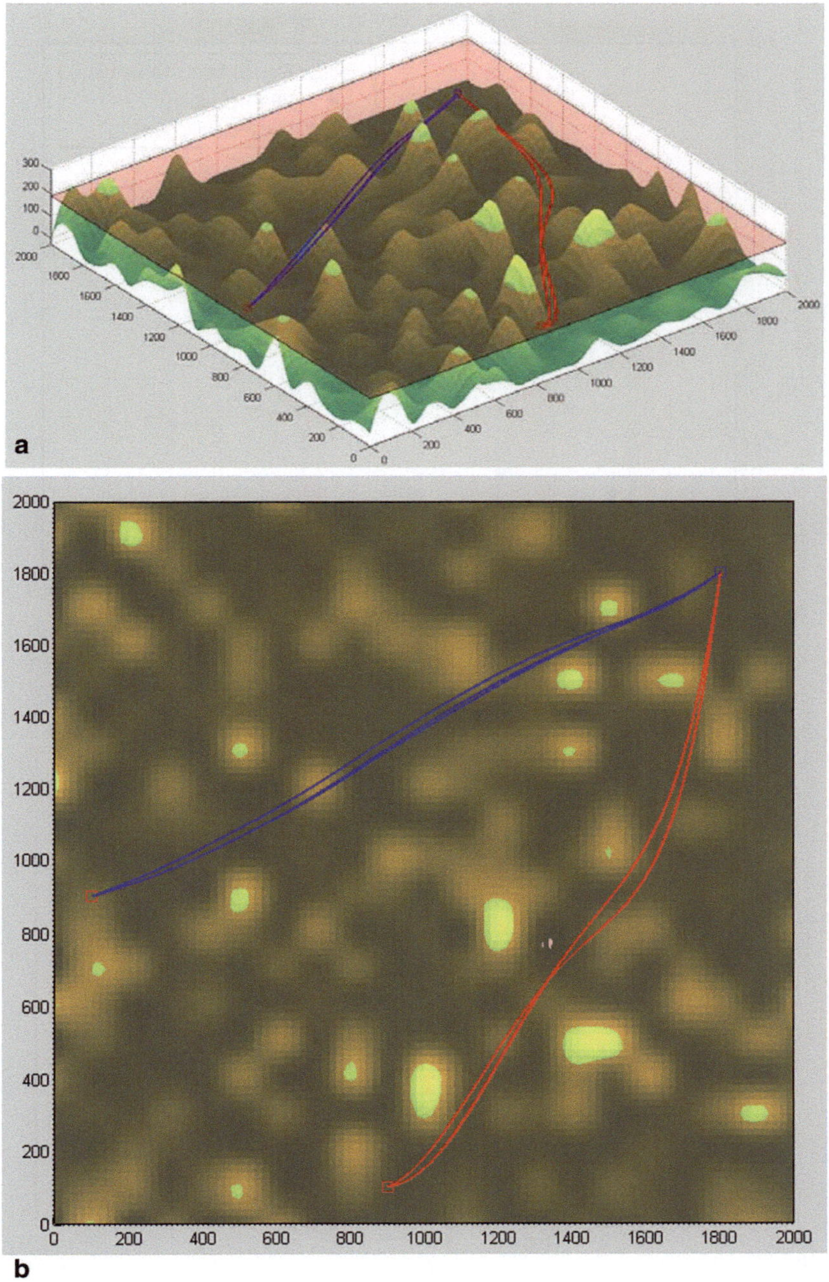

Fig. 8.42 **a** Trajectories obtained by Hybridized Games on NSGA-II (T2: 20 members). **b** Trajectories obtained by Hybridized Games coupled with NSGA-II (T2)

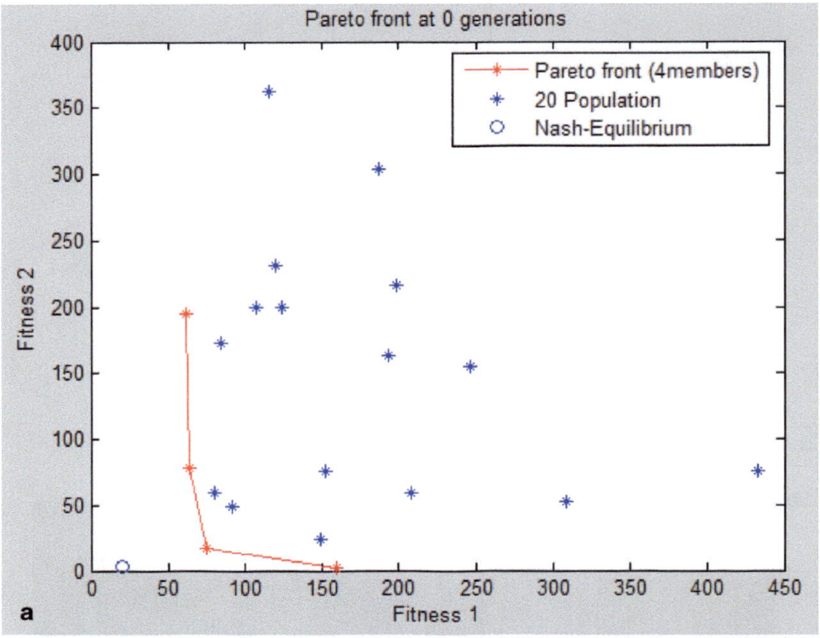

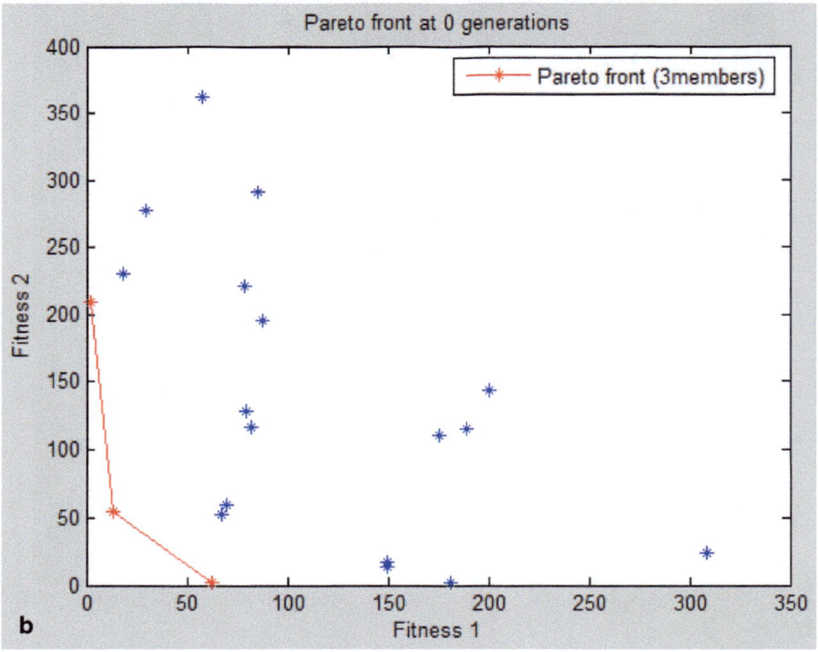

Fig. 8.43 **a** Population of Hybridized Games on NSGA-II at generation 0. **b** Population of NSGA-II at generation 0

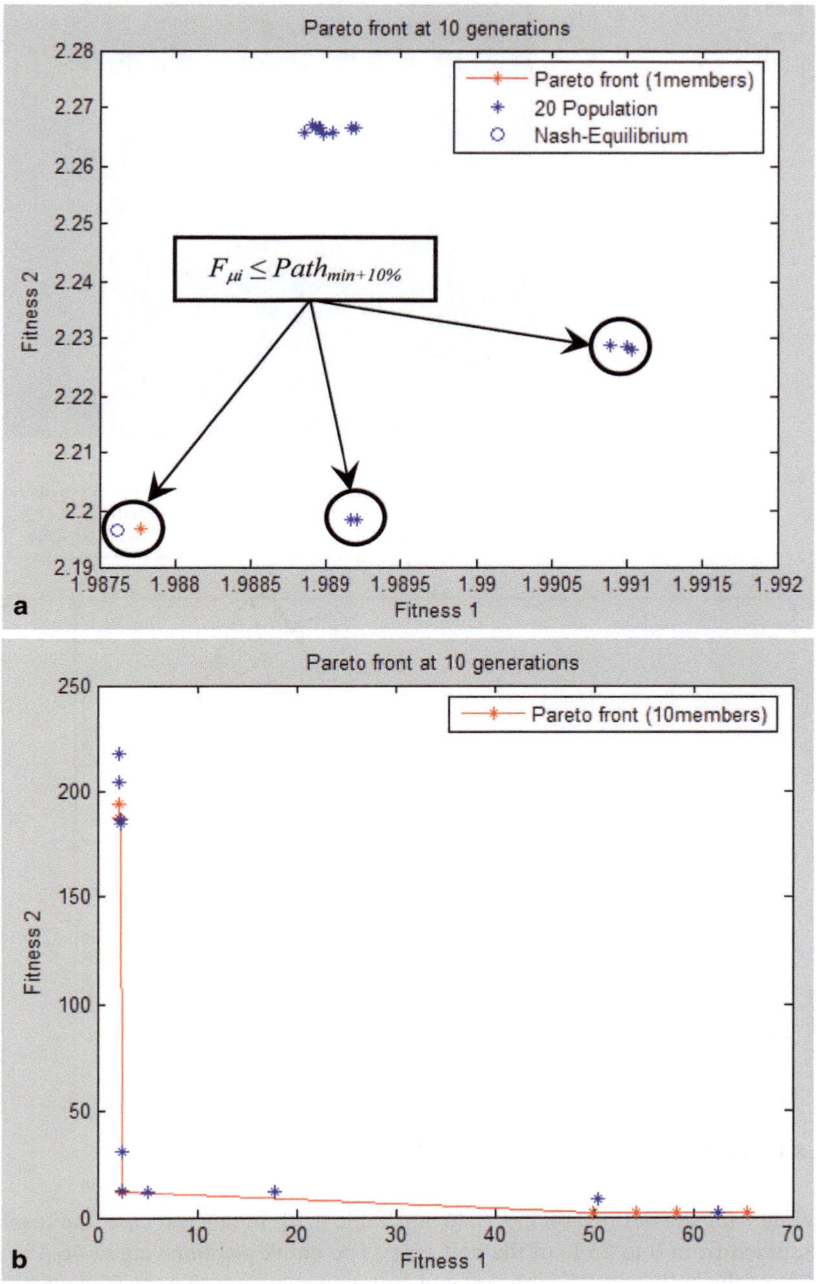

Fig. 8.44 **a** Population of Hybridized Games on NSGA-II at generation 10. **b** Population of NSGA-II at generation 10

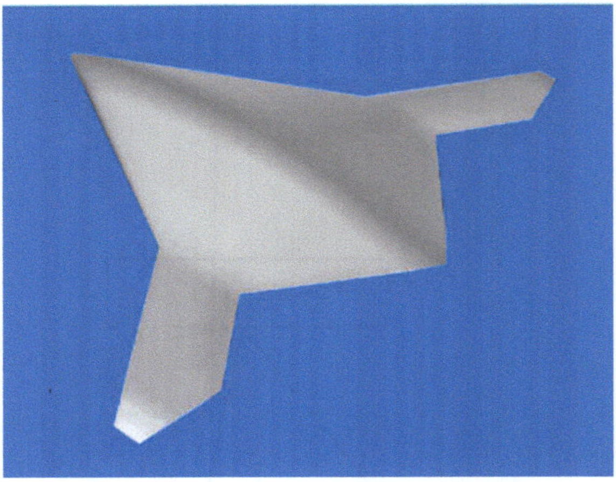

Fig. 8.45 Baseline design in 3D view

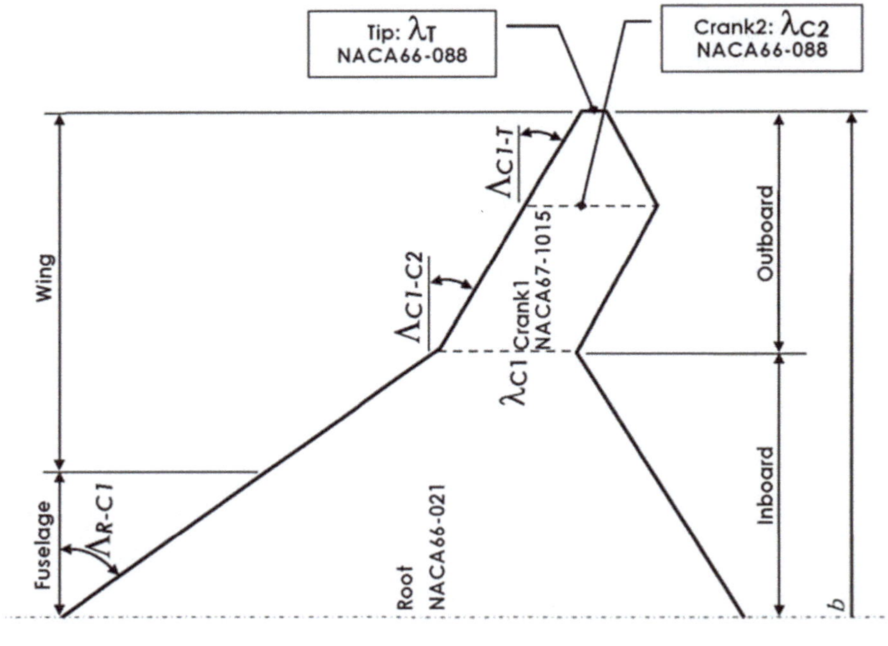

Fig. 8.46 The UCAV baseline configuration

baseline wing are defined on Fig. 8.46 and Table 8.10. In this test case, the fuselage is assumed from 0 to 25 % of the half span. The crank positions are at 46.4 % and 75.5 % of the half span. The inboard and outboard sweep angles are 55 degrees and 29 degrees respectively. Inboard and outboard taper ratios are 20 % and 2 % of the root chord respectively.

Table 8.10 Baseline UCAV wing configurations

AR	b	Λ_{R-C1}	Λ_{C1-C2}	Λ_{C2-T}	λ_{C1}	λ_{C2}	λ_{T}	$\Gamma_{Overall}$
4.377	18.9 m	55°	29°	29°	20	20	2	0°

Note: λ is given with respect to percentage of root chord length

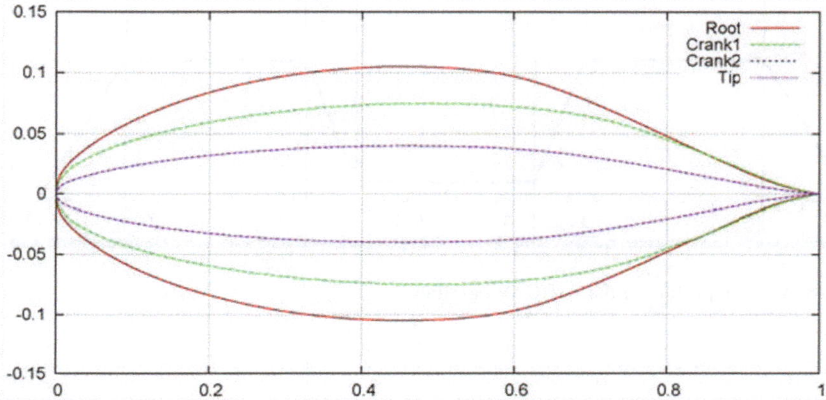

Fig. 8.47 UCAV baseline air foil sections

It is assumed that the baseline design contains three types of airfoils at root, crank 1, crank 2 and tip section as shown in Fig. 8.46; NACA 66–021 and NACA 67–1015 are at inboard and for outboard NACA 66–021 and NACA 67–008 are employed. These airfoils are shown in Fig. 8.47. The maximum thickness at root is 21 % of the chord that is approximately 3 % thicker than X-47B due to increment of payload and fuel capacity.

The mission profile of UCAV includes eight (8) sectors: Reconnaissance, Intelligence, Surveillance and Target Acquisition (RISTA) as illustrated in Fig. 8.48 where it is divided into eight Sectors.

Where

Sector1: T/O & Climb	; Sector 2: start Cruise	Sector 3: transition dash	; Sector 4: Ingress
Sector 5: Target strike	;Sector 6: start; R-Cruise	Sector7: end R-Cruise	Sector 8: Decent & Land

where R-Cruise represents the returning cruise

Figure 8.49 shows the weight distribution along the mission profile (Sector 1~ Sector 8). The weight between Sector 4 and Sector 5 is significantly reduced since 80 % of ammunitions weight is used for target strike.

In this section, flight conditions for Sector 2 to Sector 4 are considered and minimum lift coefficients ($C_{L_{Minimum}}$) are 0.296 and 0.04 for Sector 2 and Sector 4 respectively as shown in Fig. 8.50. The baseline design produces 30 % higher lift

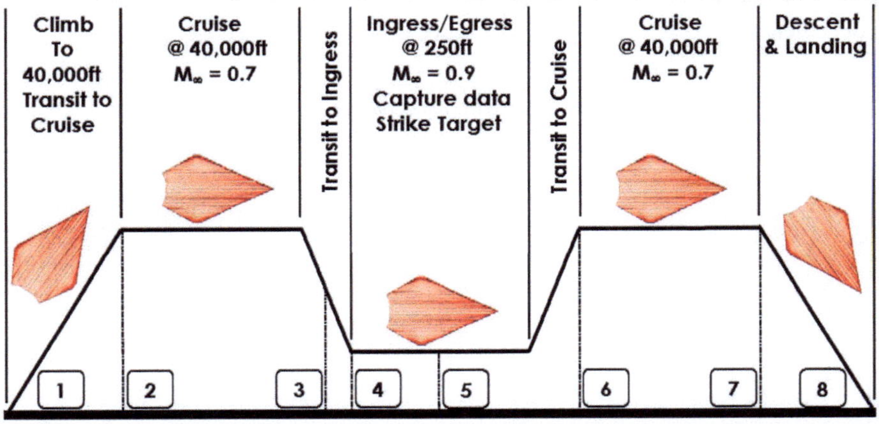

Fig. 8.48 Mission profile for the J –UCAV baseline

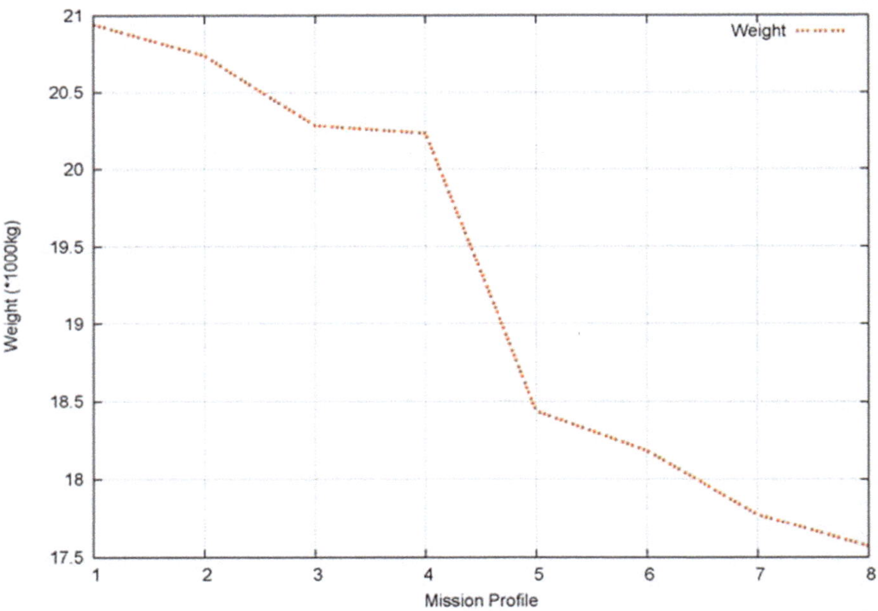

Fig. 8.49 Weight distribution corresponding to Sectors

coefficient at Sector 2 when compared to $C_{L_{Minimum}}$ while only 7 % higher at Sector 4. The aim of optimisation is the improvement of aerodynamic performances at Sector 4 while maintaining aerodynamic performance at Sector 2.

Problem Definition This test case considers the design optimisation of a UCAV wing airfoil sections and plan form geometry. The objectives are to maximize both

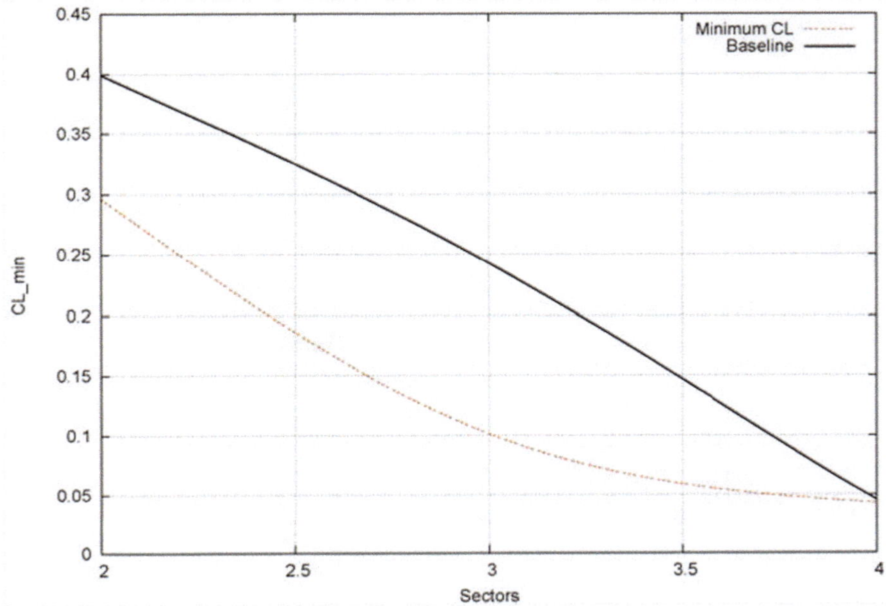

Fig. 8.50 Minimum lift coefficient (CL) required for Sector 2–Sector 4

mean values of lift coefficient $\overline{(C_L)}$ and lift to drag ratio $\overline{(L/D)}$. Fitness functions and flight conditions are as follows;

$$fitness(f_1) = \min\left(1/(\overline{C_L})\right) \tag{8.22}$$

$$fitness(f_2) = \min\left(1/(\overline{L/D})\right) \tag{8.23}$$

at

Sector 2: $M_\infty = 0.7, \alpha = 6.05^\circ$ and altitude 40,000 ft
Sector 4: $M_\infty = 0.9, \alpha = 0.5^\circ$ and altitude 250 ft

where $\overline{C_L} = \frac{1}{2}(C_{L_{Mission2}} + C_{L_{Mission4}})$ and $\overline{L/D} = \frac{1}{2}(L/D_{Mission2} + L/D_{Mission4})$
This optimisation uses Algorithm 4 in Sect. 6.5.5 of Chap. 6.

Design Variables In this test case, the airfoil geometry is represented using Bezier splines with a combination of a mean line and thickness distribution as shown in Chap. 7. The upper and lower bounds for mean and thickness control points at root, crank 1, crank 2 and tip sections are given in Figs. 8.51a, b, c, d.

The wing plan form shape is parameterized with nine (9) design variables including three (3) sectional areas, three (3) sweep angles and three (3) taper ratios described in Fig. 8.52. These will alter the span length (b) and Aspect Ratio (AR).

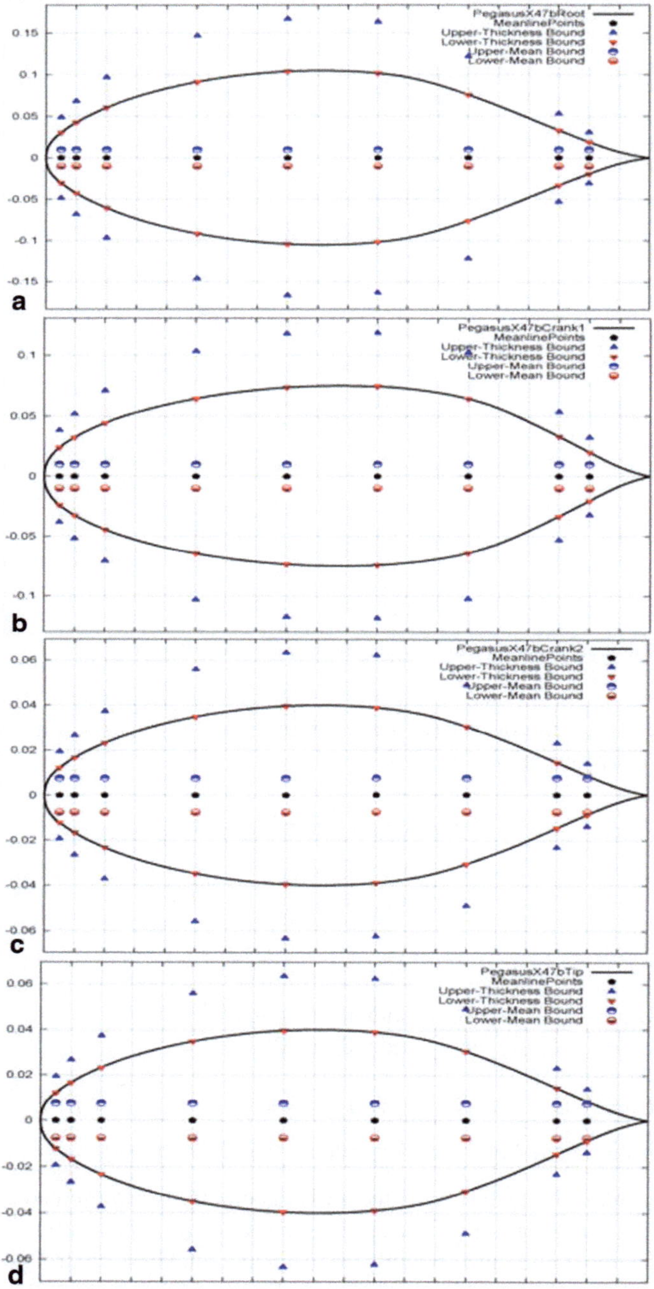

Fig. 8.51 **a** Root mean and thickness control points. **b** Crank1 mean and thickness control points. **c** Crank 2 mean and thickness control points. **d** Tip mean and thickness control points

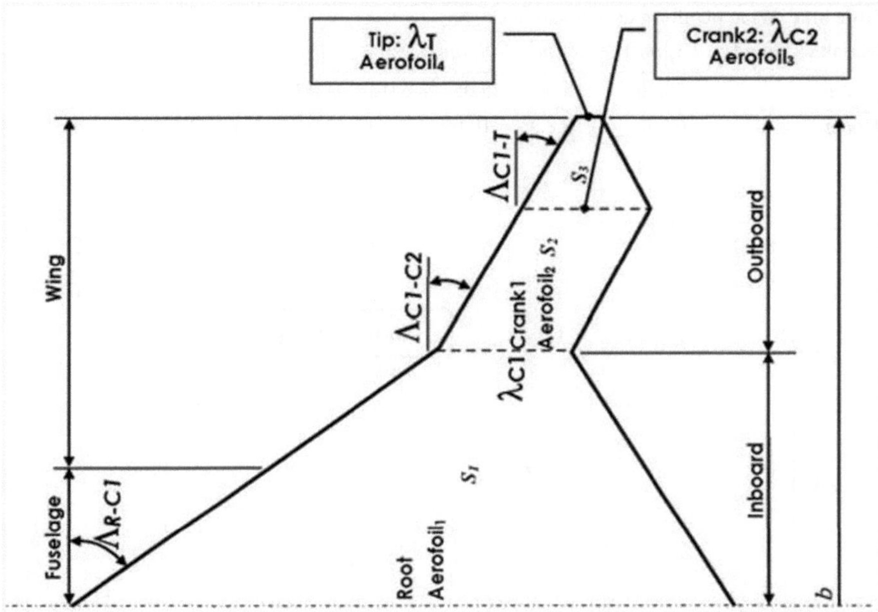

Fig. 8.52 Wing plan form design variables

The taper ratio at crank 2 will not be higher than the taper ratio at crank 1 i.e., $(\lambda_{C2} \leq \lambda_{C1})$.

The design variables are for the airfoil sections located at four (4) span wise stations as described in Figs. 8.51a, b, c, d. Lower design bounds of thickness distributions are the same as the baseline design to increase the fuel capacity, avionics and missile payloads. Eight (8) design variables that define wing configurations are considered (λ_{C1}, λ_{C2}, $\Lambda_{R\text{-}C1}$, $\Lambda_{C1\text{-}C2}$, $\Lambda_{C2\text{-}T}$, S_1, S_2 and S_3). The taper ratio (λ_T) at tip is fixed as 0.2 % of the root chord. Two crank positions (C_1 and C_2) and wing span (b) are recalculated without washout or dihedral effects. The upper and lower design bounds values are provided on Table 8.11. Seventy six (76) design variables are considered in total.

Implementation The FLO22 and FRICTION CFD solvers are utilised and the following specific parameters are considered for HAPMOEA.

1st Layer	Population size of 15 and intermediate recombination Computational grid of 96 × 12 × 16 cells used by FLO22 solver
2nd Layer	Population size of 20 and intermediate recombination Computational grid of 68 × 12 × 16 cells used by FLO22 solver
3rd Layer	Population size of 30 and intermediate recombination Computational grid of 48 × 12 × 16 cells used by FLO22 solver

Table 8.11 Wing planform design variables

Variables bounds	$S_1\ (m^2)$	$S_2\ (m^2)$	$S_3\ (m^2)$	λ_{C1}	λ_{C2}	$\Lambda_{R\text{-}C1}$	$\Lambda_{C1\text{-}C2}$	$\Lambda_{C2\text{-}T}$
Lower	50.46	10.09	5.05	0.15	0.15	49.5°	25°	25°
Upper	63.92	16.82	10.09	0.45	0.45	60.5°	35°	35°

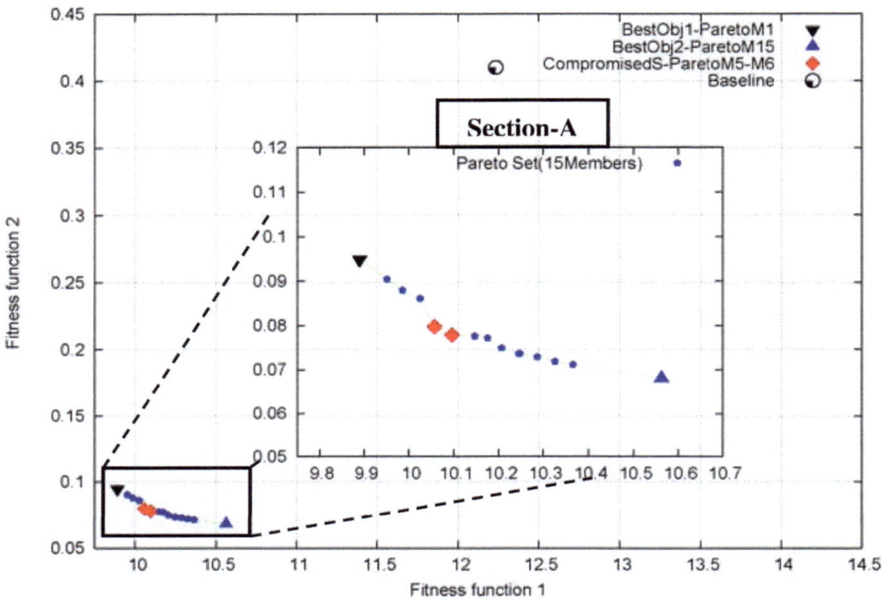

Fig. 8.53 Pareto optimal front

Numerical Results The algorithm was allowed to run approximately 6666 function evaluations and took 200 h on two 2.4 GHz processors. The resulting Pareto set is shown in Fig. 8.53 where the *black inverse triangle* represents the best solution for fitness function 1. The *blue triangle* indicates the best solutions for the fitness function 2. *Red squares* represent compromised solutions. A convex Pareto front can be seen between the first and the second objective as shown in Section-A.

Table 8.12 compares the fitness values obtained by the baseline and Pareto members (1, 5, 6 and 15). It can be seen that all non-dominated solutions produce higher $\overline{C_L}$ and $\overline{L/D}$. A 17.5 % of $\overline{C_L}$ and 80 % $\overline{L/D}$ improvement can be noticed when compared to the baseline design performances.

The wing plan form shapes corresponding to each member of the Pareto front are represented on Fig. 8.54 and Table 8.13. It can be noticed that all Pareto members have a longer span when compared to the baseline design due to considering aerodynamics only for fitness 1 and fitness 2. The aspect ratio of Pareto members are twice higher compared to the one of the baseline design. There was more than 40 % span length expansion while a reduction of 20 % taper ratios at crank 1 and

Table 8.12 Comparison of objectives obtained by the baseline design and HAPMOEA software

Description	Baseline	ParetoM1 (BO1)	ParetoM5 (CS1)	ParetoM6 (CS2)	ParetoM15 (BO2)
$1/\overline{CL}$	12.232	9.890 (−19%)	10.056 (−18%)	10.096 (−17%)	10.562 (−14%)
$1/\overline{L/D}$	0.410	0.095 (−77%)	0.079 (−80%)	0.078 (−81%)	0.068 (−83%)

Note: BO represents a Best Objective and CS indicates a Compromised Solution

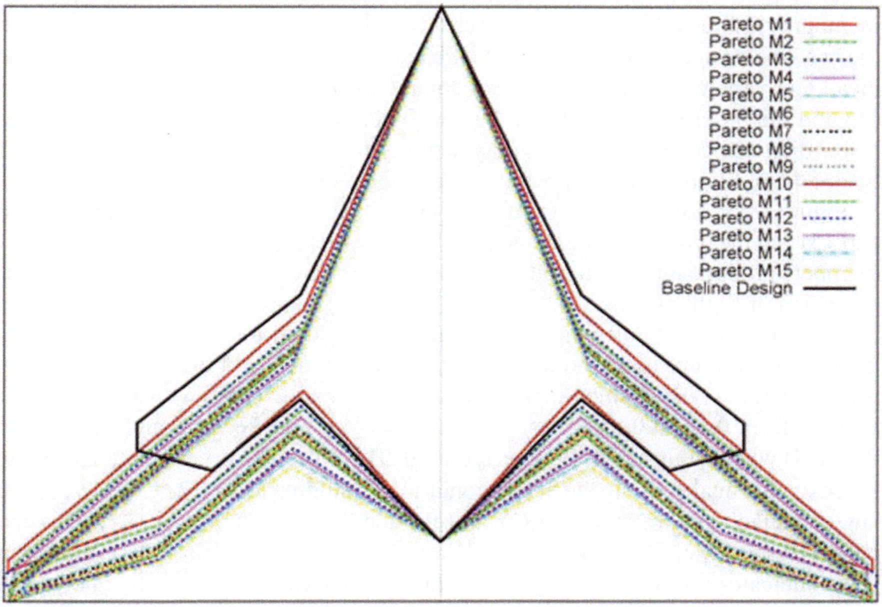

Fig. 8.54 Optimal plan form shapes

Table 8.13 Comparison of wing configurations

Models	*AR*	*B*	*λC1*	*λC2*	*ΛR-C1*	*ΛC1-C2*	*ΛC2-T*
Baseline	4.45	18.9	19.7	19.7	55°	29°	29°
Pareto M1	8.74	26.89	15.3	15.3	56.9°	31.3°	29.8°
Pareto M5	8.38	27.18	15.8	15.8	58.6°	30.3°	29.8°
Pareto M6	8.34	26.81	15.9	15.9	58.5°	30.7°	29.8°
Pareto M15	7.49	25.68	16.7	16.7	59.8°	26.6°	29.7°

crank 2 sections. It can also be observed that there was higher leading edge sweep angle between root to crank 1 section. These wing configurations result in higher $\overline{C_L}$ and $\overline{L/D}$.

The Sector sweep is plotted with lift and drag coefficient in Figs. 8.55a and b. The range of Sector sweep is $M_{\infty} = 0.7:0.9, \alpha = 6.05:0.5$ at an altitude $= 40{,}000:250$. It can be observed that all Pareto members (1, 5, 6 and 15) produce higher *CL* when compared to the baseline and minimum *CL* requirement. Pareto member 1 (Best objective 1 for $1/\overline{CL}$) indicates higher lift coefficient along the sectors when compared to other solutions while Pareto member 15 (Best objective 2 for $1/\overline{L/D}$) has lower drag coefficient when Mach number is higher than 0.775. In addition, Pareto members (1, 5, 6 and 15) produce lower *CD* without fluctuation when compared to the baseline design as shown in Fig. 8.55b and Table 8.14. Table 8.14 compares the mean and variance of drag coefficient obtained from Pareto members (1, 5, 6 and 15) and the baseline design. It can be concluded that all Pareto members produce lower drag coefficient with sensitivity/stability by more than 50 % at Sector 2 and Sector 4 conditions when compared to the baseline design.

Figure 8.55c compares the lift to drag ratio obtained by Pareto members (1, 5, 6 and 15) and the baseline design along the Sector sweep. It can be seen that all Pareto members produce higher lift to drag ratios along the Sector sweep which means an extension of flight range.

Even though the multi-objective optimisation method found useful Pareto non-dominated solutions and produced aerodynamic improvement in Sector 2 and Sector 4, there is a considerable fluctuation of lift to drag ratio (L/D) with Mach number ranging from 0.75 to 0.85 (transition points: Sector 2 to Sector 3 and Sector 3 to Sector 4) where a high dash flight is required. Therefore it is necessary to check the aerodynamic quality along the Sector conditions including Mach ($M_{\infty} = 0.75:0.85$), angle of attack ($\alpha_{\infty i} \in 4.662:1.887$) and altitude ($ATI_{\infty i} \in 30062:10187$). This can be expressed in terms of mean and variance of lift to drag ratio (L/D): the mean value indicates the scalar of objective while the variance value can be interpreted as the stability or sensitivity of objective. Table 8.15 compares the quality of lift to drag ratio (L/D) obtained from the Pareto members (1, 5, 6 and 15) with the baseline design. The variances of lift to drag ratio (L/D) obtained by Pareto members are higher than the baseline design. This means Pareto members are over-optimised solutions to maximise an aerodynamic performance at design conditions.

This fluctuation can lead to structural or control failure at transition points: Sector 2 to Sector 3 and Sector 3 to Sector 4. This fluctuation can be avoided by using an uncertainty robust design technique during optimization. However, particular care is required for deciding variability of flight conditions. For instance, the variable operating conditions are considered between blue centre lines in Fig. 8.55c then the variance (Line-B) of the baseline is higher than Pareto member 1 (Line-A) even though the baseline is more stable (low variance) from Sector 2 to Sector 4 conditions. The introduction of uncertainty with effective variability of operating conditions is implemented in the next application to produce stable solutions on both drag coefficient and lift to drag ratio. This test case will be considered with uncertainty design techniques implemented in Sect. 9.4 of Chap. 9.

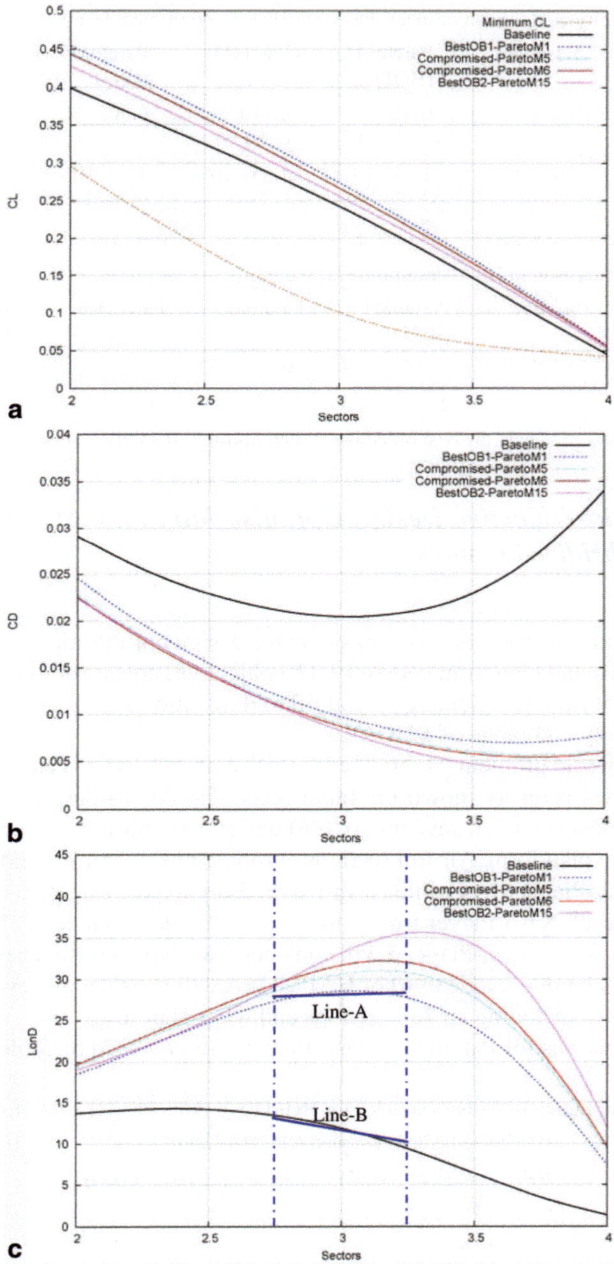

Fig. 8.55 **a** Lift coefficient (*CL*) vs. Sector sweep. **b** Drag coefficient (*CD*) vs. Sector sweep. **c** Lift to drag ratio (*L/D*) vs. Sector sweep

Table 8.14 Quality comparison of mean and variance drag coefficient (CD)

Description	Baseline	ParetoM1 (BO1)	ParetoM5 (CS1)	ParetoM6 (CS2)	ParetoM15 (OB2)
$\overline{C_D}$	0.025	0.011 (−56%)	0.010 (−60%)	0.009 (−64%)	0.009 (−64%)
δC_D	5.49×10^{-5}	1.49×10^{-5}	1.54×10^{-5}	$1.56' 10^{-5}$ (−81%)	2.11×10^{-5}

Table 8.15 Comparison of lift to drag ratio (L/D) quality

Objective	Baseline	ParetoM1 (BO1)	ParetoM5 (CS1)	ParetoM6 (CS2)	ParetoM15 (BO2)
$\overline{L/D}$	10.525	27.62	30.03	31.05	33.222
$\delta L/D$	8.25	23.53	42.08	50.19	127.10

Note: BO and CS represent the best objective and compromised solutions respectively

8.6.1 Multi-Objective Design Optimisation of UCAV Using Hybridized Games

Problem Definition This test case considers the design optimisation of UCAV wing air foil sections and plan form geometry. The objectives consist of maximizing both mean values of lift coefficient $(\overline{C_L})$ and lift to drag ratio $(\overline{L/D})$ and maximising a manoeuvrability and range of a UCAV.

For a real world problem, Hybridized Games employs five (5) Nash-Players and one (1) Pareto-Player as shown on Table 8.16. The Pareto-Player of Hybridized Games solely considers all seventy six (76) design variables including air foil sections and wing plan form. Air foil sections at root, crank 1, crank 2 and tip are optimized by Nash-Players 1–4 while Nash-Player 5 optimises wing plan form only. In other words, each Nash-Player from 1 to 4 will optimize seventeen (17) air foil design variables while Nash-Player 5 will consider 8 wing plan form design variables. In contrast, each node (Node 0- 6) of HAPMOEA will consider all seventy six (76) design variables including air foil sections and wing plan form.

Fitness functions for Pareto and Nash-Players are indicated in Table 8.17.

Table 8.16 Distribution of design variables for Hybridized Games and HAPMOEA

Design variables	Hybridized games coupled with HAPMOEA						HAPMOEA
	N-P1	N-P2	N-P3	N-P4	N-P5	P-Player	
Air foil $_{Root}$	√					√	√
Air foil $_{Crank1}$		√				√	√
Air foil $_{Crank2}$			√			√	√
Air foil $_{Tip}$				√		√	√
Wing Plan form					√	√	√

$N\text{-}P_i$ represents ith Nash-Player and P-Player the Pareto-Player

Table 8.17 Fitness functions for each player in Hybridized games

Player	Fitness function	Optimization criteria
Pareto-Player	$f_{PP1} = \min\left(1/\left(\overline{C_L}\right)\right)$ $f_{PP2} = \min\left(1/\left(\overline{L/D}\right)\right)$.	Optimize wing plan form and airfoil sections at root, crank 1, crank 2 and tip to maximize $\overline{C_L}$ and $\overline{L/D}$
Nash-Player 1	$f_{NP1} = \min\left(1/\left(\overline{C_L}\right)\right)$	Maximize total wing $\overline{C_L}$ using design variables for air foil Root only, all other design variables are fixed
Nash-Player 2	$f_{NP2} = \min\left(1/\left(\overline{C_L}\right)\right)$	Maximize total wing $\overline{C_L}$ using design variables for air foil Crank 1 only, all other design variables are fixed
Nash-Player 3	$f_{NP3} = \min\left(1/\left(\overline{C_L}\right)\right)$	Maximize total wing $\overline{C_L}$ using design variables for air foil Crank 2 only, all other design variables are fixed
Nash-Player 4	$f_{NP4} = \min\left(1/\left(\overline{C_L}\right)\right)$	Maximize $\overline{C_L}$ using design variables for air foil Tip only, other design variables are fixed
Nash-Player 5	$f_{NP5} = \min\left(1/\left(\overline{L/D}\right)\right)$	Maximize total wing $\overline{L/D}$ using design variables for wing plan form only, all other design variables are fixed

Note: $\overline{C_L} = \frac{1}{2}\left(C_{L_{Sector2}} + C_{L_{Sector4}}\right)$ and $\overline{L/D} = \frac{1}{2}\left(L/D_{Sector2} + L/D_{Sector4}\right)$

Flight conditions are:
Sector 2: $M_\infty = 0.7$, $\alpha = 6.05°$ and altitude 40,000 *ft*
Sector 4: $M_\infty = 0.9$, $\alpha = 0.5°$ and altitude 250 *ft*
This optimisation uses Algorithm 4 described in Sect. 6.5.5 of Chap. 6.

Design Variables The airfoil geometry is represented using Bezier curves with a combination of a mean line and thickness distribution control points. Four (4) air foils at root, crank 1, crank 2 and tip sections are considered to optimize and the control points for air foil design are sixty eight (68: 4 sections × 17 control points). The wing plan form shape is parameterized by considering eight design variables including three wing sectional areas, three sweep angles and two taper ratios and upper and lower bounds values of these variables are described on Table 8.11. These lead to the different span length (b) and Aspect Ratio (AR).

Numerical Results Both HAPMOEA and Hybridized Games are run on two 2.4 GHz processors. The HAPMOEA algorithm was allowed to run approximately 6667 function evaluations and took two hundred (200) CPU hours. The Hybrid-Game (Hybrid Nash-HAPEA) algorithm was run approximately for 1300 function evaluations and took fifty (50) hours which represents 25 % computation cost of HAPMOEA. The Pareto fronts obtained by HAPMOEA and Hybridized Games are

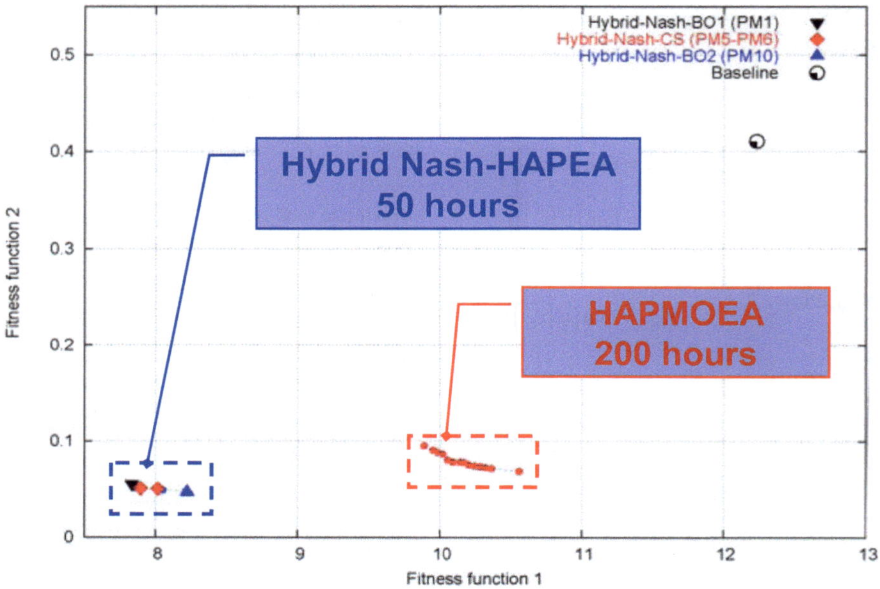

Fig. 8.56 Comparison of Pareto front obtained by Hybridized Games and HAPMOEA

Table 8.18 Comparison of fitness values obtained by HAPMOEA and Hybridized games

Objective	Baseline design	HAPMOEA (200 h)			Hybrid-Game (50 h)		
		PM1 (BO1)	PM6 (CS)	PM15 (BO2)	PM1 (BO1)	PM6 (CS)	PM10 (BO2)
$1/(\overline{C_L})$	12.232	9.890 (−19 %)	10.096 (−17 %)	10.562 (−14 %)	7.836 (−36 %)	8.017 (−34 %)	8.223 (−32 %)
$1/(\overline{L/D})$	0.410	0.095 (−77 %)	0.078 (−81 %)	0.068 (−83 %)	0.054 (−87 %)	0.050 (−88 %)	0.046 (−89 %)

Note: BO represents the best objective solution and CS stands for the compromised solution

compared to the baseline design in Fig. 8.56. It can be seen that Hybridized Games produce much better solutions when compared to HAPMOEA solutions.

Table 8.18 compares the Pareto optimal solutions obtained by HAPMOEA software and Hybridized Games. It can be seen that Hybridized Games software produces twice the value of the lift coefficient (*CL*) while slightly better for inverse mean lift to drag ratio (*L/D*) when compared to Pareto members obtained by HAPMOEA software.

Table 8.19 compares the quality of drag coefficient obtained by HAPMOEA software and Hybridized Games using the uncertainty mean and variance statistical formulas. It can be seen that Pareto members of Hybridized Games produce lower drag at [Sector 2:Sector 4] while Pareto members from HAPMOEA software produce stable drag.

Table 8.19 Comparison of total drag coefficient quality ($C_{D_{Quality}}$) obtained by HAPMOEA and Hybridized Games

Objective	Baseline design	HAPMOEA (200 h)			Hybrid-Game (50 h)		
		PM1 (BO1)	PM6 (CS)	PM15 (BO2)	PM1 (BO1)	PM6 (CS)	PM10 (BO2)
$\overline{C_D}$	12.232	0.025	0.011 (−56%)	0.009 (−64%)	0.009 (−64%)	0.009 (−64%)	0.0089 (−64%)
δC_D	0.410	5.49×10^{-5}	1.49×10^{-5}	1.56×10^{-5}	2.11×10^{-5}	2.29×10^{-5}	2.24×10^{-5}

Note: Quality is represented by mean (performance) and variance (sensitivity/stability)

The mission sweep is plotted with the lift coefficient and lift to drag ratio (L/D) as shown in Figs. 8.57a, b. The range of Sector sweep is $M\infty \in [0.7{:}0.9]$, $\alpha \in [6.05°{:}\ 0.5°]$ and altitude (*feet*) ∈ [40,000: 250]. Pareto fronts obtained from HAPMOEA software and Hybridized Games software produce higher *CL* and *L/D* when compared the baseline design. Pareto members from both optimisation techniques produce similar results for *CL* and *L/D* at Sector 2 however Pareto non-dominated solutions from Hybridized Games produce better *CL* and *L/D* at Sector 4.

The top, side, front and 3D view of trade off model from HAPMOEA software (Pareto member 5) and Hybridized Games (Pareto member 6) are shown in Figs. 8.58a and b. Even though Hybridized Games spent less computational time when compared to HAPMOEA, both compromised solutions look geometrically similar.

8.7 Aerostructural Optimisation of a Medium Alitude Long Endurance (Male) UAS

Problem Definition This problem considers the design optimisation of a MALE UAV wing similar to the one shown in Fig. 8.59. The two objectives are the maximization of the lift to drag ratio (*L/D*), and the minimization of wing weight. The cruise Mach number and altitude are 0.3663 and 12680 m respectively and the wing area is set to 29.21 m². The two fitness functions to be optimised are defined as maximization of lift to drag ratio (*L/D*) and minimization of wing weight (*W*). For penalty function in fitness function, (2.4) and (2.5) expressions defined in Chap. 2 are applied. This optimization uses Algorithm 5 in Sect. 6.5.6, Chap. 6.

$$f_1 = \min\left(\frac{1}{(L/D)} + penalty\right) \tag{8.24}$$

$$f_2 = \min(W + penalty) \tag{8.25}$$

Design Variables Design variables for the external and internal wing geometry were considered. Figure 8.60 illustrates the external geometry variables and Fig. 8.61 illustrates the internal geometry variables.

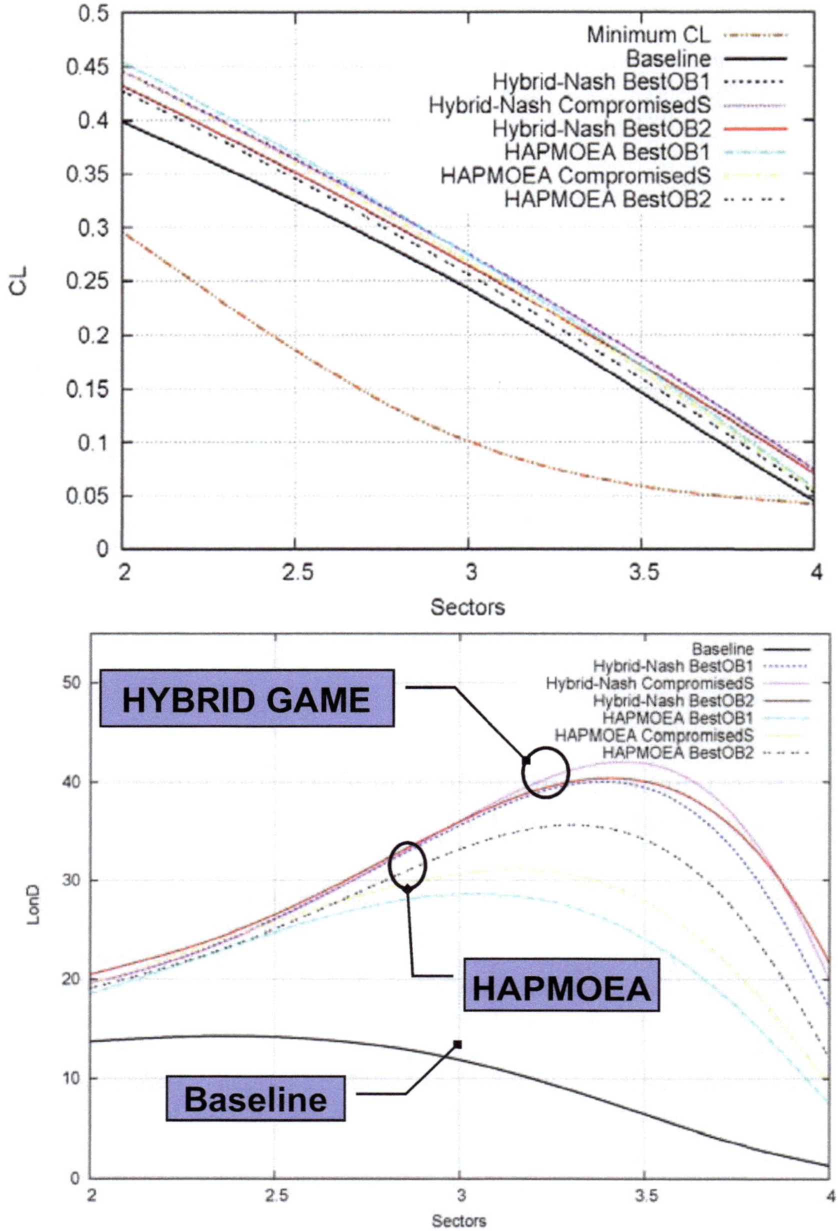

Fig. 8.57 **a** Lift coefficient (*CL*) vs. Sectors: Hybridized -Nash represents Hybridized -Games. **b** Lift to drag ratio (*L/D*) vs. Sectors: Hybrid-Nash represents Hybridized Games

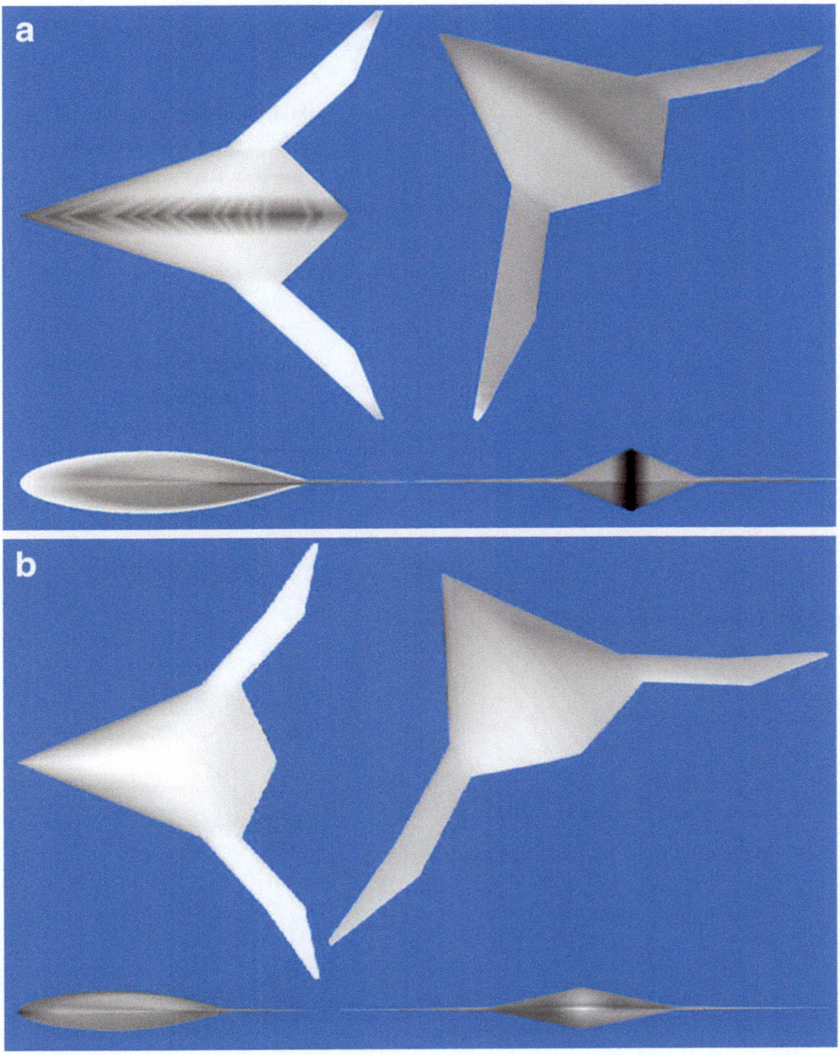

Fig. 8.58 **a** Pareto Member 5 obtained by HAPMOEA software. **b** Pareto Member 6 obtained by Hybridized Games software

Fig. 8.59 MALE UAV

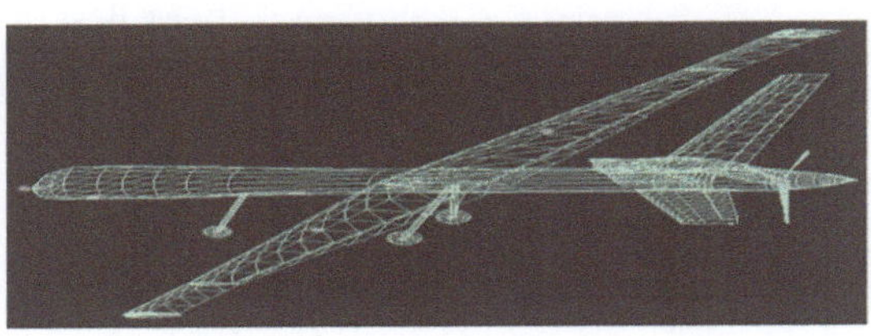

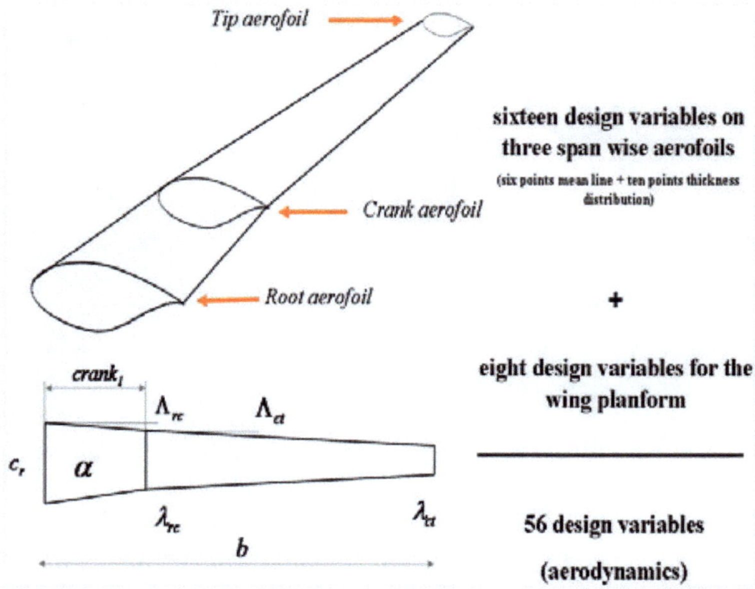

Fig. 8.60 External geometry variables

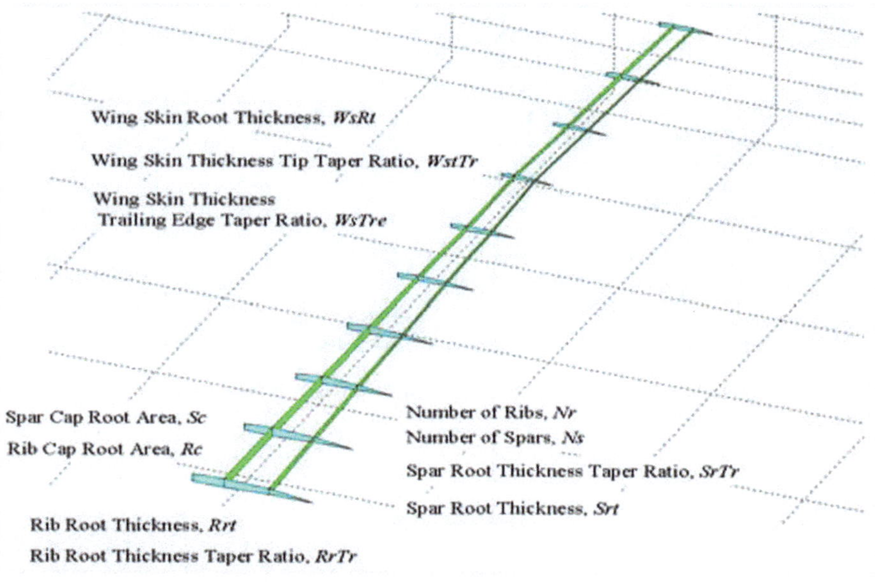

Fig. 8.61 Internal geometry variables

The following constraints, based on the baseline design, are considered:

- The pitching moment must not be greater than −0.1768;
- The calculated coefficient of lift must be greater than 0.64;
- The thickness of each aerofoil must exceed 12 % ($t/c \geq 0.12$).
- No internal or external structural sections can buckle or fail due to excessive strains;
- The wing tip may not twist more than one (1°) degree;
- The wing may not deflect more than twenty per cent (20 %) of the span.

All the abovementioned constraints are applied by penalizing either or both fitness values via an exponential penalty method. In addition, any airfoil generated outside the thickness bounds of 10–15 % is rejected immediately before any analysis being performed on the wing.

Implementation This optimization uses Algorithm 5 in Sect. 6.5.6 of Chap. 6. An aero-structural solver is developed integrating two analysis tools for FEA and CFD namely MSC. NASTRAN and the high order panel method PANAIR. The entire aero-structural program is controlled through a Mat lab® script file. This allows for an easy coupling of the different required programs since a continuous Mat lab® data structure can be utilized to define all the information passed between programs. Figure 8.62 illustrates the procedure.

Two approaches were compared, one using a single population approach and one using a hierarchical/multi-fidelity approach. Details on each approach are:

Single-Expensive –High Fidelity Approach EA with CMA/Pareto tournament selection with a population size of 20 and a fine computational grid between 1904 and 2992 aerodynamic panels.

Multi-fidelity- Hierarchical Approach

1st Layer	Population size of 20 and intermediate recombination with a computational grid between 1904 and 2992 aerodynamic panels
2nd Layer	Population size of 40 and intermediate recombination with a computational grid between 1428 and 2244 aerodynamic panels
3rd Layer	Population size of 60 and intermediate recombination with a computational grid between 952 and 1496 aerodynamic panels

Numerical Results The algorithm was run for 500 function evaluations in both cases and took approximately 36 h to compute on a cluster of three machines. Figure 8.63 illustrates a well-distributed Pareto front in both cases. It can be seen that the use of a multi-fidelity (hierarchical) approach gives an overall lower front as compared to a single model approach. This problem demonstrates use of the framework for UAV wing design and optimisation. Results indicate a computational gain and better results by using a hierarchical topology of fidelity models.

Figure 8.64 shows some of the wings in the Pareto front and Fig. 8.65 illustrates the airfoil geometries for each of these Pareto members. The designer will have the

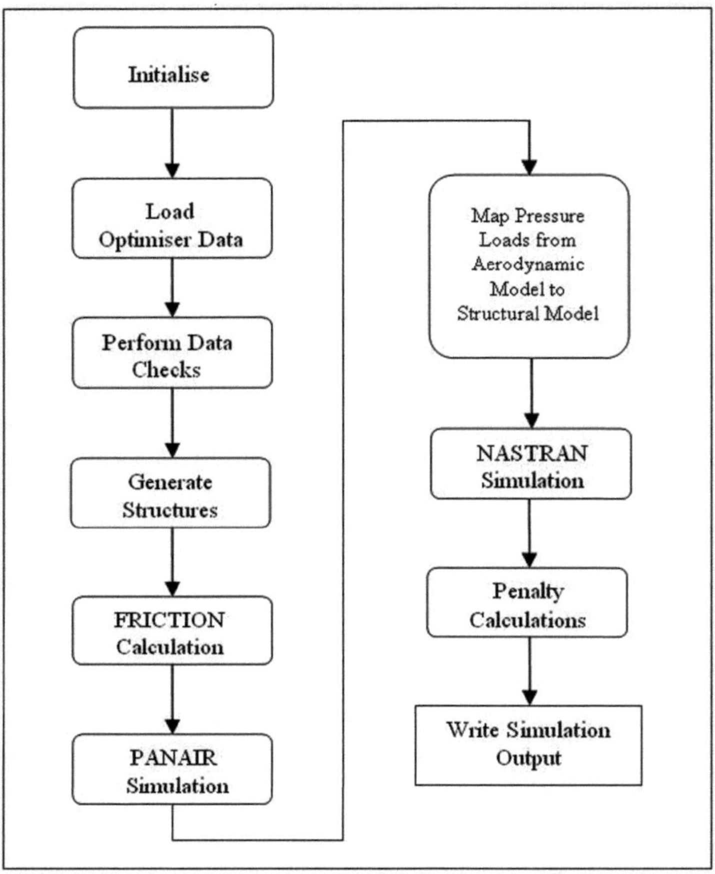

Fig. 8.62 Aero-structural analysis

flexibility to choose amongst these solutions depending on preference and objectives of the design.

Figure 8.66 compares the coefficient of lift versus coefficient of drag for selected Pareto members and baseline/benchmark. A better performance of Pareto members can be observed.

Table 8.20 compares some of the aerodynamic and structural characteristics of the designs. Again benefits are clearly shown on the methodology as it produces optimal designs with higher *L/D* and lower structural weight.

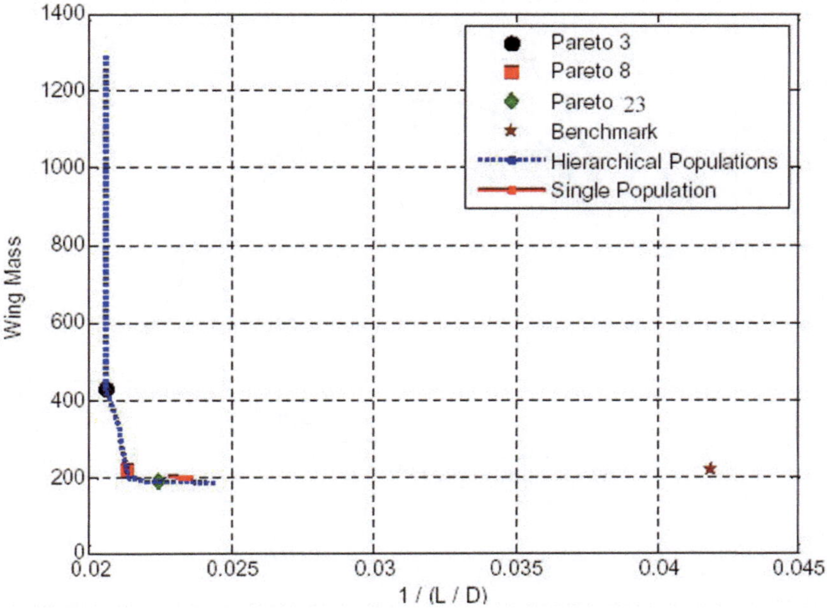

Fig. 8.63 Internal geometry variables

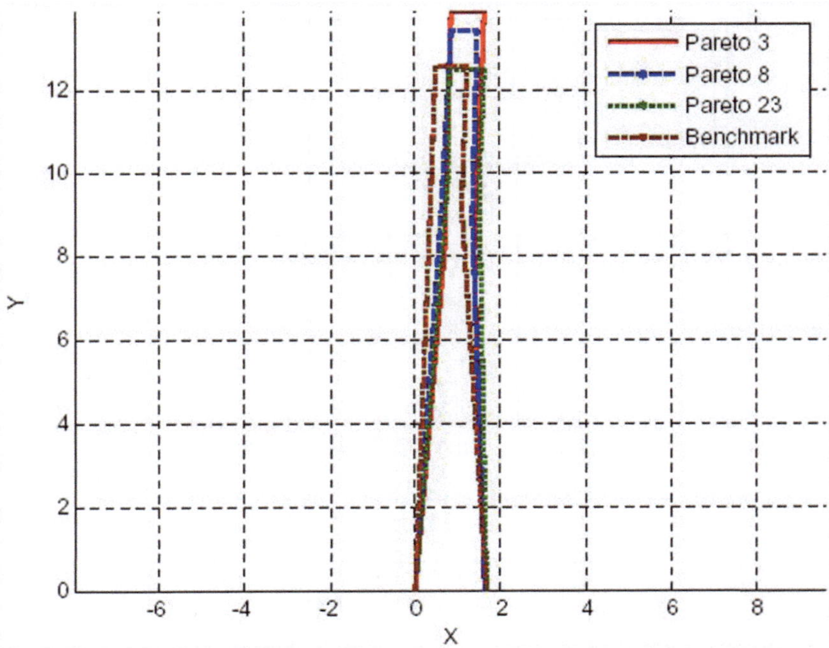

Fig. 8.64 Top views of selected Pareto members

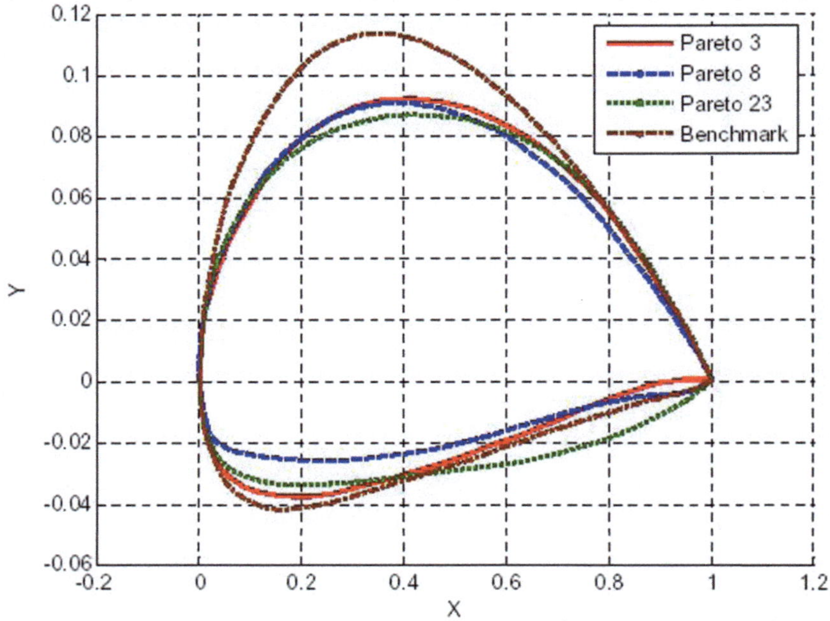

Fig. 8.65 Airfoil section of selected Pareto members

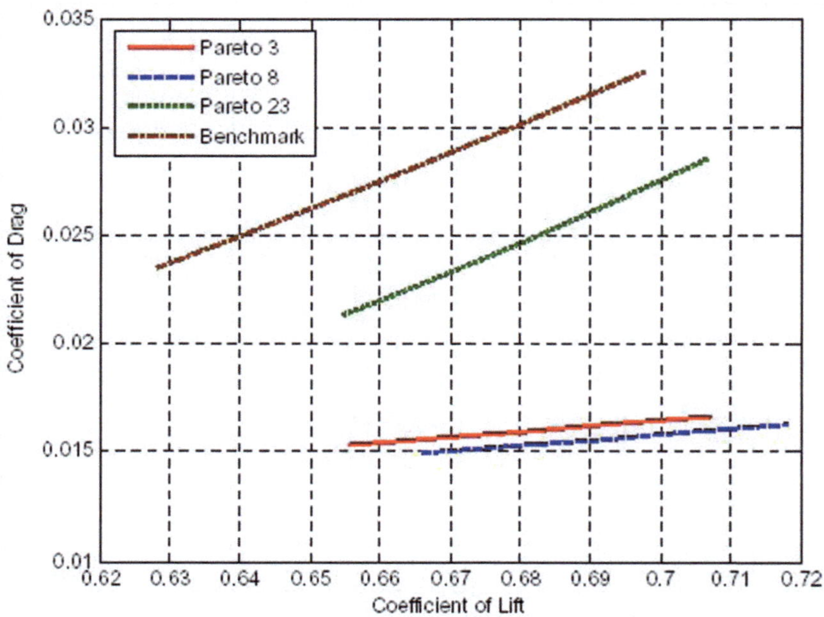

Fig. 8.66 Lift coefficient (*CL*) vs. drag coefficient (*CD*) for selected MALE Pareto members and Benchmark

Table 8.20 Aerodynamic and structural characteristics of the optimal designs

	Pareto 3	Pareto 8	Pareto 23	Benchmark
Deflection (%)	16	17.3	13.8	12.5
Maximum strain	0.000396	0.00052	0.000485	0.000403
Mass (kg)	233.735	217.873	185.1	220.9
L/D	43.367	44.4	44.25	23.9
Mass penalty (kg)	40	0	0	0
Drag penalty	0.0027	0	0	0

8.8 Aero-Electromagnetic Optimization of a UAS

Problem Definition This test case considers both mono and bi-static radar analysis to produce a stealth model targeting enemy aircraft and ground radar. The same flight conditions as shown on Table 8.5 of Chap. 8.5 are applied for the design and optimisation of a UCAV wing air foil sections and plan form geometry operating with a variability of Mach numbers. First and second objectives are minimization of the mono and bi-static RCS. A third fitness function evaluates aerodynamic stability. The value of variance being normally quite small, there is little effect from mean value being log scaled. This optimization uses Algorithm 5 shown in Sect. 6.5.6 of Chap. 6. Fitness functions are defined in (8.26), (8.27) and (8.28):

$$f_1 = \min\left(\ln\left(RCS_{Mono}\right)\right) = \ln\left[\sum_{i=0}^{q}\left(\exp(RCS_{\theta i}) + \exp(RCS_{\phi i})\right)\right] \tag{8.26}$$

where $\theta = [0^\circ : 3^\circ : 360^\circ]$ and $\phi = [0^\circ : 0^\circ : 0^\circ]$ (Mono-static)

$$f_2 = \min\left(\ln\left(RCS_{Bi}\right)\right) = \ln\left[\sum_{i=0}^{q}\left(\exp(RCS_{\theta i}) + \exp(RCS_{\phi i})\right)\right] \tag{8.27}$$

where Incident angles $\theta = 135^\circ$, $\phi = 90^\circ$ at $\theta = [0^\circ : 3^\circ : 360^\circ]$, $\phi = [0^\circ : 0^\circ : 0^\circ]$ (Bi-static)

$$f_3 = \min\left(\frac{-1}{\ln\left(\overline{1/(L/D)}\right)} + \frac{-1}{\ln\left(\delta(1/(L/D))\right)}\right) \tag{8.28}$$

Design Variables The design variables for the air foil sections at four span-wise stations are described in Figs. 8.20a, b, c. Two thickness constraints are considered as 20 and 80 %c for structural concern. For the wing plan form design, nine (9) design variables are considered including $\lambda_{R\text{-}B1}$, $\lambda_{B1\text{-}B2}$, $\lambda_{B2\text{-}T}$, $\Lambda_{R\text{-}B1}$, $\Lambda_{B1\text{-}B2}$ and $\Lambda_{B2\text{-}T}$. The break point positions are recalculated without twist and dihedral factors by

Table 8.21 Wing plan form design variables

Variables	S_1 (ft^2)	S_2 (ft^2)	S_3 (ft^2)	$\lambda_{R\text{-}B1}$	$\lambda_{B1\text{-}B2}$	$\lambda_{B2\text{-}T}$	$\Lambda_{R\text{-}B1}$	$\Lambda_{B1\text{-}B2}$	$\Lambda_{B2\text{-}T}$
Lower bounds	66.88	54.07	58.87	0.6	80 (%$\lambda_{R\text{-}B1}$)	0.002	45°	45°	45°
Upper bounds	81.74	66.09	71.95	0.9	100 (%$\lambda_{R\text{-}B1}$)	0.1	65°	65°	65°

obtaining inboard and outboard areas. The upper and lower design bounds values are given on Table 8.21.

Implementation The Euler FLO22 solver (A. Jameson) is utilized and the following specific parameters are considered for the evolutionary optimiser using a hierarchical topology concept.

POFACETS [36] developed at the Naval Postgraduate School (NPS) is an implementation of the physical optics approximation for predicting the radar cross section (RCS) of complex objects. It represents the constituent parts by triangular facets. POFACETS can calculate mono or bi static RCS of the object.

1st Layer	Population size of 15 and intermediate recombination computational grid of 96 × 12 × 16 cells used by FLO22 solver
2nd Layer	Population size of 40 and intermediate recombination computational grid of 82 × 12 × 16 cells used by FLO22 solver
3rd Layer	Population size of 60 and intermediate recombination computational grid of 68 × 12 × 16 cells used by FLO22 solver

Numerical Results The algorithm was allowed to run approximately 1100 function evaluations and took 150 CPU hours on two single 1.8 GHz. The resulting Pareto set is shown in Fig. 8.67. Pareto Members 1, 5, 6 and 10 are compared to the quality of models in Table 8.22 where Pareto Member 1 made a 27 % reduction of mono-static RCS when compared to baseline while Pareto Member 10 reduces 33 % of bi-static RCS. The shapes of Pareto Members 5 and 6 produce 25 % lower observability in both mono and bi-static radar systems. Pareto Members 5 and 6 also produce up to 10 % better aerodynamic model quality. Pareto Members 5 and 6 are selected for further evaluation.

Wing plan form shapes corresponding to the Pareto front are illustrated in Fig. 8.68 and the configuration of baseline is compared to Pareto Members 5 and 6 in Table 8.23. The aspect ratios of Pareto Members 5 and 6 are increased about 20 and 7 % increment at breakpoint 1 taper ratio while 23 % reduction at breakpoint 2 taper ratio. It is also observed that there were similar sweep angles at inboard and outboard when compared to baseline.

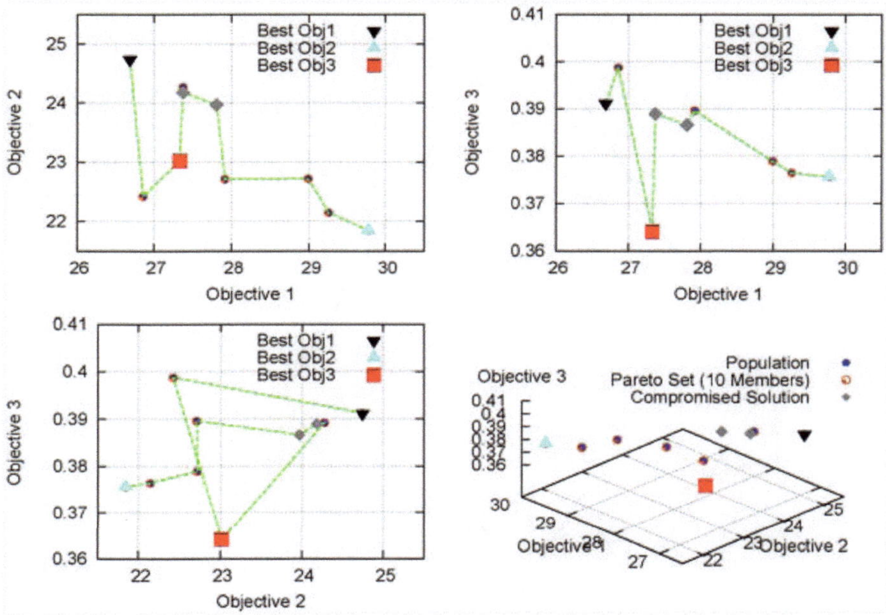

Fig. 8.67 Pareto optimal fronts for UCAV wing airfoil sections/plan form

Table 8.22 Comparison of objectives obtained by the baseline design and Pareto optimal solutions (Pareto Members 1, 5, 6 and 10)

Description	Baseline	Pareto M1	Pareto M5	Pareto M6	Pareto M10
RCS_{Mono}	36.390	26.69 (−27%)	*27.37 (−25%)*	*27.81 (−24%)*	29.78 (−18%)
RCS_{Bi}	32.702	24.74 (−24%)	*24.18 (−26%)*	*23.97 (−27%)*	21.84 (−33%)
AQ	0.430	0.390 (−9.3%)	*0.389 (−9.5%)*	*0.386 (−10.2%)*	0.375 (−12.8%)

AQ represents aerodynamic quality (lower is better)

The airfoil sections are compared on Table 8.24 and Fig. 8.69 where it can be seen that air foil sections have less max-thickness compared to baseline design. The thickness ratio at root and break point1 is reduced by 3% when compared to the baseline design while the camber at all sections is increased.

The span wise *Cp* distribution between compromised solutions and baseline design at standard design conditions as illustrated in Figs. 8.70.

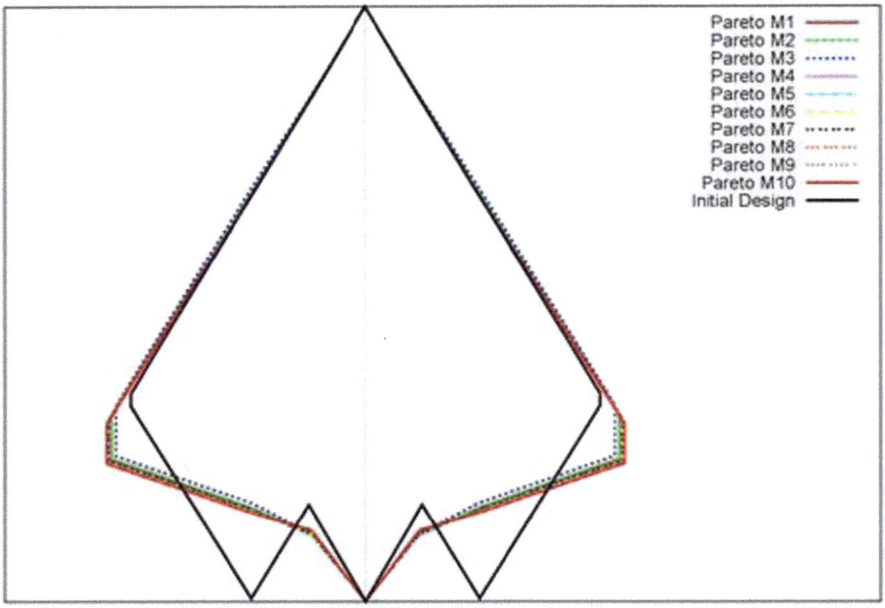

Fig. 8.68 Optimal wing plan form shapes

Table 8.23 Comparison of wing configurations

Models	AR	B	$\lambda_{R\text{-}B1}$	$\lambda_{B1\text{-}B2}$	$\lambda_{B2\text{-}T}$	$\Lambda_{R\text{-}B1}$	$\Lambda_{B1\text{-}B2}$	$\Lambda_{B2\text{-}T}$
Baseline	1.66	18.22	0.68	0.68	0.02	55°	55°	55°
ParetoM5	2.0	19.86	0.73	0.520	0.070	55.56°	54.75°	53.71°
ParetoM6	2.0	19.89	0.73	0.520	0.069	55.47°	54.78°	53.73°

Figures 8.71 and 8.72 show the comparison of bi-static RCS between Pareto Member 5, 6 and baseline in linear plot. The RCS of baseline peak at 27°, 147°, 207° and 333° where the baseline produces higher RCS than Pareto Member 5 and 6 marked with circles. Figure 8.73 show the comparison between Pareto Members 5, 6, the non-dominated solution (Pareto Member 1) from mono-static test and baseline design. Pareto Members 5 and 6 produce similar value as mono-static compromised solution (Pareto Member 1) that is lower by 25 % when compared to baseline. Top, side and three-dimensional views for optimal wing geometries (Pareto member 5) are illustrated in Fig. 8.74. From a designer point of view, these Pareto Members can be applied to stealthy air battle or and reconnaissance of UCAVs.

Table 8.24 Comparison of airfoil sections configurations

Description	$Thick_{Root}$	$Camber_{Root}$	$Thick_{BP1}$	$Camber_{BP1}$	$Thick_{BP2}$	$Camber_{BP2}$	$Thick_{Tip}$	$Camber_{Tip}$
Baseline design	15 % @50 %c	0.0	15 % @50 %c	0.0	8 % @45 %c	0.0	8 % @45 %c	0.0
Pareto M5	*12.30 % @47 %c*	*0.04 % @17.3 %c*	*12.05 % @47.0 %c*	*0.07 % @4.9 %c*	*6.31 % @45.3 %c*	*0.02 % @77.6 %c*	*6.25 % @44.2 %c*	*−0.02 % @56.1 %c*
Pareto M6	*12.30 % @46 %c*	*0.04 % @18.9 %c*	*12.0 % @47.0 %c*	*0.07 % @70 %c*	*6.3 % @45.3 %c*	*0.03 % @77.3 %c*	*6.25 % @44.2 %c*	*−0.02 % @54.0 %c*

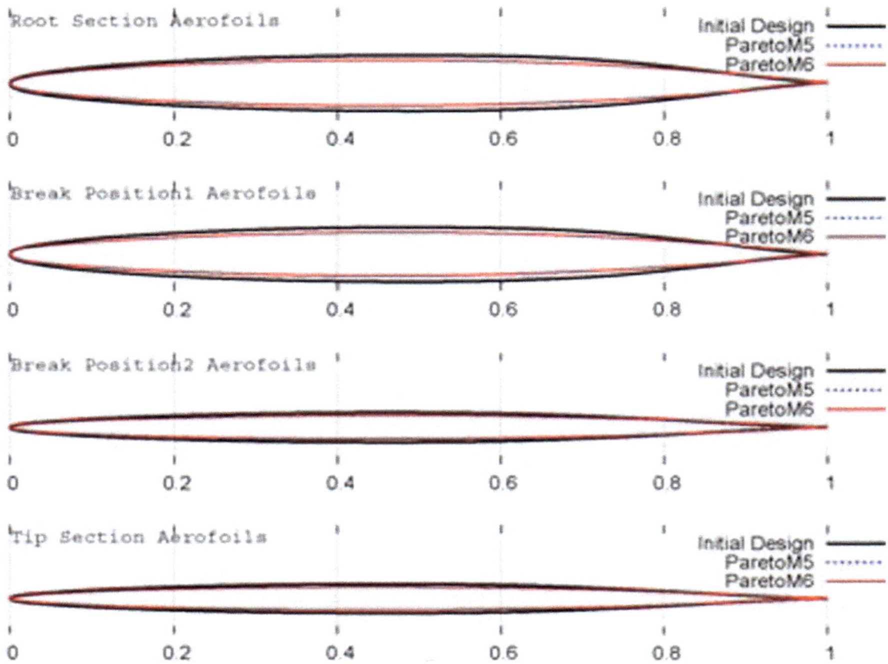

Fig. 8.69 Airfoils sections for the non-dominated solutions (Pareto Members 5 and 6)

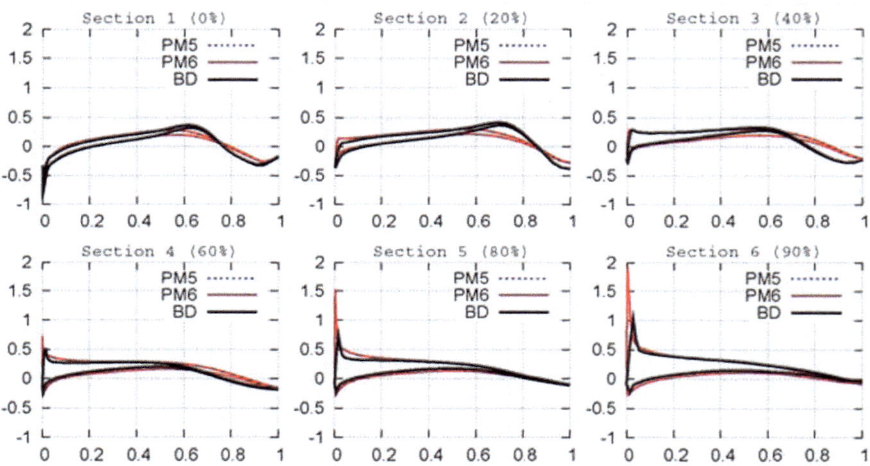

Fig. 8.70 Pressure coefficient (Cp) distribution along the span at standard design point

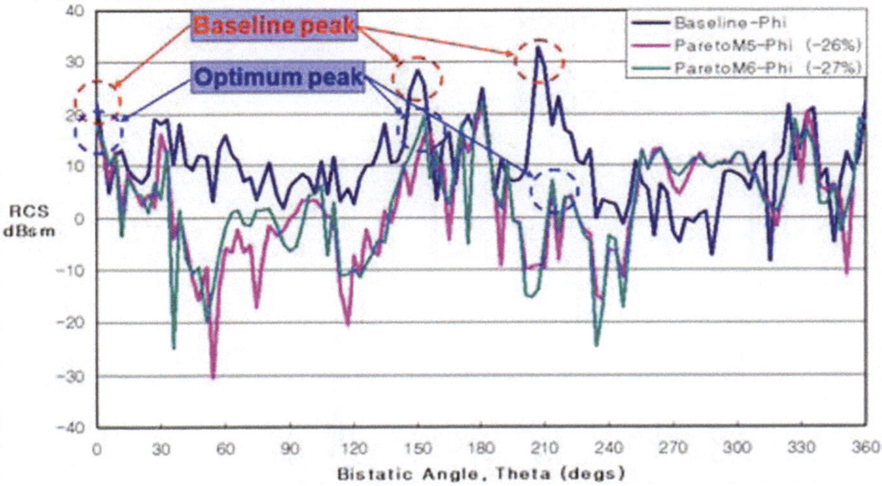

Fig. 8.71 RCS-Phi comparison between Pareto Member 5, 6 and baseline design. [solid: theta, dashed: phi phi=0°, wave (m)=0.3]

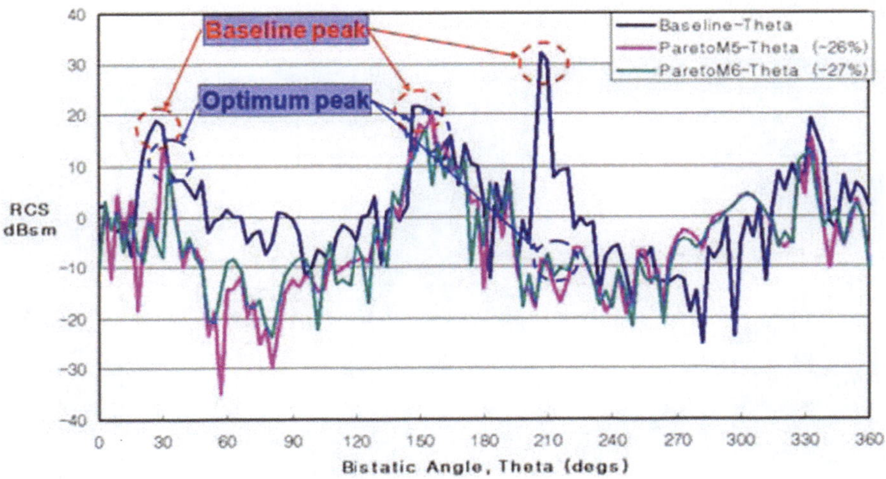

Fig. 8.72 RCS-Theta (linear) comparison between Pareto Member 5, 6 and baseline design. [solid: theta, dashed: phi phi=0°, wave (m)=0.3]

8.9 Summary of Chapter 8

This chapter introduced several applications using advanced EAs and MOEAS to solve multi-objective and multi-disciplinary problems in aeronautics. A detailed optimization mechanism coupled with flow solvers, a definition of design variables, fitness functions as well as analysis of optimized performance efficiency and design quality are implemented, results presented and discussed.

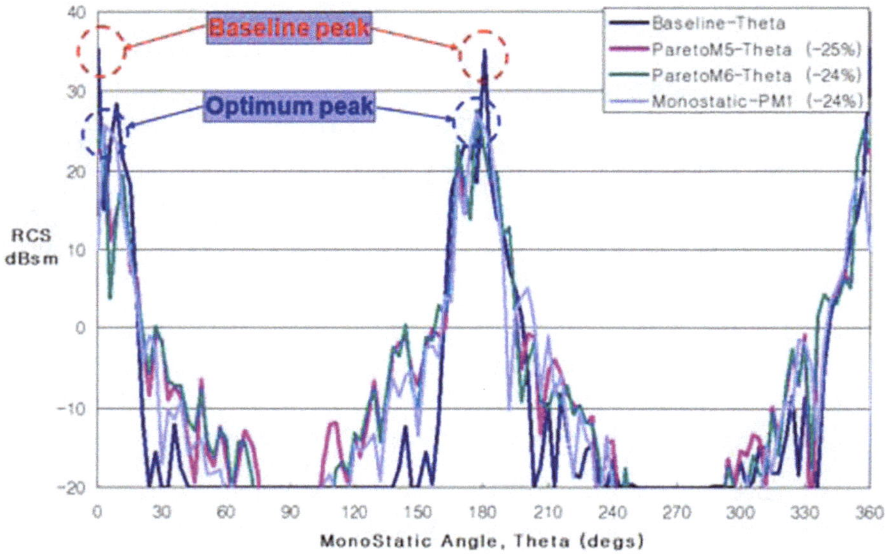

Fig. 8.73 Mono-static RCS-Theta comparison between Pareto Member 5, 6, Mono-Pareto member 1 (Test1-mono) and baseline design

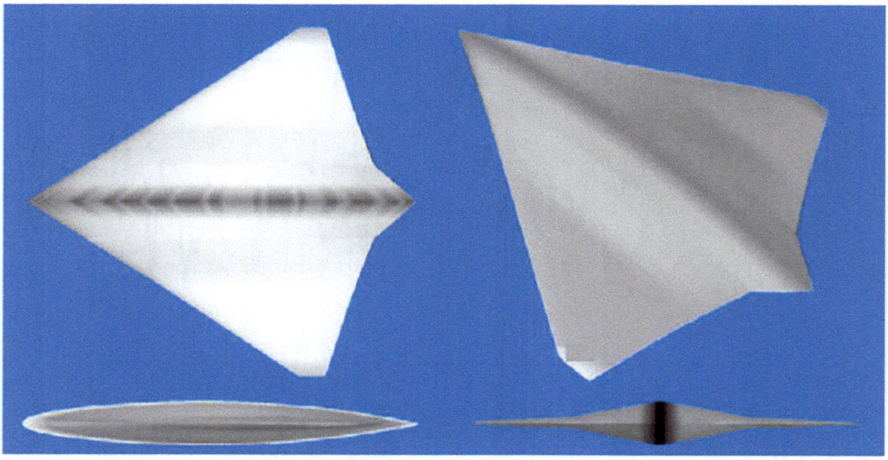

Fig. 8.74 Top, side, front and 3D view for compromised solution (Pareto Member 5)

References

1. Jameson A (2001) A perspective on computational algorithms for aerodynamic analysis and design. Prog Aerosp Sci 37(2):197–243(47)
2. Kim HJ, Rho OH (1997) Dual-point design of transonic airfoils using the hybrid inverse optimization method. J Aircr 34(5):612–618
3. Obayashi S (1997). Aerodynamic optimization with evolutionary algorithms. In: von Karman Institute for Fluid Dynamics-Lecture Series. Belgium.
4. Obayashi S (1998). Multidisciplinary Design Optimization of Aircraft Wing Plan form Based on Evolutionary Algorithms. In: Proceedings of the 1998 IEEE International Conference on Systems, Man, and Cybernetics, La Jolla, California.
5. Obayashi S, Takanashi S (1996) Genetic optimization of target pressure distributions for inverse design methods. AIAA J 34(5):881–886
6. Zetina AM, Shinkyu J, Shigeru O (2013) Airfoil aerodynamic optimization for a high-altitude long-endurance aircraft using multi-objective genetic-algorithms. IEEE C Evol Computat: 2314–2320
7. Oyama A, Liou MS, Obayashi S (2002) Transonic axial-flow blade shape optimization using evolutionary algorithm and three-dimensional navier-stokes solver, 9th AIAA/ISSMO symposium on multidisciplinary analysis and optimization, Georgia, Atlanta
8. Abott H, Von Doenhoff AE. Theory of wing sections. Dover
9. Geist A, Beguelin A, Dongarra J, Jiang W, Manchek R, Sunderam V. PVM: parallel virtual machine. A User's guide and tutorial for networked parallel computing. Massachusetts Institute of Technology
10. Schaffer JD (1985) Multiple objective optimization with vector evaluated genetic algorithms. Lawrence Erlbaum Associates, Pittsburgh, pp 93–100
11. Srinivas N, Deb K (1994) Multi objective optimisation using non-dominated sorting in genetic algorithms. Evolut Comput 2(3):221–248
12. Drela M (1993) A user's guide to MSES V2.3.
13. Reneaux J (2004) Overview on drag reduction technologies for civil transport aircraft. ECCOMAS 24–28 July
14. Wagner S, Sommerer A, Lutz A (2000) Numerical optimisation of adaptive transonic airfoils with variable camber. In: Proceedings 22nd ICAS Congress, number ICA2.111, Harrogate, United Kingdom, p 1–10
15. Quagliarella D, Vicini A (1999) Designing high-lift airfoils using genetic algorithms. In: Kaisa M, Marko M, Pekka N, Jacques P (eds) Proceedings of EUROGEN'99, University of Jyváskylä, Finland
16. Quagliarella D, Vicini A. A multiobjective approach to transonic wing design by means of genetic algorithms. In: NATO RTO AVT symposium on aerodynamic design and optimization, number RTO-MP—35, Ottawa, Canada.
17. Chen HQ, Mantel B, Periaux J, Sefrioui M (1998) Solution of some non linear fluid dynamics problems by means of genetic algorithms. In: Chetverushkin B, Desideri J-A, Kuznetsov YA, Muzafarov KA, Periaux J, Pironneau O (eds) Experimentation, modeling and computation in flow, turbulence and combustion, vol 2. Computational methods in Applied Sciences. Wiley, Chichester
18. Gano S, Renaud J. Optimized unmanned aerial vehicle with wing morphing for extended range and endurance. 9th AIAA/ISSMO Symposium and Exhibit on multidisciplinary analysis and optimization, Atlanta GA
19. Vickers M, Robert M (December 2001) Future warfare 20XX Wargame series: lessons learned report', Center for Strategic and Budgetary Assessments (CSBA)
20. Warner CG (1999) Implementing Joint Vision 2010: A revolution in military affairs for strategic air campaigns. Air University Press, Maxwell Air Force Base

21. Geer H, Bolkcom C. Unmanned aerial vehicles: background and Issues for Congress. CRC report for congress, Congressional Research Service, Library of Congress, Washington DC, USA.
22. Air-Attack. X-47 Pegasus Unmanned Combat Aerial Vehicle-Naval Version (UCAV-N), http://www.air-attack.com/page/28/X-47-Pegasus-UCAV-N.html
23. Directory of U.S. Military Rockets and Missiles: Northrop Grumman X-47 Pegasus http://www.designation-systems.net/dusrm/app4/x-47.htm
24. Samli U, Cotuk N (2003) Shaping considerations in the RCS reduction. 81504 Pendik Istanbul Turkey, AFCEA
25. Stimon G Introduction to airborne radar. SciTech Publishing, Inc.
26. UCAVs Chapter 16 of 19 (1998) http://www.vectorsite.net/twuav_16.html
27. Kok J, Gonzalez LF (2013) N.A Kelson FPGA implementation of an evolutionary algorithm for autonomous unmanned aerial vehicle on-board path planning. IEEE Trans Evol Comput 17(2):272–281
28. Gonzalez F, Narayan P, Castro MPG, Zeller L (2011) Development of an autonomous unmanned aerial system to collect time-stamped samples from the atmosphere and localize potential pathogen sources. J Field Robot 28(6):961–976
29. Lee D, Gonzalez LF, Periaux J, UAS Mission Path Planning System, (MPPS) using hybrid-game coupled to multi-objective optimizer. J Dyn Sys Meas and Control 132(4)
30. Bertola A, Gonzalez LF (February 2013) Adaptive dynamic path re-planning RRT algorithms with game theory for UAVs. In 15th Australian International Aerospace Congress (AIAC15), 25-28, Melbourne Converntion Centre, Melbourne, VIC
31. Florent C, Gonzalez LF, Motta N (February 2013) Coordination of UAV path planning in 2D and 3D in environmental/bio sensing applications. In 15th Australian International Aerospace Congress (AIAC15), 25–28, Melbourne Convention Centre, Melbourne, VIC
32. Kok J, Bruggemann T, Gonzalez LF (2013) An evolutionary computation approach to three-dimensional path planning for unmanned aerial vehicles with tactical and kinematic constraints. In Proceedings of the 15th Australian International Aerospace Congress, Melbourne Convention Centre, Melbourne, VIC.
33. Rappa G, Gonzalez LF, Kok J, Quagliotti F Hybrid game evolutionary algorithm for mission path planning of aerial survey tasks. In Grant I (Ed.) Proceedings of the 28th International Congress of the Aeronautical Sciences, Optimage Ltd., Brisbane Convention & Exhibition Centre, Brisbane, QLD, pp. 1–13.
34. Northrop Grumman-Defining The Future. X-47B UCAS, http://www.is.northropgrumman.com/systems/nucasx47b.html
35. Global Security: Military X-47 (A/B) Pegasus UCAV http://www.globalsecurity.org/military/systems/aircraft/x-47-specs.htm
36. Chatzigeorgiadis F, Jenn DC (2004) A MATLAB physical-optics RCS prediction code. IEEE Antennas Propag Mag 46(4):136–139

Chapter 9
Robust Multi-Objective and Multi-Disciplinary Model Optimization Test Cases

9.1 Overview

All previous chapters considered solutions of design optimisation 'at fixed design' conditions. However as discussed in Chap. 4, there are many situations in engineering where a design does not operate at 'fixed design' points due to uncertainties in values of manufacturing or operational parameters. In this chapter we illustrate the use of advanced EAs for five (5) robust design optimisation problems taking into account uncertainties.

9.2 Robust Active Flow Control Design Optimization

Active flow control is desirable to reduce drag at cruise tran-sonic conditions. In the following test cases the natural laminar flow RAE 5243 airfoil is selected as a baseline design. The airfoil has a maximum thickness of 0.14 at 41 % of the chord from the leading edge and a maximum camber of 0.0186 at 54.0 % of the chord. The baseline design is tested at flow conditions M_∞=0.68, C_l=0.82, Re = 19.0×10^6 with a Boundary Layer Transition (BLT) at 45 % of the chord from the leading edge. Figure 9.1 shows the pressure coefficient (*Cp*) contours. It can be seen that there is a strong normal shock on the suction side of the baseline design at 45 % of the chord BLT conditions.

The shock occurs at 60.0 % of chord for the BLT position at 45 % of the chord. This baseline design will be compared to two (2) SCB design optimisations; the first test case considers minimisation (Sect. 9.2.2) of total drag (Cd_{Total}) at flow conditions M_∞=0.68, C_l=0.82, Re = 19.0×10^6 with Boundary Layer Transition (BLT) at 45 % of the chord. The second test case (Sect. 9.2.3) considers finding a robust Shock Control Bump (SCB) at flow conditions M_∞=0.68 and Re = 19.0×10^6. Two uncertainty parameters, C_l and three Boundary Layer Transi-tion (BLT) positions are considered i.e., C_l=[0.7, 0.82], BLT = [25 %c, 37.5 %c, 50 %c].

J. Periaux et al., *Evolutionary Optimization and Game Strategies for Advanced Multi-Disciplinary Design,* Intelligent Systems, Control and Automation: Science and Engineering 75, DOI 10.1007/978-94-017-9520-3_9

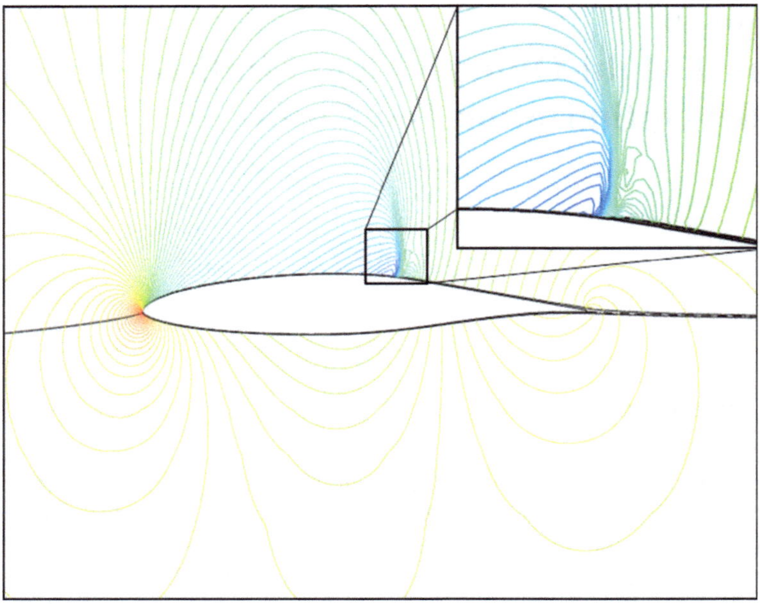

Fig. 9.1 Pressure coefficient (Cp) contour obtained by RAE 5243 airfoil

Table 9.1 Shock Control Bump (SCB) design variables and bounds for SCB geometry

Design variables	Lower bound	Upper bound
SCB_L	0	30
SCB_H	0	5
SCB_P	0	100

Peak position is in % of SCB length (SCB_L)

9.2.1 SCB Shape Design Optimisation at 45 % of the chord Boundary Layer Transition

Problem Definition This test case considers a single objective SCB design optimisation on the upper surface of the RAE 5243 air foil to minimise the total drag at flow conditions M_∞=0.68, C_l=0.82, Re=19.0×10^6 with BLT at 45 %c from the leading edge. This optimisation uses Algorithm 3 in Sect. 6.5.4 Chap. 6. The fitness function is shown in (9.1).

$$fitness(f) = \min(Cd_{Total}) = \min(Cd_{Viscous} + Cd_{Wave}). \tag{9.1}$$

Design Variables The design variable bounds for the SCB geometry are shown on Table 9.1. The centre of the SCB (50 % of SCB length) is located at the shock

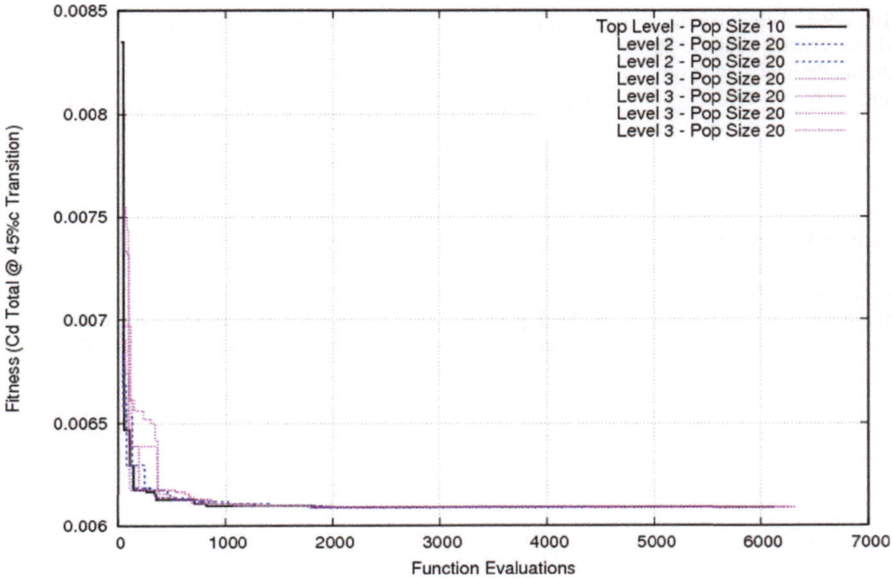

Fig. 9.2 Convergence history for SCB design at 45 % of the chord BLT

where the flow speed transits from supersonic to subsonic (60 % of chord). The SCB spans from approximately 45 % of the chord to 75 % of the chord if SCB length is assigned as 30 % of the chord.

The maximum length of SCB is limited to 30 %c since SCB will not be located over flap and aileron control surfaces which are located from 75 % of the chord to trailing edge (100 % of the chord).

Implementation The optimization was computed using HAPMOEA optimizer coupled to MSES [1] flow analyzer at the following details on multi-resolution/ population hierarchical populations.

1st Layer	Population size of 15 and intermediate recombination with a Computational grid of 36 × 16 cells
2nd Layer	Population size of 40 and intermediate recombination with a Computational grid of 24 × 16 cells
3rd Layer	Population size of 60 and intermediate recombination with a Computational grid of 36 × 16 cells

These grid conditions produce less than 5 % accuracy error compared to precise model at the 1st layer (Node 0)

Numerical Results As illustrated in Fig. 9.2, the algorithm was allowed to run 24 h and for 6135 function evaluations using a single 4 × 2.8 GHz processor. Convergence occurred after 1826 function evaluations which is equivalent to the CPU time of seven and half (7.5) hours.

Table 9.2 Aerodynamic characteristics obtained by the single-objective design approach

Aerofoil	Cd_{Total}	Cd_{Wave}	L/D
Baseline (RAE 5243)	0.01003	0.0032	81.72
with optimal SCB	0.00609 (−40 %)	0.00014 (−95 %)	134.56 (+65 %)

C_l value is fixed to 0.82

Table 9.3 Optimal SCB design parameters obtained by the single-objective design approach

Variables	SCB_L (%c)	SCB_H (%c)	SCB_P (%SCB_L)
SCB	29.22	1.04	67.7

Peak position (SCB_P) is in % of SCB length. The SCB starts from x and y coordinates (0.4516, 0.0858) to (0.7440, 0.0475)

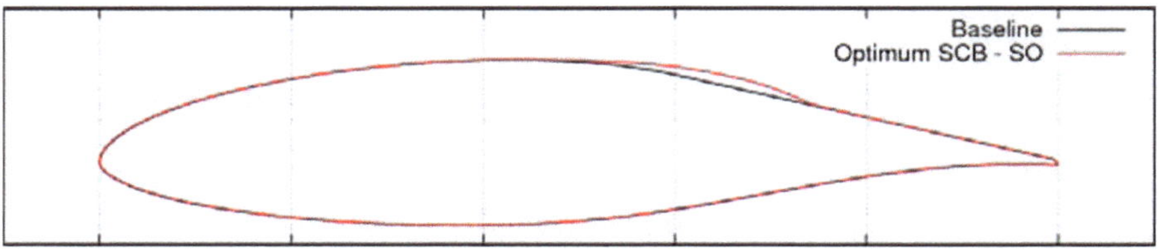

Fig. 9.3 Baseline design with the optimal SCB at 45 %c BLT

Table 9.2 compares the aerodynamic characteristics obtained by the baseline design (RAE 5243) and the baseline design with upper SCB. It can be seen that applying SCB on the upper surface of RAE 5243 aerofoil reduces the wave drag by 95 % which leads to 40 % total drag reduction. This optimal SCB improves L/D by 65 %.

The optimal shape of SCB is shown in Table 9.3. Figure 9.3 compares the geometry of the baseline design and baseline with optimal SCB. The baseline (RAE 5243) design with optimal SCB has a maximum thickness of 0.14 (t/c_{max} = 0.14) at 41 %c from the leading edge while a maximum camber is 0.0215 at 63.1 %c. Adding the optimal SCB increases the maximum camber by 0.003, and the camber position is moved towards the trailing edge by 9 %c while keeping the same maximum thickness as the baseline design.

Figure 9.4 shows the Cp contours obtained by the baseline design with optimal SCB. It can be seen that the strong shock on the baseline design shown in Fig. 9.1 is now 95 % weaker by adding the SCB with optimal position and shape (Fig. 9.4).

Figures 9.5a and b compare total drag (Cd_{Total}) and wave drag (Cd_{Wave}) distributions obtained by the baseline design and one with the optimal SCB along the Mach range i.e. $M_\infty \in [0.5{:}0.75]$ with Cl_{Fixed} = 0.82, Re = 19.0×10^6 with a BLT at 45 % of the chord. The baseline design with SCB starts to produce lower total drag when Mach number is higher than 0.67. By adding the optimal SCB, the baseline design reduces its total drag by 40 % and its wave drag by 95 % at the standard flight condition marked with the dashed line shown in Fig. 9.5. The critical Mach number (M_C = 0.64) for baseline design is extended to 0.66 due to the optimal SCB.

Figures 9.6a and b compare total drag distributions obtained by the baseline design and one with the optimal SCB for a C_l range i.e., $C_l \in [0.1{:}1.1]$ with M_∞ = 0.68,

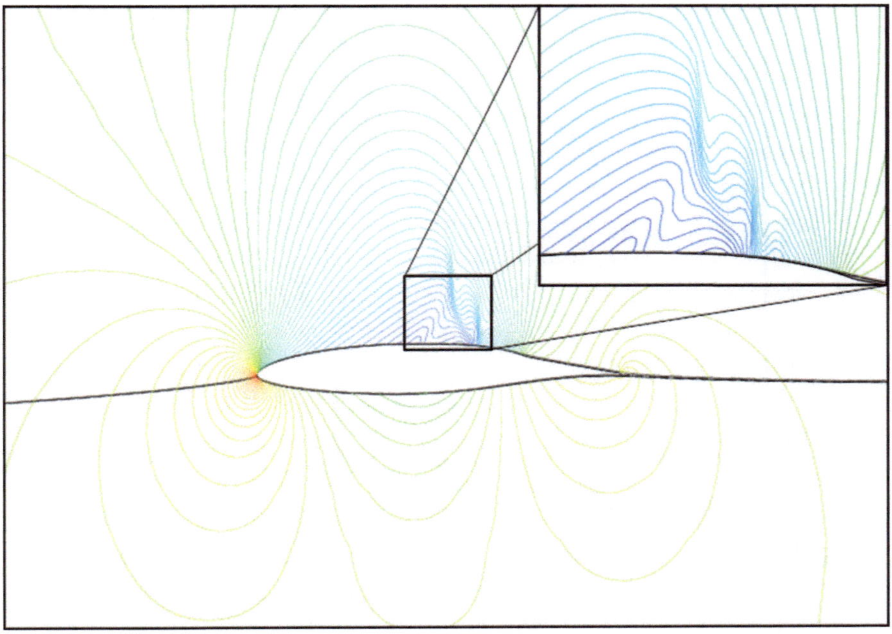

Fig. 9.4 Pressure coefficient (Cp) contour RAE 5243 with the optimal SCB at 45 %c BLT

Re = 19.0×10^6 with a BLT at 45 %c. The baseline design with SCB starts to produce lower total drag when its C_l number is higher than 0.65. The critical C_l number (C_{lc} = 0.25) for the baseline design is extended to 0.39 by applying optimal SCB on the suction side of the baseline design.

Even though good results were obtained by using a single-objective design approach at the standard flow conditions, the optimal SCB produces Cd_{Total} fluctuation at a range of C_l from 0.6 to 0.82 as shown in Figs. 9.6a and b. This optimal solution is an over-optimised solution which fails before reaching the standard flow/flight conditions. Such a fluctuation should be treated as uncertainty in the design parameters and the design engineer should take this into account during the optimisation. Therefore, it is necessary to use an uncertainty design technique to produce a set of solutions which have both low mean and sensitivity (no fluctuation and stable) Cd_{Total} under the considered variable C_l numbers and BLT positions.

9.2.2 *Robust SCB Shape Design Optimization with Uncertainty Boundary Layer Transitions*

Problem Definition This test case considers a robust multi-objective SCB design optimisation on the upper surface of the RAE 5243 airfoil to minimize mean and standard deviation of total drag ($\overline{Cd_{Total}}, \sigma Cd_{Total}$) at flow conditions M_∞ = 0.68 and Re = 19.0×10^6. Two C_l and three Boundary Layer Transition (BLT) positions are

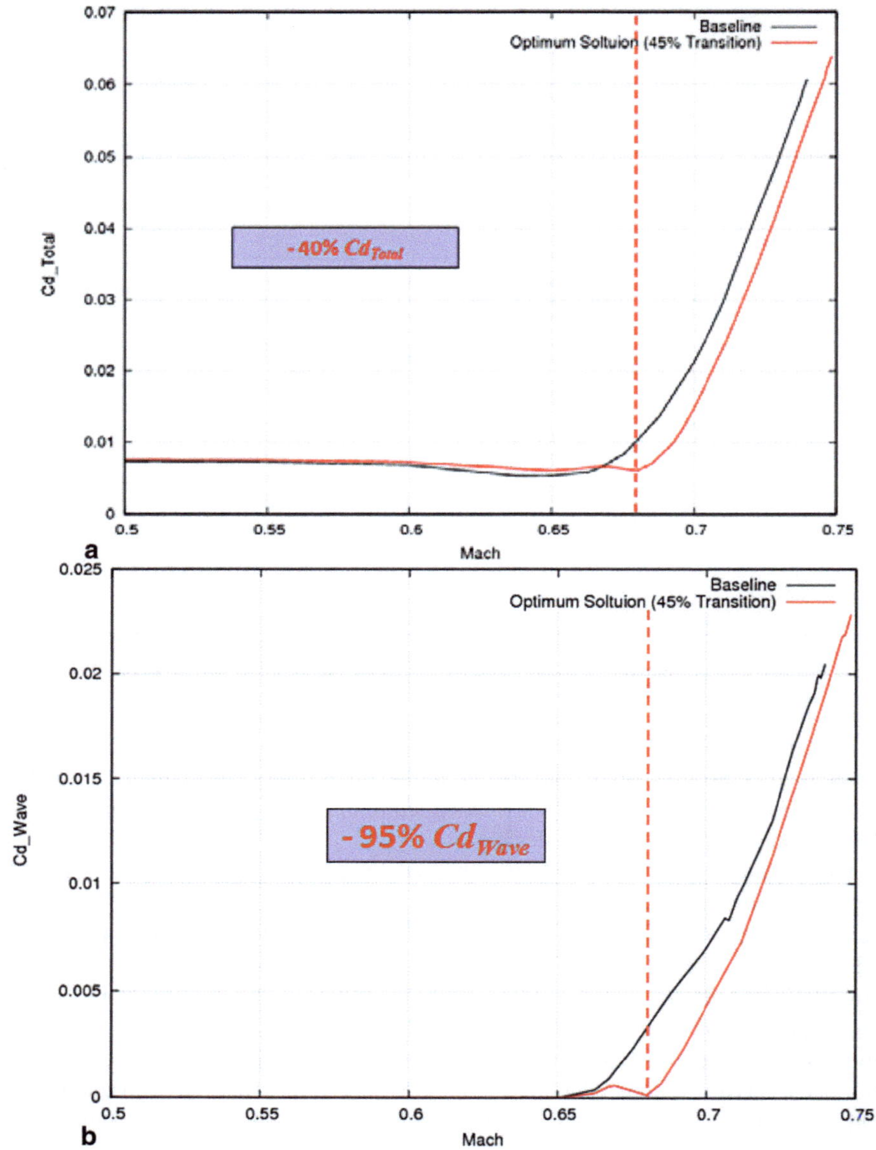

Fig. 9.5 Total drag coefficient (Cd_{Total}) **a** & wave drag coefficient (Cd_{Wave}) **b** vs. Mach at 45 %c BLT

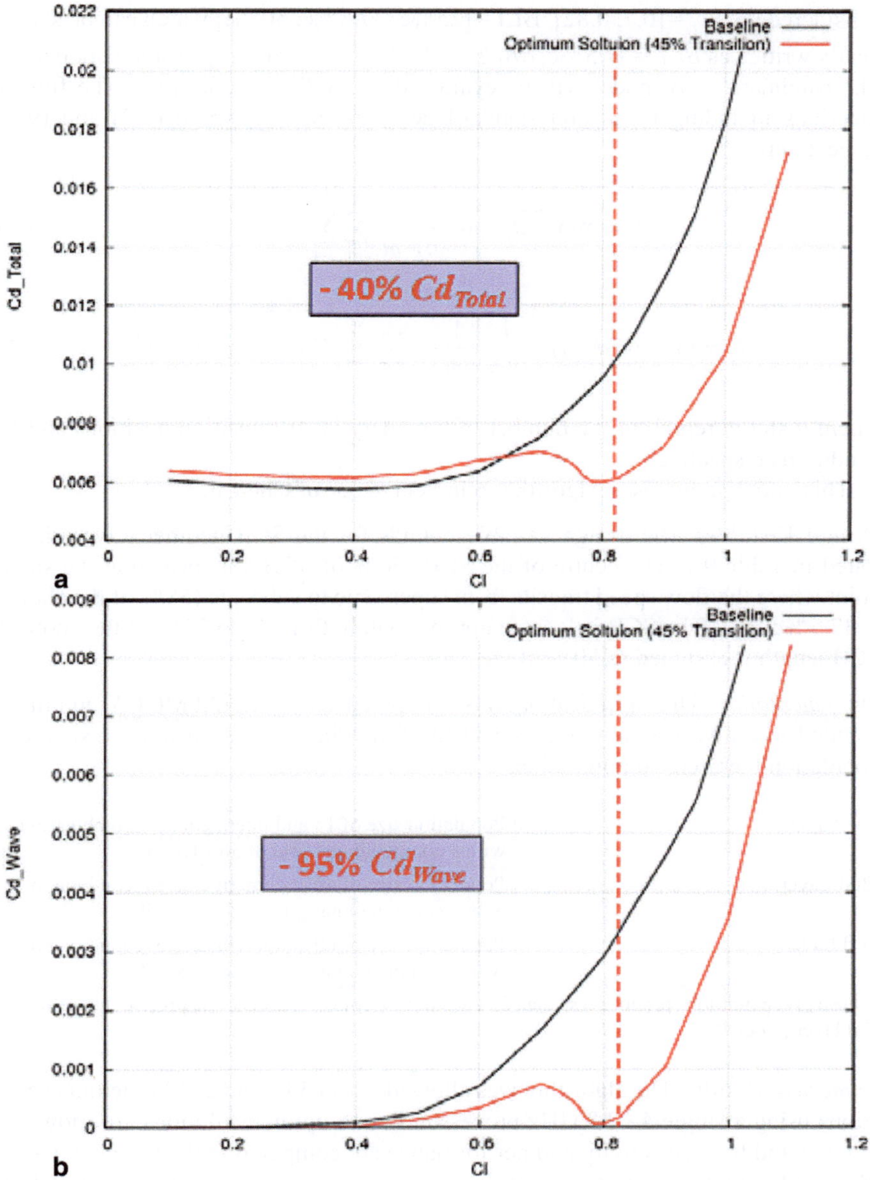

Fig. 9.6 Total drag coefficient (Cd_{Total}) **a** and wave drag coefficient (Cd_{Wave}) **b** vs. Cl at 45% of the chord BLT

considered i.e., $C_l = [0.7, 0.82]$, BLT=[25 %c, 37.5 %c, 50 %c] which can be statistically written as $\overline{BLT} = 37.5$ %c, $\sigma BLT = 12.5$ %c for uncertainty design parameters. The candidate SCB model will be evaluated at six flight conditions. The fitness functions including mean and standard deviation are shown in (9.2) and (9.3) respectively.

$$f_1 = \min\left(\overline{Cd_{Total}}\right) = \frac{1}{n \times m} \sum_{i=1}^{n} \sum_{j=1}^{m} Cd_{Total_{ij}} \tag{9.2}$$

$$f_2 = \min(\sigma Cd_{Total}) = \sqrt{\frac{1}{n \times m - 1} \sum_{i=1}^{n} \sum_{j=1}^{m} \left(Cd_{Total_{ij}} - \overline{Cd_{Total}}\right)^2} \tag{9.3}$$

where n and m represent the number of boundary layer transition positions and C_l conditions respectively.

This optimisation uses Algorithm 6 in Sect. 6.5.7 of Chap. 6.

Design Variables The design variable bounds for the SCB geometry were illustrated in Table 9.1. The centre of the SCB (50 % of SCB_L) is located at the sonic point where the flow speed transits from supersonic to subsonic (58 % of the chord) at 37.5 %c BLT. The SCB is located approximately from 43 to 73 % of the chord if SCB length is assigned as 30 % of the chord.

Implementation The optimisation was computed using a HAPMOEA optimizer coupled to an MSES flow analyser at the following details on multi-resolution/population hierarchical populations.

1st Layer	Population size of 15 and intermediate recombination with a computational grid of 36 × 16 cells
2nd Layer	Population size of 40 and intermediate recombination with a computational grid of 24 × 16 cells
3rd Layer	Population size of 60 and intermediate recombination with a computational grid of 36 × 16 cells

These grid conditions produce less than 5 % accuracy error compared to the precise model at the 1st layer (Node 0)

Numerical Results The algorithm was allowed to run 50 h and 2450 function evaluations using a single 4 × 2.8 GHz processor. Pareto optimal solutions are shown in Fig. 9.7 and their sensitivity and per formance are compared to the baseline design (Sect. 9.2.1) and the optimal design from a single-objective approach (Sect. 9.2.2). It can be seen that all Pareto members dominate the baseline for the second fitness function (standard deviation of Cd_{Total} : σCd_{Total}) while Pareto Members 1–9 dominate the baseline design in terms of mean and standard deviation of Cd_{Total}. Pareto Members 1–3 dominate the optimal design from Sect. 102.2. Pareto Members 1 and 4 are selected to compare aerodynamic performance to the baseline design and optimal solution from Sect. 9.2.2.

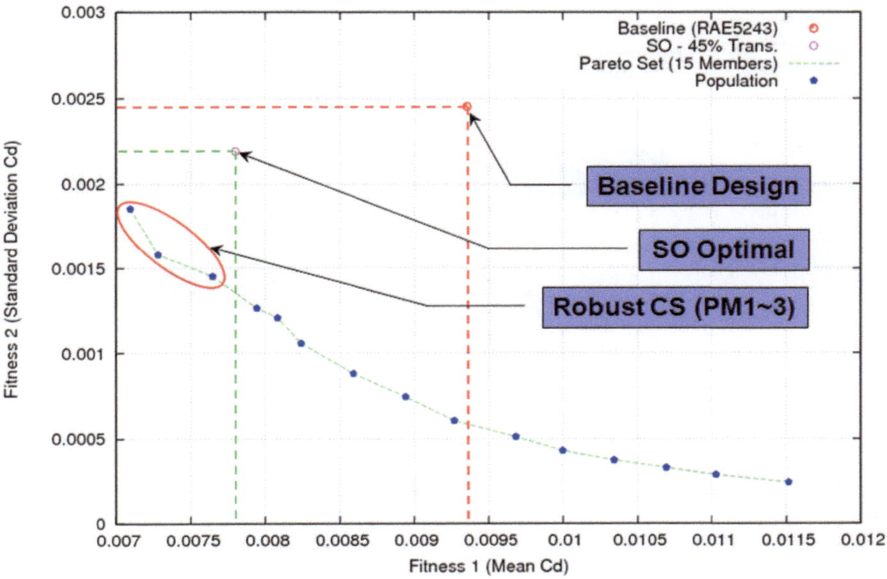

Fig. 9.7 Pareto optimal front for robust SCB design at 45 % of the chord BLT (Remark: SO and CS represent the single-objective approach and compromised solutions respectively)

Table 9.4 Comparison of fitness values obtained by the robust design approach

Aerofoil	$\overline{Cd_{Total}}$	σCd_{Total}	$\overline{L/D}$
Baseline	0.00935	0.00245	87.70
Single-Objective	0.00780 (−17 %)	0.00219 (−11 %)	105.13 (+20 %)
with PM1-SCB	0.00709 (−24 %)	0.00185 (−24 %)	115.66 (+32 %)
with PM2-SCB	0.00728 (−22 %)	0.00158 (−35 %)	112.64 (+28 %)
with PM3-SCB	0.00764 (−18 %)	0.00145 (−41 %)	107.33 (+22 %)
with PM4-SCB	0.00794 (−15 %)	0.00126 (−48 %)	103.27 (+18 %)

PM_i represents Pareto optimal member obtained by robust design optimisation

Table 9.4 compares the aerodynamic characteristics obtained by the baseline design (RAE 5243 airfoil) and the baseline design with SCB obtained by Pareto members 1 and 4. It can be seen that applying the optimal SCB obtained by Pareto member 1 on the suction side of RAE 5243 airfoil reduces the mean total drag by 24 %. This optimal SCB improves *L/D* by 32.0 %.

The mean and standard deviations obtained by the baseline design, single-objective and robust Pareto members can be compared using Cumulative Distribution Function (CDF) and Probability Density Function (PDF). Figure 9.8a shows the CDF obtained by the baseline design, the optimal from single-objective (marked as SO optimal) and the robust compromise Pareto solutions (marked as Robust CS). It can be seen that all the solutions obtained by single-objective and robust design

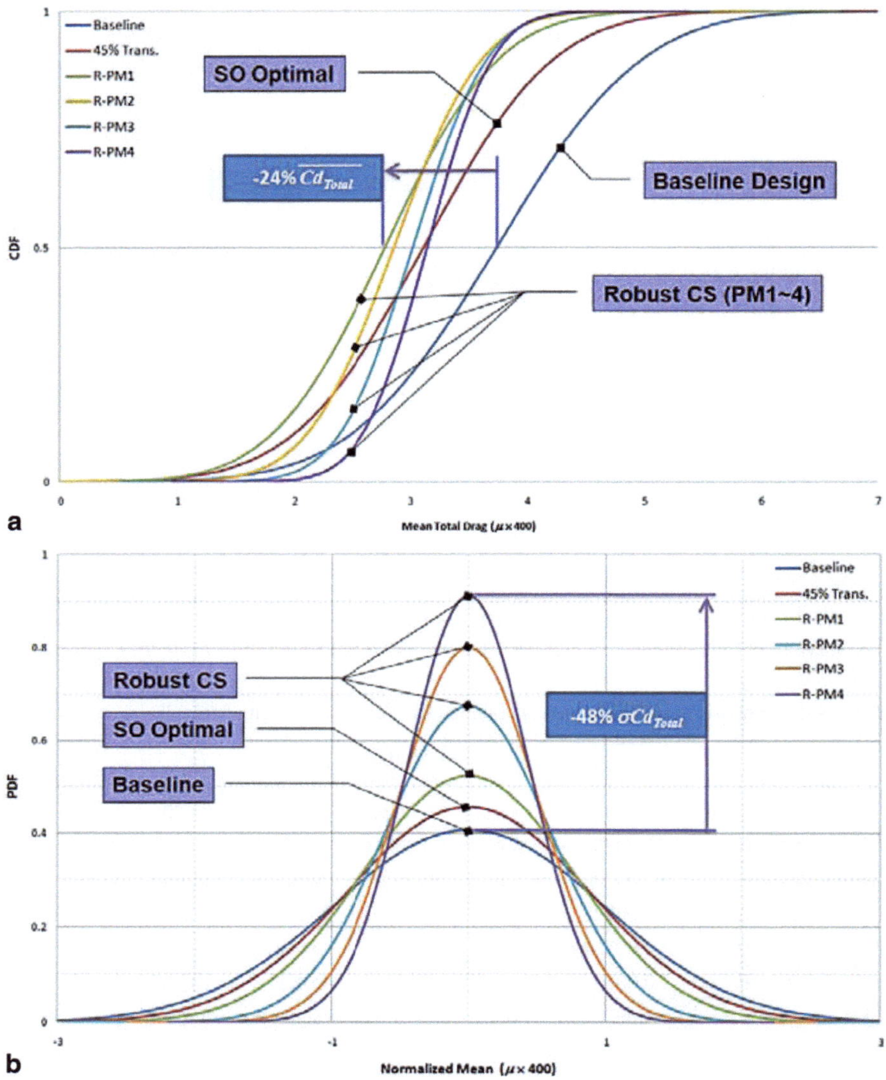

Fig. 9.8 **a** Mean total drag comparison using CDF. **b** Total drag sensitivity comparison using PDF

methods have lower mean total drag when compared to the baseline design. Pareto Member 1 reduces the mean total drag by 24% when compared to the baseline while the optimal obtained by the single-objective approach reduces total drag by 17%. The standard deviation (sensitivity) can be represented by evaluating gradient of the lines to the CDF value of 0.5 or 1 (steep gradient means low sensitivity). Herein PDF is plotted in Fig. 9.8b to have a clear sensitivity comparison between the baseline design, single-objective, robust design method. It can be seen that all the solutions obtained by the single-objective and robust design methods have lower

Table 9.5 Comparison of mean and standard deviation of drag obtained by the baseline design, single-objective optimal solution and Pareto members 1 to 3 under considering five hundred uncertainty flight conditions

Aerofoil	$\overline{Cd_{Total}}$	σCd_{Total}	$\overline{L/D}$
Baseline	0. 00919	0.000893	83.31
Optimal solution (Section V.A)	0.00705 (−23 %)	0.000756 (−15 %)	109.59 (+31 %)
Pareto Member 1	*0.00678 (−26 %)*	*0.000696 (−22 %)*	*113.17 (+36 %)*
Pareto Member 2	0.00720 (−22 %)	0.000591 (−34 %)	106.38 (+28 %)
Pareto Member 3	0.00757 (−18 %)	0.000498 (−44 %)	100.93 (+21 %)

sensitivity (narrow bell curve). Pareto Member 4 obtained by the robust design method has 48 % total drag sensitivity reduction when compared to the baseline design while the optimal obtained by single-objective approach reduces total drag by only 11 %. In other words, the robust design method has capabilities to produce a set of solutions which have better performance and sensitivity when compared to the single-objective optimization method.

For the detailed sensitivity analysis, the baseline design, the optimal solution obtained in Section V.A and Pareto members 1–3 are selected and their $\overline{Cd_{Total}}$ and σCd_{Total} values in (9.2) and (9.3) respectively are recomputed at the variability of $\overline{BLT}=37.49\,\%c$, $\sigma BLT=0.0729\,\%c$; [25.0 %c− 50.0 %c] and the variability of $\overline{Cl}=0.761$, $\sigma Cl=0.0385$, [0.7:0.82]. In total, five hundred non-uniformly distributed flight conditions (50 BLT positions × 10 C_l values) obtained by Latin Hypercube Sampling[19] are considered.

Table 9.5 compares the aerodynamic characteristics obtained by the baseline design and the optimal solution obtained in Sect. 9.2.2 and Pareto Members 1–3. It can be seen that both optimal solutions produce lower total mean drag and drag sensitivity respect to five hundred uncertain design conditions (boundary layer transition location and lift coefficient). Applying the optimal SCB obtained by Pareto member 1 produces 26 % lower total drag while lowering the drag sensitivity by 22 % when compared to the baseline design. The sensitivity obtained by the baseline design and all solutions is more than two and half times lower due to increment of number of uncertainty design conditions from six to five hundred. In addition, Fig. 9.9 compares mean and standard deviation using CDF and PDF.

It can be noticed that $\overline{Cd_{Total}}$ and σCd_{Total} behaviours obtained by the CDF and PDF (shown in Fig. 9.8) considering six uncertain flight conditions are similar to the $\overline{Cd_{Total}}$ and σCd_{Total} obtained by the CDF and PDF (shown in Figs. 9.9a and b) considering five hundred flight conditions between the baseline design and Pareto Members 1–3. In other words, the simplified robust method with six uncertain flight conditions still produces both lower total mean drag and drag sensitivity with respect to the variability of Boundary Layer Transition positions and lift coefficient.

The design parameters of the SCB obtained by Pareto Members 1–4 are shown in Table 9.6. It can be seen that the length and height of SCB are reduced by approximately 5 and 0.4 %c respectively when compared to the optimal solution

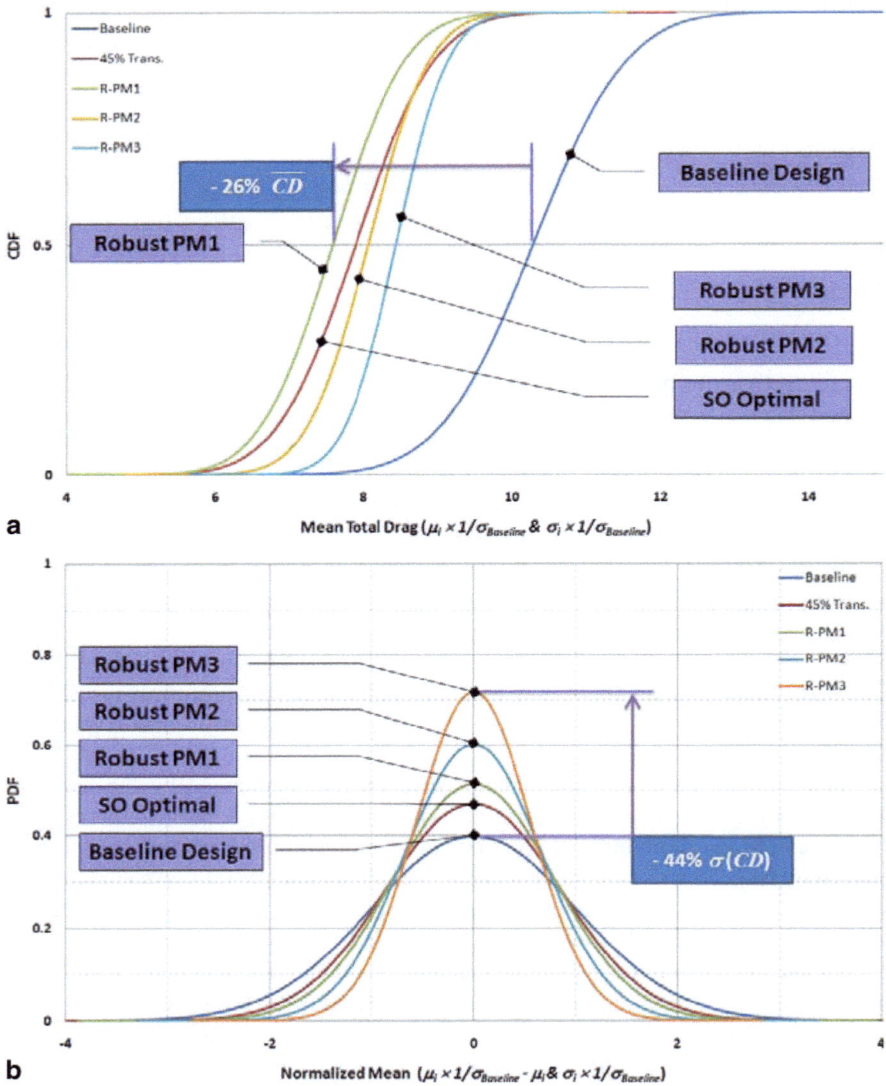

Fig. 9.9 **a** Total mean drag comparison using CDF with five hundred uncertainty flight conditions. **b** Standard deviation (Sensitivity) of total drag comparison using PDF with five hundred uncertainty flight conditions

from Sect. 9.2.2 (Table 9.3). Figure 9.10 compares the geometry of the baseline design and baseline with optimal SCBs. The baseline (RAE 5243) design with SCB obtained by Pareto Member 2 (highest SCB) has a maximum thickness 0.14 (t/c_{max}=0.14) at 41 %c from the leading edge while a maximum camber 0.0212 at 63.0 %c. Adding a SCB obtained by Pareto Member 2 increases the maximum camber by 0.0026 and its position is moved towards the trailing edge by 9 %c while keeping the same maximum thickness as the baseline design.

Table 9.6 Pareto optimal SCB design parameters

Variables	SCB_L (%c)	SCB_H (%c)	SCB_P (%SCB_L)
PM1-SCB	25.93	0.76	69.3
PM2-SCB	27.12	0.85	76.2
PM3-SCB	26.45	0.75	81.3
PM4-SCB	25.56	0.69	84.9

Peak position is in % of SCB length. The PM1-SCB starts from x and y coordinates (0.4490, 0.0859) to (0.7083, 0.0540)

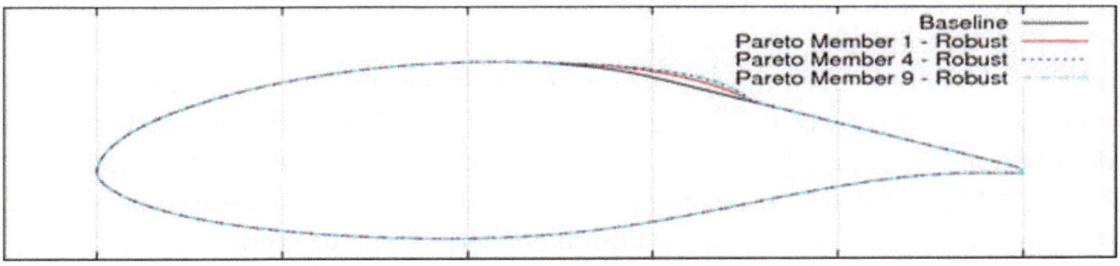

Fig. 9.10 Baseline design with the optimal SCB at 45 %c BLT

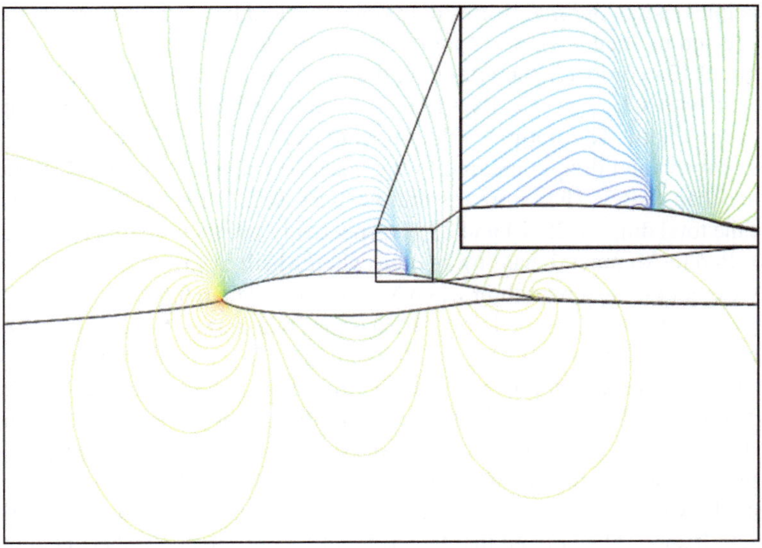

Fig. 9.11 Pressure coefficient (Cp) contours of the RAE 5243 airfoil with the PM1-SCB at 45 % of the chord BLT

Figure 9.11 shows the pressure contour obtained by Pareto Member 1 at 45 % of the chord Boundary Layer Transition. It can be seen that the strong shock on the baseline design shown in Fig. 9.1 is 77 % weaker by adding SCB on the suction side. The SCB for Pareto Member 1 reduces the total drag by 35 % and improves the lift to drag ratio by 53 %.

Figure 9.12a compares the total drag (Cd_{Total}) distributions obtained by the baseline design, optimal solution from Sect. 5.1 and Pareto Members 1 and 4. The flight conditions are; $M_\infty \in [0.6:0.72]$ with $Cl_{Fixed}=0.82$, $Re=19.0\times10^6$ and the boundary layer transition at 45 % of the chord. It can be seen that all solutions obtained by single-objective and robust design methods have lower total drag when Mach number is higher than 0.67. The optimal solution from Sect. 9.2.2 produces lower total drag (−40 %) at the Mach 0.68 and Pareto Members 1 and 4 reduce the total drag by 35 and 27 % respectively. Pareto Member 1 has lower drag compared to other solutions when Mach is lower than 0.6775 while Pareto Member 4 produces lower drag when Mach is higher than 0.685.

Figure 9.12b compares total drag (Cd_{Total}) distributions at the flow conditions; $C_l \in [0.1:1.1]$ with $M_{Fixed}=0.68$, $Re=19.0\times10^6$ and the boundary layer transition at 45 % of the chord. Even though the optimal solution obtained by single-objective method produces lower total drag, Pareto Members 1–4 have a stable total drag distribution at C_l range [0.6:0.82] without fluctuation due to the stable wave drag. Figure 9.12c compares wave drag (Cd_{Wave}) distributions obtained by the baseline design, single-objective and robust design methods. It can be seen that Pareto Members 1–4 produce stable wave drag when compared to the baseline design and the single-objective approach.

One thing should be noticed is that the critical lift coefficient numbers (Cl_C) for the baseline design, optimal from single-objective and Pareto members obtained by robust design method are 0.3, 0.4 and 0.5 respectively.

Figure 9.13a compares the total drag distributions at a range of Mach i.e. $M_\infty \in [0.5:0.75]$ with $Cl_{Fixed}=0.82$, $Re=19.0\times10^6$ and the boundary layer transition at 25 % of the chord. The optimal solution from Sect. 5.1, and Pareto members 1–4 produce lower total drag compared to the baseline design when the Mach number is higher than 0.665. The optimal solution (Sect. 9.2.2), Pareto Members 1 and 4 reduce the total drag by 25, 24 and 23 % respectively when compared to the baseline design. Pareto Member 1 has lower drag when Mach is lower than 0.6775 while Pareto Member 4 has lower drag when Mach is lower than 0.685.

Figure 9.13b compares total drag distributions at a range of C_l i.e., $C_l \in [0.1:1.1]$ with $M_{Fixed}=0.68$, $Re=19.0\times10^6$ and the boundary layer transition at 25 % of the chord. The optimal solution from Sect. 5.1 fluctuates at the range of $C_l=[0.6: 0.82]$ while Pareto members 1 and 4 have a stable C_l distribution. Figure 9.13c compares wave drag (Cd_{Wave}) distributions obtained by the baseline design, single-objective and robust design methods. It can be clearly seen that Pareto Members 1–4 produce stable wave drag when compared to the baseline design and single-objective approach. It should be noticed that the critical lift coefficient numbers (Cl_C) for the baseline design, optimal from single-objective and Pareto Members obtained by robust design method are 0.3, 0.4 and 0.5 respectively.

The baseline design with the optimal SCB obtained by the single-objective and robust design methods are also tested at six (6) normal flight conditions shown in Table 9.7.

The histogram shown in Fig. 9.14a compares the total drag obtained by the baseline design, the single-objective (Sect. 9.2.2) and the robust design approaches. It can be seen that the baseline design with the optimal SCB—robust

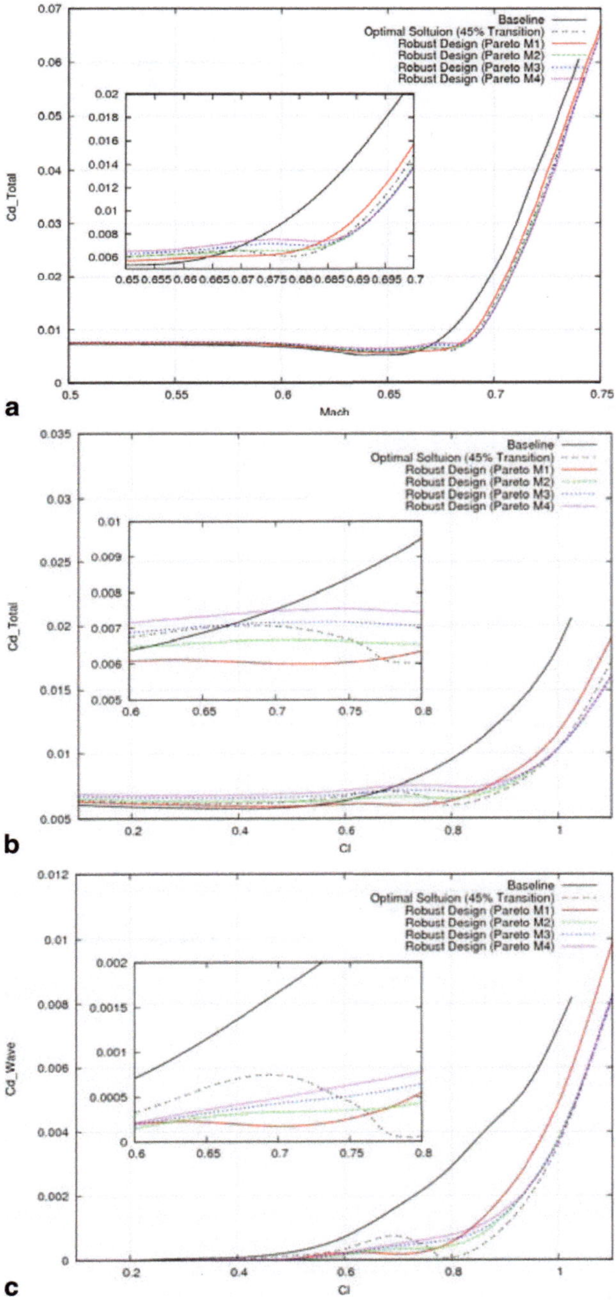

Fig. 9.12 **a** Total drag coefficient (Cd_{Total}) vs. Mach at 45 % of the chord BLT. **b** Total drag coefficient (Cd_{Total}) vs. Cl at 45 %c BLT. **c** Wave drag coefficient Cd_{Wave} vs. Cl at 45 % of the chord BLT

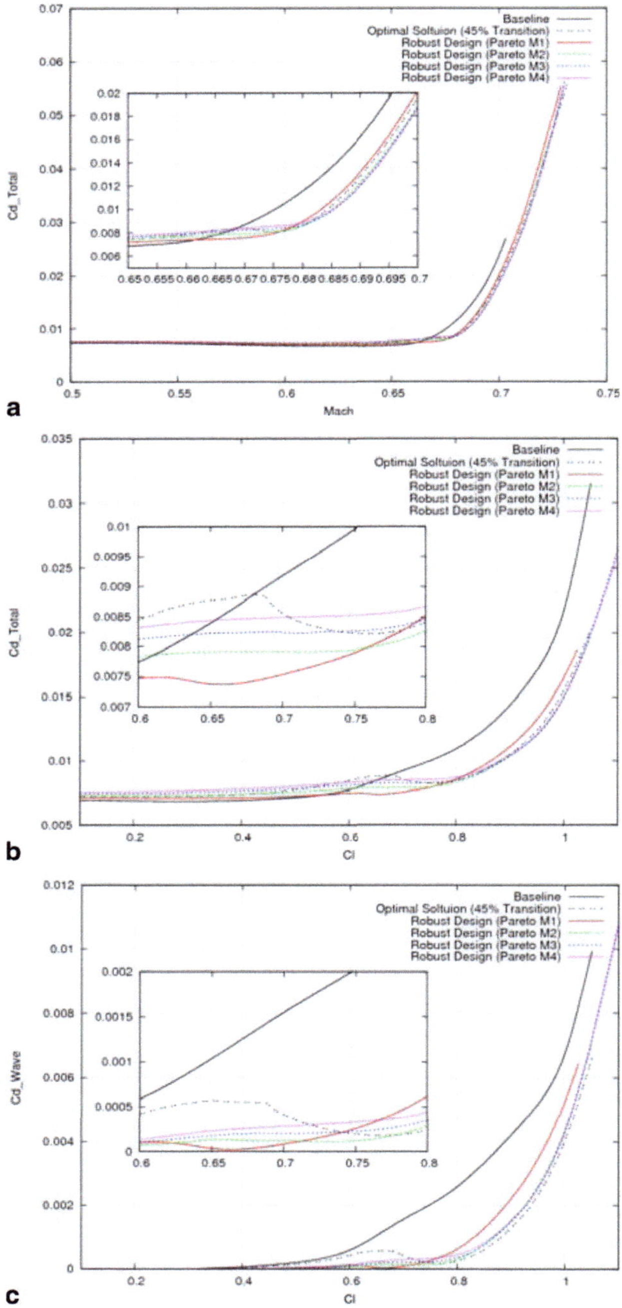

Fig. 9.13 **a** Total drag coefficient (Cd_{Total}) vs. Mach at 25 %c BLT. **b** Total drag coefficient (Cd_{Total}) vs. Cl at 25 % of the chord BLT. **c** Wave drag coefficient (Cd_{Wave}) vs. $C\,l$ at 25 % of the chord BLT

Table 9.7 Six flight conditions

Conditions	M_∞	Re	C l	X_{Trans}—Upper	X_{Trans}—Lower
Flight condition 1	0.69	11.7×10^6	0.54	0.51	0.54
Flight condition 2	0.69	11.7×10^6	0.69	0.51	0.54
Flight condition 3	0.70	11.7×10^6	0.38	0.51	0.54
Flight condition 4	0.70	11.7×10^6	0.50	0.51	0.54
Flight condition 5	0.73	11.7×10^6	0.22	0.51	0.54
Flight condition 6	0.73	11.7×10^6	0.34	0.51	0.54

design produces lower total drag when compared to the baseline design and the optimal SCB obtained by the single-objective method. The optimal SCB—Pareto Member 1 reduces the total mean drag by 36 % while lowering the total drag sensitivity by 77 %. The optimal SCB obtained in Sect. 9.2.2 reduces the total mean drag and sensitivity by 9.5 and 48 % respectively even though it produces higher drag at the flight conditions 1, 3 and 5. Figure 9.14b compares the lift to drag ratio obtained by the baseline design, the optimal SCB from Sect. 9.2.2 and Pareto Member 1. It can be seen that Pareto Member 1 has the biggest lift to drag improvement at the flight condition 4.

One example (Cond 4) is shown in Fig. 9.15 where the pressure contours obtained by the baseline design and Pareto Member 1 are illustrated. Even though the SCB obtained by Pareto member 1 is optimized at the critical flight condition, the optimal SCB—Pareto Member 1 reduces the wave drag ($Cd_{Wave}=0.0025$) obtained by the baseline design by 99.5 % ($Cd_{Wave}=0.00001$) and reduces the total drag by 40 % which leads to 65 % improvement of *L/D* when compared to the baseline design.

To summarize the design test cases (Sects. 9.2.2 and 9.2.3), the design engineer can choose one of the solutions (Pareto Members 1–4) obtained by robust multi-objective design optimization due to two main reasons: the first reason is that even though the optimal solution from Sect. 9.2.2 produces lower total drag at the standard flight conditions (45 % of the chord Boundary Layer Transition), Pareto Members 1–4 have lower sensitivity (stable means no fluctuation) at the variability of C_l and boundary layer transition positions [25–50 % of the chord]. The second reason is that Pareto members 1–4 have a smaller SCB than the optimal solution obtained by the single-objective method. In other words, the manufacturer can reduce significantly the cost of the material and lower the modification on the manufacturing system.

9.3 Robust Multi-Objective Generic Aircraft Wing Optimization

Problem Definition This test case considers the design and optimisation of wing airfoil sections of a generic aircraft using robust (uncertainty) design method (Sect. 6.5.7 Chap. 6). Instead of designing a wing at a single design point, an

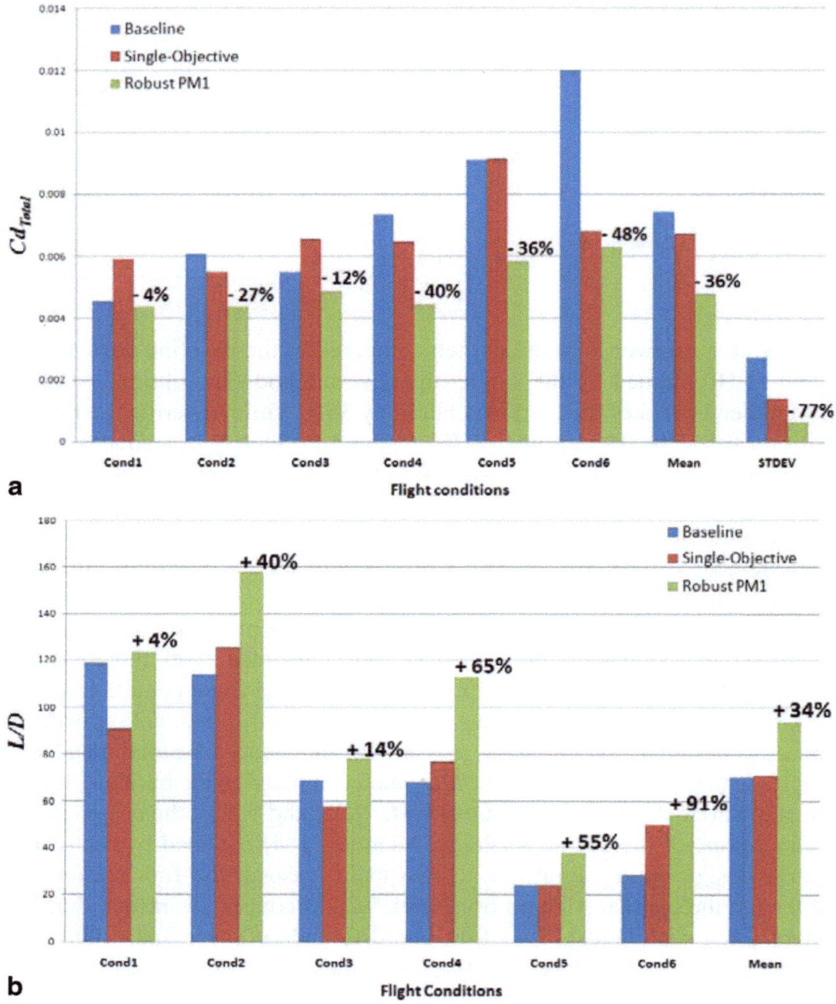

Fig. 9.14 **a** Drag reduction obtained by the baseline design, the single-objective solution and robust Pareto member 1 (Remark: Robust PM1 represents Pareto Member 1 obtained by the robust design optimisation. Mean and STDEV represent the mean and standard deviation of Cd_{Total} at six flight conditions). **b** Drag reduction obtained by the baseline design, the single-objective solution and robust Pareto Member 1

uncertainty method uses statistic formulations and designs aerodynamically stable models at off-design conditions. In this problem, five off-set Mach numbers are considered;

$$\overline{M_\infty} = 0.82 \; and \; \sigma M_\infty = 0.01581$$

$$\text{or} \quad M_{\infty i} \in [0.8, 0.81, M_S = 0.82, 0.83, 0.84]$$

and

$$C_{L_S} = 0.691, \left({}^{L}\!/\!{}_{D} \right)_{Min@Ms} \geq 8.46.$$

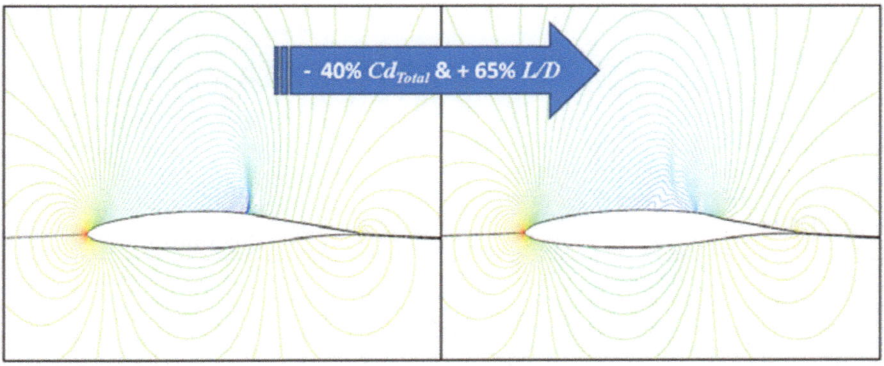

Fig. 9.15 *Cp* contours comparison obtained by the baseline design (*left*) and Pareto Member 1 (*right*) at the flight condition 4

The problem considers two (2) objectives where the fitness functions are minimization of discrete averaged (9.4) and variance (9.5) of the inverse of lift on drag:

$$f_1 = \min\left(\overline{1/(L/D)}\right) = 1/K \sum_{i=1}^{K} \left(1/(L/D_i)\right) \frac{M_{\infty i}^2}{M_S^2} + Penalty \tag{9.4}$$

$$\begin{aligned} f_2 &= \min\left(\delta\left(1/(L/D)\right)\right) \\ &= \frac{1}{(K-1)} \sum_{i=1}^{K} \left(1/(L/D_i) \frac{M_{\infty i}^2}{M_S^2} - \overline{1/(L/D)}\right)^2 + Penalty \end{aligned} \tag{9.5}$$

where the penalty term in Chap. 2.7 is computed and add to the fitness functions; if the thickness (t/c) is higher than 20% and less than 10.0% of the chord, the thickness ratio should be $10\% \leq t/c \leq 20\%$. If the ratio of lift to drag at flight conditions is lower than 8.46, the lift to drag ratio should be $L/D_{min} \geq 8.46$. If the *mean* and the *variance* of inverse lift to drag ratio are higher than 0.1008 then $\left(\overline{1/(L/D)} \leq 0.1008\right)$ and 3.0×10^{-4} $\left(\delta(1/(L/D)) \leq 3.0 \times 10^{-4}\right)$.

Design Variables The external wing geometry is fixed as shown in Fig. 7.40 and Table 7.12 Sect. 7.5.1 Chap. 7. Four aerofoil sections at root, crank 1, crank 2 and tip are considered as the design variables. There are eighty eight (88) design variables (11 mean line and 11 thickness at each section) in total. The control points of mean and thickness distribution are illustrated in Figs. 7.43a–d.

Implementation FLO22 and FRICTION solvers are software utilised and the following specific parameters below are considered for the HAPMOEA optimizer. The grid sizes of model are changed from 82 to 68 for second layer and from 68 to 48 for third layer to make a fast search.

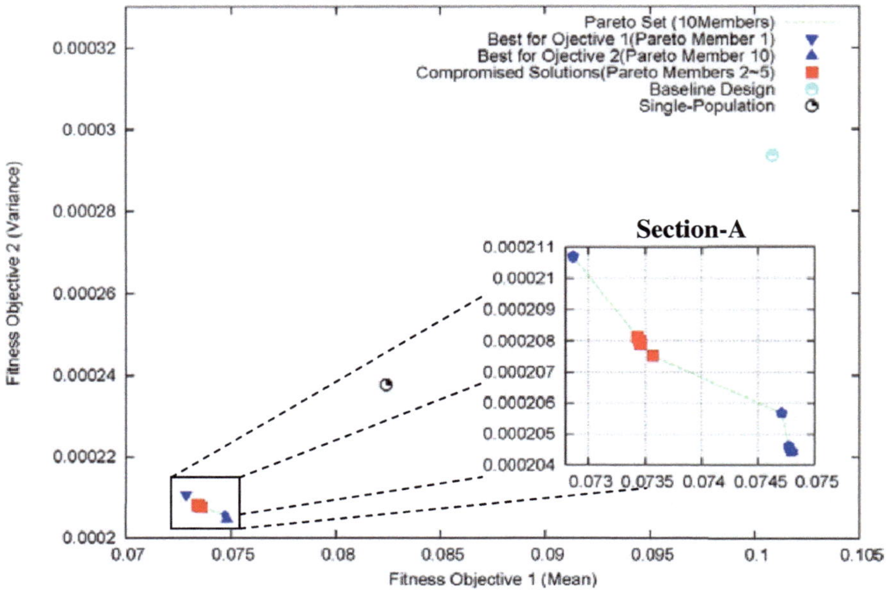

Fig. 9.16 Pareto optimal fronts for a generic aircraft wing with airfoil sections

1st Layer	Population size of 10 and intermediate recombination with a computational grid of 96 × 12 × 16 cells on the FLO22 solver
2nd Layer	Population size of 40 and intermediate recombination with a computational grid of 68 × 12 × 16 cells on the FLO22 solver
3rd Layer	Population size of 60 and intermediate recombination with a computational grid of 48 × 12 × 16 cells on the FLO22 solver

Numerical Results The algorithm was allowed to run approximately 1200 function evaluations and took 150 h on two 1.8 GHz processors. The Pareto optimal fronts is shown in Fig. 9.16 where it can be seen that the baseline and single-objective result from Sect. 7.5.4 are computed in terms of mean and variance of inverse lift to drag ratio. The Pareto fronts are zoomed in Section-A. It can be seen that Pareto (non-dominated) solutions produce lower *mean* and *variance* of inverse lift to drag ratio. Pareto members provide improvement of both aerodynamic efficiency and stability at variability of the flight conditions.

Five Pareto Members 2, 3, 4 and 5 are selected for further evaluation and Pareto member 4 and 5 are compared to the baseline and single-objective optimality from Sect. 7.5.4 Chap. 7. Pareto Members 4 and 5 produce 27 % improvement in mean inverse lift to drag ratio $\left(\overline{1/(L/D)}\right)$ with lower sensitivity $\left(\delta(1/(L/D))\right)$. Table 9.8 compares the drag coefficient at the standard design point. Pareto members 4 and 5 produce 28 % lower drag when compared to the baseline design.

Table 9.8 Comparison of the aerodynamic performance obtained by the baseline design, single-objective solution and compromised solutions (Pareto members 4 and 5)

Description	Baseline	SO optimal	Pareto M4	Pareto M5
$\overline{1/(L/D)}$	0.1008	0.0824 (−18%)	0.0734 (−27%)	0.0735 (−27%)
$\delta(1/(L/D))$	3.0×10^{-4}	2.4×10^{-4}	2.078×10^{-4}	2.075×10^{-4}
$C_{D_{@M_S}}$	0.0695	0.0565 (−18.7%)	0.050 (−28%)	0.050 (−28%)

SO represents Single-Objective

The *mean and standard* deviations at a range of Mach (M_∞=[0.8:0.01:0.84] or $\overline{M_\infty}=0.82$ and σ M_∞=0.01581) obtained by the baseline design, single-objective and robust Pareto members can be compared using Cumulative Distribution Function (CDF) and Probability Density Function (PDF). Figure 9.17a shows the CDF obtained by the baseline design, the optimal from single-objective (marked as SO optimal) and the robust compromise Pareto solutions (marked as Robust CS). It can be seen that all solutions obtained by single-objective and robust design methods have lower mean inverse *L/D* when compared to the baseline design. Pareto Members 4 and 5 improve the mean *L/D* by 27% when compared to the baseline while the optimal obtained by the single-objective approach improves *L/D* by 18%. The standard deviation (sensitivity) can be represented by evaluating the gradient of the line to the CDF value of 0.5 or 1 (steep gradient = low sensitivity). Herein PDF is plotted in Fig. 9.17b to have a clear sensitivity comparison between the baseline design, single-objective, robust design method. It can be seen that all solution obtained by the single-objective and robust design methods have lower sensitivity (narrow bell curve). Pareto members 4 and 5 obtained by the robust design method has 17% *L/D* sensitivity reduction when compared to the baseline design while the optimal obtained by single-objective approach reduces total drag by only 10%. In other words, the robust design method has capabilities to produce a set of solutions which have better performance and sensitivity when compared to the single-objective optimisation method.

Figure 9.18 shows the set of airfoils non—dominated solutions or Pareto fronts. It can be seen that classical aerodynamic shapes have been evolved, even considering that the optimisation was started completely from random and the evolution algorithm had no problem specific knowledge of appropriate solution types for uncertainty.

The airfoil sections characteristics for selected Pareto Members (4 and 5) are compared to the baseline design in Fig. 9.19 and Table 9.9. The airfoil sections of the selected Pareto Members 4 and 5 are 10% thinner than the baseline design. Negative camber is observed at the crank 2 airfoil section.

Figures 9.20a, b, and c show the pressure coefficient distributions (0, 20, 40, 60, 80 and 90% of the span). It can be seen that the optimal wing has a pressure improvement at the leading edge marked with blue circles from 40~90% of the span when compared to the baseline design. In the figures @ means at that design condition.

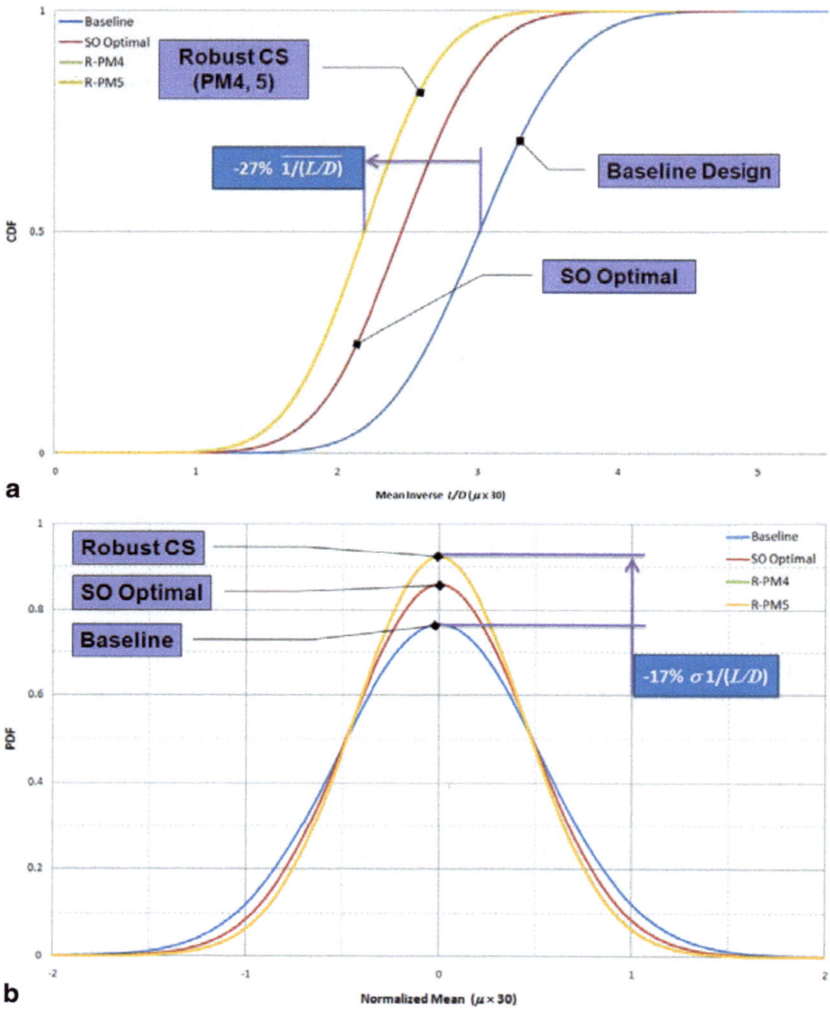

Fig. 9.17 **a** Mean inverse lift to drag ratio comparison using CDF. **b** Inverse lift to drag ratio (L/D) sensitivity comparison using PDF

Figures 9.21a and b compare the Mach sweeps on the drag coefficient and the lift to drag ratio obtained by uncertainty approach, single-objective (Sect. 7.5.4) and the baseline design. The two approaches including single-objective and uncertainty provide an improvement of the lift to drag ratio (L/D) when compared to the baseline wing. The results obtained using the uncertainty concept demonstrate better aerodynamic performances at all Mach numbers with less sensitivity. Table 9.10 compares the aerodynamic quality between Pareto members 4 and 5, single-objective solution and the baseline design. The Pareto members 4 and 5 are not only

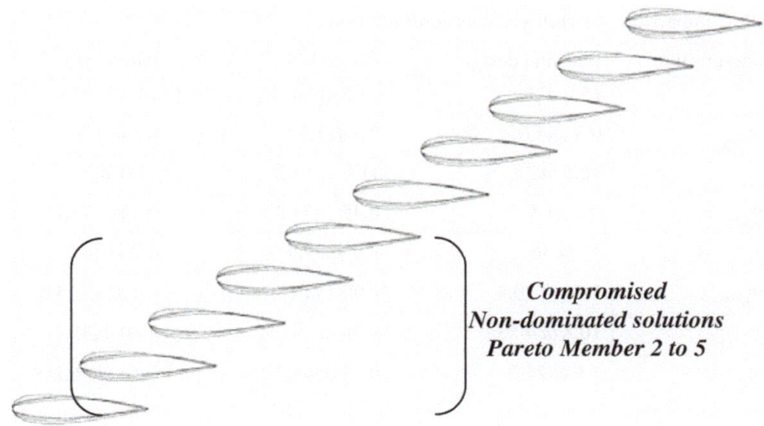

Fig. 9.18 Optimal airfoil sections

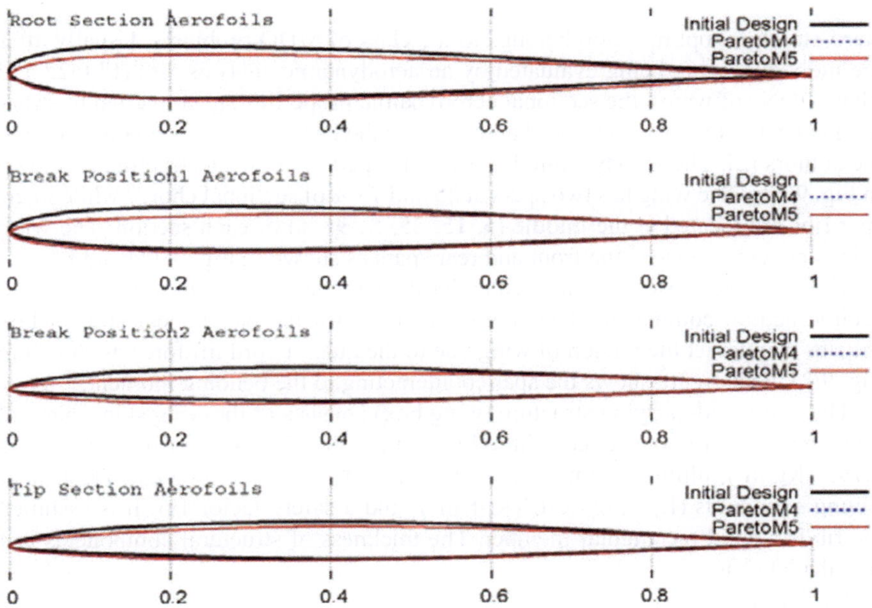

Fig. 9.19 Comparison of Airfoils sections between Pareto members 4, 5 and the baseline

stable at standard design point but also at other flight conditions when compared to the single optimal and baseline design. They also produce 8208 and 8196 km range (R) which have about 35 % range extension compared to the baseline and 12 % compared to the result obtained by the single-objective optimisation approach. The benefit of taking into account of uncertainty in the design is clearly demonstrated from the results.

Table 9.9 Comparison of airfoil sections configurations

Design description	Baseline design	Pareto M4	Pareto M5
t/c_{Root}	15.4 @19	13.5 @26	13.47 @26
$Camber_{Root}$	0.3 @4.6	0.5 @1.3	0.5 @1.3
t/c_{Crank1}	12.5 @28	11.0 @33.2	11.0 @33.6
$Camber_{BP1}$	0.7 @10.3	0.45 @74.2	0.44 @74.2
t/c_{Crank2}	10 @38	9.0 @39.4	8.7 @38.5
$Camber_{BP2}$	1.44 @20.4	−0.51 @15.7	−0.51 @15.4
t/c_{Tip}	10.8 @40.3	9.78 @39.7	9.80 @39.7
$Camber_{Tip}$	1.6 @21.2	0.55 @44.7	0.55 @46.2

9.4 Robust Aero-Structural Generic Aircraft Wing Optimization

Aero-structural optimisation belongs to the class of MDO problems. Usually after the individual wing being evaluated by an aerodynamic analysis tool (FLO22 and FRICTION software), the sectional aerodynamic properties are converted to structural concentrated loads. In this chapter we use the PSEC code developed by one of the authors [2]. The aerodynamic loads are mapped to the wing structure as shown in Fig. 9.22a. The wing has two spars at 25 and 75 % of sectional chords while inter-spar ribs are located at the middle (5, 15, 25, … 95 %) of each section. The wing skin is covered between the front and rear spars as shown in Fig. 9.22a and b.

The converted loads are checked under the structural constraints and applied to each structural component. The skin counteracts for the panel shear stress while the ribs counteract the torsion of wing due to the quarter cord lift force as shown in Fig. 9.23. This figure shows the spar counteracting to the bending moment.

The wing load-carrying structure (wing box) consists of the wing skin, spar and inter-spar ribs are considered with Aluminium Alloy 2024-T351: density (ρ) is 2,795 (Kg/m^3), ultimate compressive strength ($\sigma_{ultimate}$) is 37.965×10^6 (Kgf/m^2), Young's modulus (E) 7.523×10^9 (Kgf/m^2), and a safety factor 1.5. It is assumed the rib is a solid rectangular member. The thickness of structural components can be calculated as:

For the skin:

$$F_{Buckling_i} = KE\left(\frac{t_{skin_i}}{b}\right)^2 \rightarrow W_{skin} = \sum_{i=0}^{N} W_{skin_i} = \sum_{i=0}^{N} 2\cdot\rho\cdot t_{skin_i}\cdot A_i \tag{9.6}$$

where i represents the section number and b denotes the width of the structural section. K is the panel buckling coefficient (NACA TN3781), E is the Young's modulus and A_i is the section panel area.

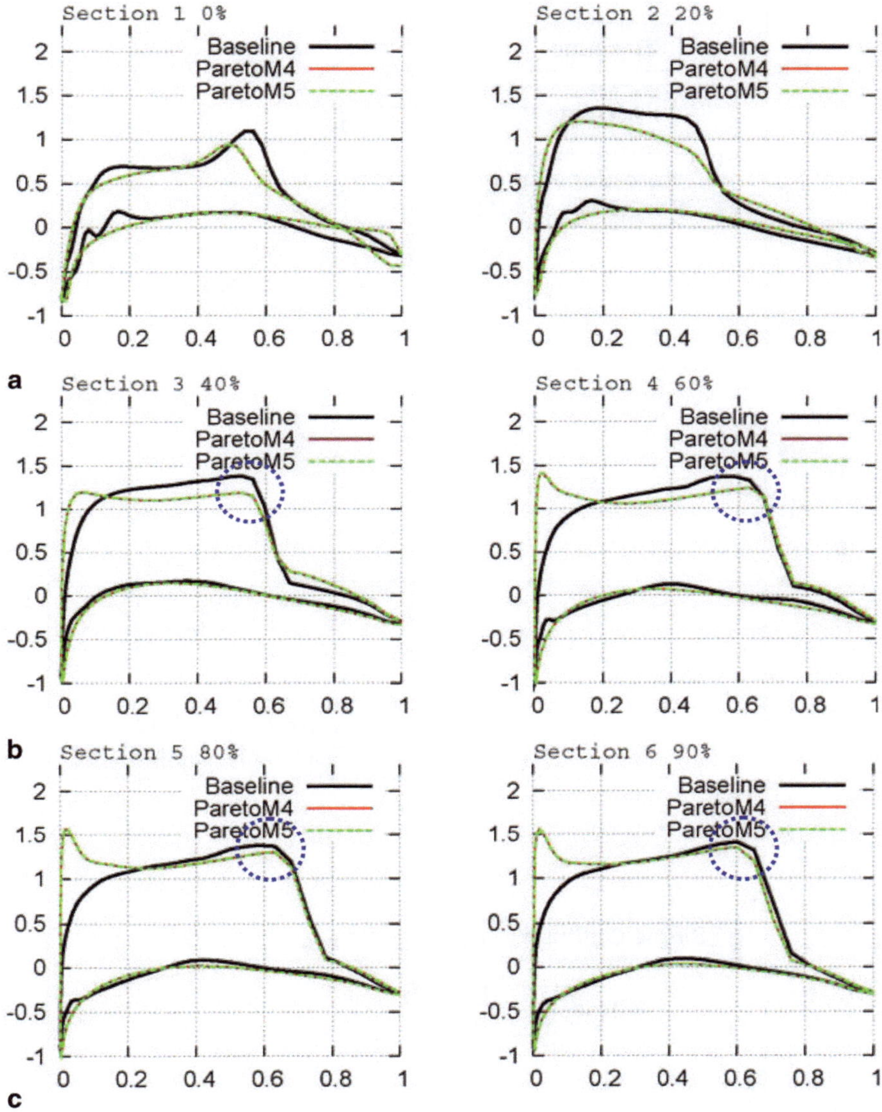

Fig. 9.20 **a** Comparison of pressure coefficient (*Cp*) distributions at 0 and 20 % of span. **b** Comparison of pressure coefficient (*Cp*) distributions at 40 and 60 % of span. **c** Comparison of pressure coefficient (*Cp*) distributions at 80 and 90 % of span

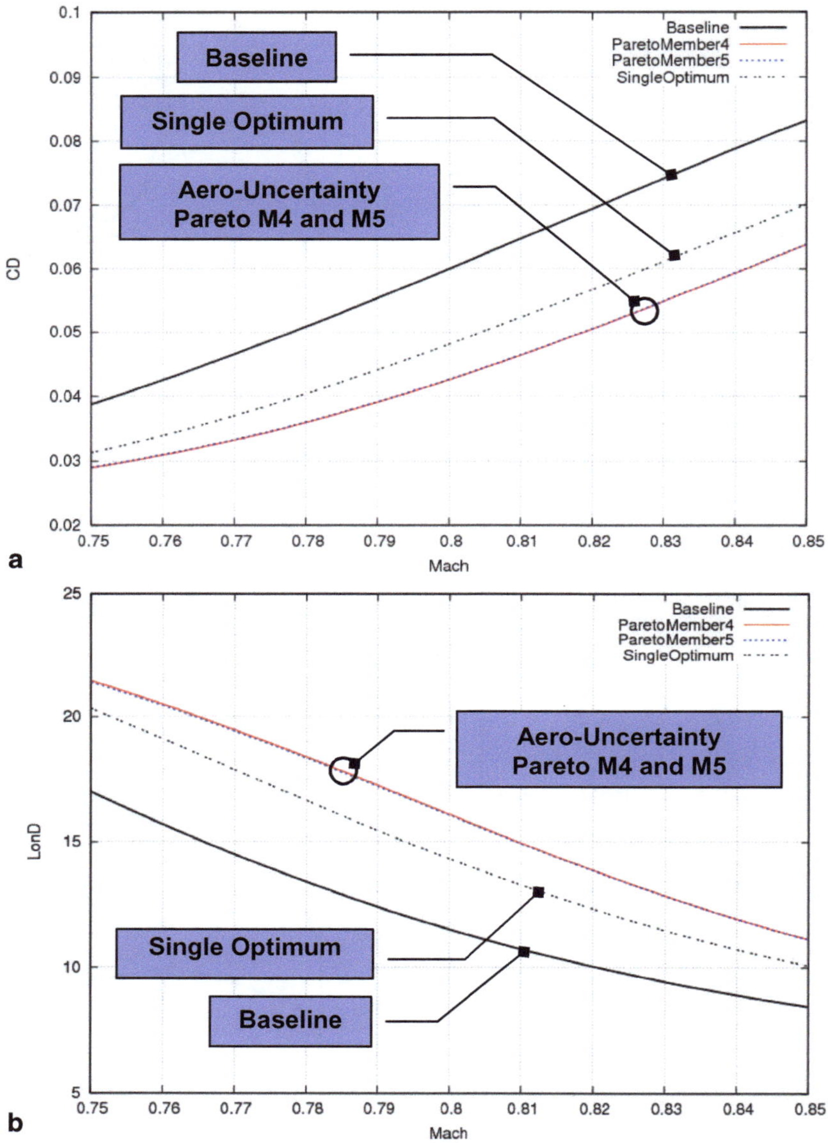

Fig. 9.21 **a** Comparison of total drag coefficient (CD_{Total}) vs. Mach number. **b** Comparison of lift to drag ratio (L/D) vs. Mach number

Table 9.10 Comparison of the aerodynamic performances

Description	Baseline	Single-objective	Pareto M4	Pareto M5
$\overline{1/(L/D)}$	0.0703	0.0582 (−18 %)	0.0532 (−24 %)	0.0533 (−24 %)
$\delta(1/(L/D))$	1.0×10^{-3}	6.0×10^{-4}	4.7×10^{-4}	4.7×10^{-4}

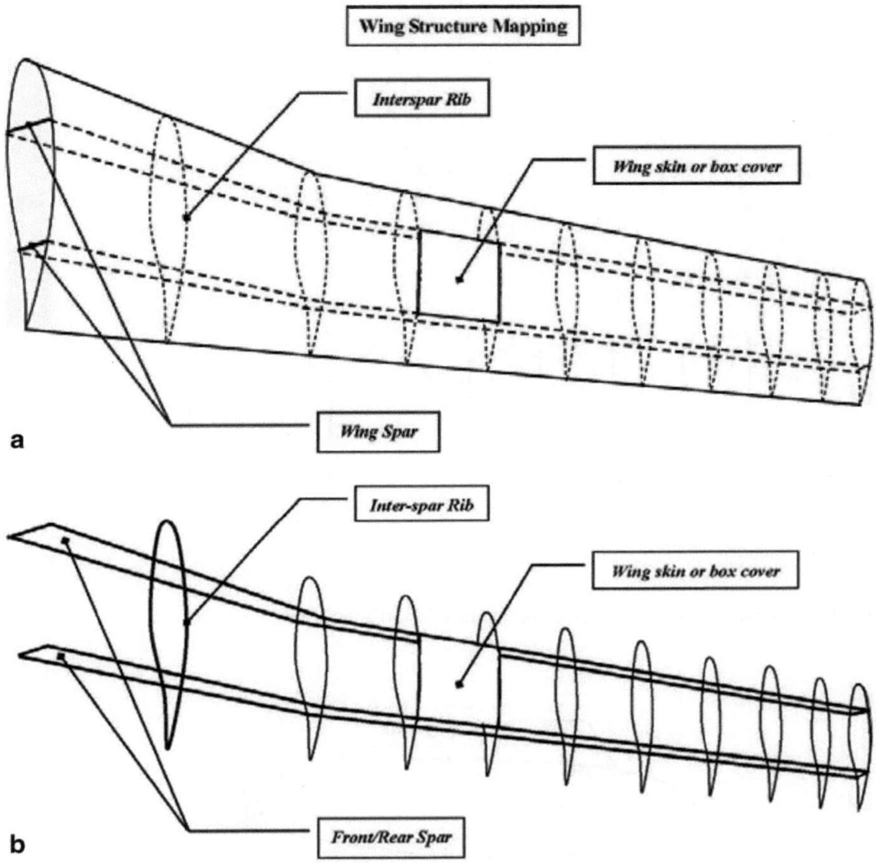

Fig. 9.22 **a** Wing load carrying structure mapping. **b** Wing load carrying structure with skin, inter-spar rib and spar

For the rib:

$$\tau_i = \frac{T_i}{\alpha b(t_{rib_i})^2} \text{ and } W_{rib} = \sum_{i=0}^{N} W_{rib_i} = \sum_{i=0}^{N} \rho t_{rib_i} \cdot c_i \cdot h_{4_i} \tag{9.7}$$

where i represents section number and b is width of structural section. α is a Torsion constants (α=0.333). c_i and h_{4_i} represent airfoil sections chord length and mean airfoil section thickness shown in Fig. 9.24.

For the spar:

$$F_{shear\,buckle_i} = KE\left(\frac{t_{spar_i}}{b}\right)^2 \text{ and } W_{spar} = \sum_{i=0}^{N} W_{spar_i} = \sum_{i=0}^{N} \rho t_{spar_i} \cdot c_i \cdot h_{4_i} \tag{9.8}$$

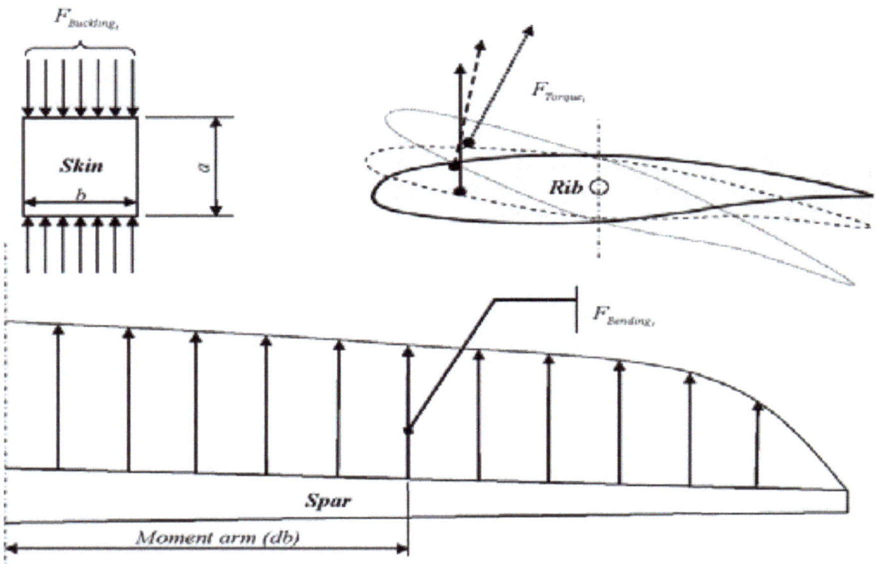

Fig. 9.23 Loads on wing structure components (skin, ribs and spar)

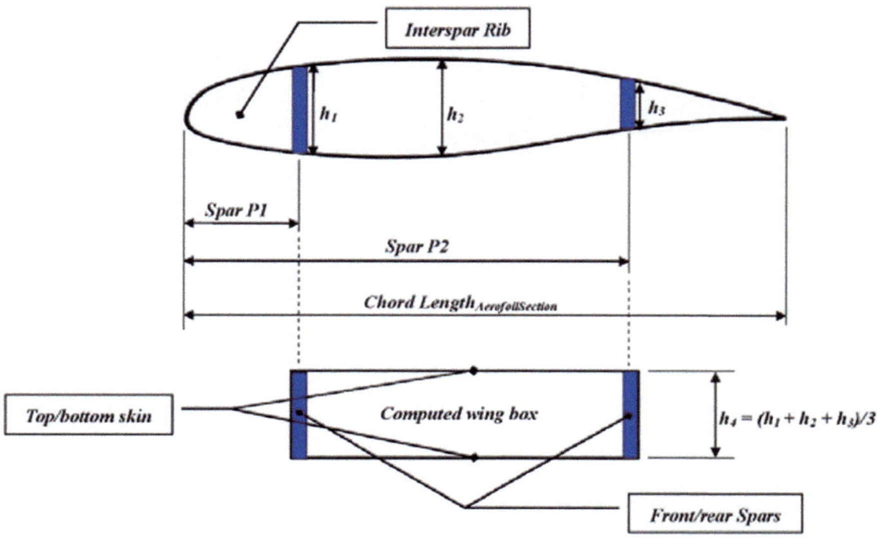

Fig. 9.24 Computed wing box

Table 9.11 Comparison of load carrying structural weight

Description	B737	PDCYL	PSEC
W_{wing} (kg)	2,455.75	2,593.18 (+5.6%)	2,687.03 (+9.4%)

PDCYL [3] is empirical weight estimation methods and parts of the General Aircraft Synthesis Program (GASP)

where i represents section number and b is the width of the structural section. K is the shear web buckling (NACA TN3781) and E the Young's modulus.

The wing structure will be evaluated by the structure analysis code (PSEC) and then the weight of wing structure is calculated as:

$$W_{wing} = 2 \cdot W_{skin} + 2 \cdot W_{rib} + 2 \cdot W_{spar} + 2 \cdot Penalty \tag{9.9}$$

where a Penalty is applied when the section stress σ_i is lower than the ultimate compressive wing strength $\sigma_{ultimate}$.

The structural weight of the baseline wing is computed and compared with the real structural weight obtained from reference [3] as shown in Table 9.11.

This structural estimation code (PSEC) [2] will be utilised in multi-disciplinary (aero-structures) design optimization to compute the structural weight of the candidate model.

Problem Definition This test case considers the multi-disciplinary design optimisation of the wing plan form and airfoil sections shapes of a generic MM-UAS using the uncertainty method. The problem is a two-objective optimisation where the fitness functions are the maximization of aerodynamic performances and the structural quality. These two fitness functions are described in terms of discrete averaged aerodynamic performance and sensitivity, averaged wing weight and weight sensitivity. The optimisation for this application is based on Algorithm 5 and Algorithm 6 described in Sect. 6.5 of Chap. 6. The expression of fitness functions for the mean and variance are normalised using a logarithm scale as follows:

$$f_{Aerodyanmic\ Quality} = Average_{Performance} + Sensitivity_{Performance} + Penalty \tag{9.10}$$

$$f_1 = \frac{-1}{\ln\left(\frac{1}{K}\sum_{i=1}^{K}\left(1/\left(L/D_i\right)\right)\frac{M_{\infty i}^2}{M_S^2}\right)} + \frac{-1}{\ln\left(\frac{1}{(K-1)}\sum_{i=1}^{K}\left(1/\left(L/D_i\right)\frac{M_{\infty i}^2}{M_S^2} - \overline{1/(L/D)}\right)^2\right)} + Penalty \tag{9.11}$$

$$f_{Structural\ Quality} = Average_{Weight} + Sensitivity_{Weight} + Penalty \tag{9.12}$$

$$f_2 = \overline{W_{wing}} + \frac{-1}{\ln\left(\frac{1}{(K-1)}\sum_{i=1}^{K}\left(W_{wing_i}\frac{M_{\infty i}^2}{M_S^2} - \overline{W_{wing}}\right)^2\right)}$$

The operating conditions are:

$$M_{\infty i} \in [0.8, 0.81, M_S = 0.82, 0.83, 0.84] \text{ and } C_{L_S} = 0.691$$

$$\text{Where } \overline{W_{wing}} = \frac{1}{K}\sum_{i=1}^{K}(W_{Wing_i})\ \frac{M_{\infty i}^2}{M_S^2}$$

A penalty (cf. Chap. 2.7) as geometry and aerodynamic performance constraints, is computed and added to fitness functions:

- if the thickness (*t/c*) of the aerofoil sections is higher than 20 % and less than 10.0 %,
- if the ratio of lift to drag at flight conditions is lower than 8.46,
- if the aerodynamic quality (*AQ*) is higher than 0.560,
- if the structural quality (*SQ*) is lower than 3.8118 that is calculated by 2.644 of *mean* ($\overline{W_{wing}} \leq 2.644$) and 0.0374 of *variance* value ($\delta(W_{wing}) \leq 0.0374$).

Design Variables Four airfoil sections at root, crank 1, crank 2 and tip are considered and there are eighty eight (88) design variables (eleven (11) mean line control points, eleven (11) thickness control points at each section) for airfoil section design. The control points of mean and thickness distribution can be found in Sect. 7.5.1 of Chap. 7. For the wing geometry design, five design variables including in and outboard sweep angles ($\Lambda_{R-C1}, \Lambda_{C1-C2}, \Lambda_{C2-T}$), taper ratios at crank1 (λ_{C1}), crank2 (λ_{C2}) and tip (λ_T) sections are considered and their design bounds are shown in Table 9.12. In total, ninety three (93) design variables are considered.

Implementation FLO22 [4] and FRICTION solvers are utilised and the following specific parameters are considered for HAPMOEA software:

1st Layer	Population size of 10 and intermediate recombination with a Computational grid of 96 × 12 × 16 cells on the FLO22 solver
2nd Layer	Population size of 40 and intermediate recombination with a Computational grid of 68 × 12 × 16 cells on the FLO22 solver
3rd Layer	Population size of 60 and intermediate recombination with a Computational grid of 48 × 12 × 16 cells on the FLO22 solver

Table 9.12 Wing geometry design bounds

Variables	Λ_{R-C1}	Λ_{C1-C2}, Λ_{C2-T}	λ_{C1}	λ_{C2}	λ_{T}
Lower	34°	19°	0.5	0.3	0.2
Upper	37°	23°	0.7	0.5	0.25

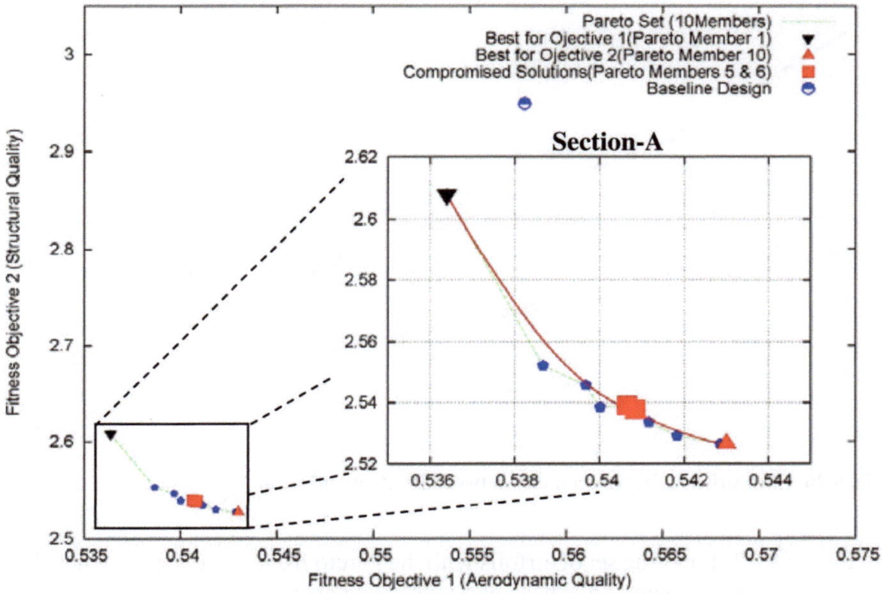

Fig. 9.25 Pareto optimal fronts for generic aircraft wing aerofoil sections

Numerical Results The algorithm was allowed to run approximately 1100 function evaluations and took 150 h of CPU time on two 1.8 GHz processors. The resulting Pareto front is shown in Fig. 9.25. The Pareto front are zoomed in Section-A. It can be seen that Pareto (non-dominated) solutions produce better aerodynamic and structural quality when compared to the values obtained with the baseline design.

Pareto Members 1 (best solution for fitness 1), 5, 6 and 10 (best solution for fitness 2) are selected and compared to the baseline design as shown in Table 9.13. Pareto Members 5 and 6 are selected for further evaluation as compromised solutions. Pareto Members 5 and 6 produce improvement on aerodynamic quality by 3 % and on structural quality by 14 %.

Figure 9.26 and Table 9.14 compare the wing geometry between Pareto Members and the baseline design. It can be seen that overall aspect ratio and span length are increased by 18 and 8 % respectively. The taper ratios at crank 2 and tip are of very similar value to the baseline design while 6.6 % decrement of taper ratio is observed at crank 1. For sweep angles, Pareto Members have 1–2 degrees swept back compared to the baseline design.

Table 9.13 Comparison of the aerodynamic and structural quality

Description	Baseline	Pareto M1	Pareto M5	Pareto M6	Pareto M10
Fitness 1 (AQ)	0.556	0.537 (−3.4%)	0.5406 (−2.7%)	0.5408 (−2.7%)	0.543 (−2.3%)
Fitness 2 (SQ)	2.949	2.608 (−11.5%)	2.539 (−14.0%)	2.537 (−14.0%)	2.526 (−14.3%)

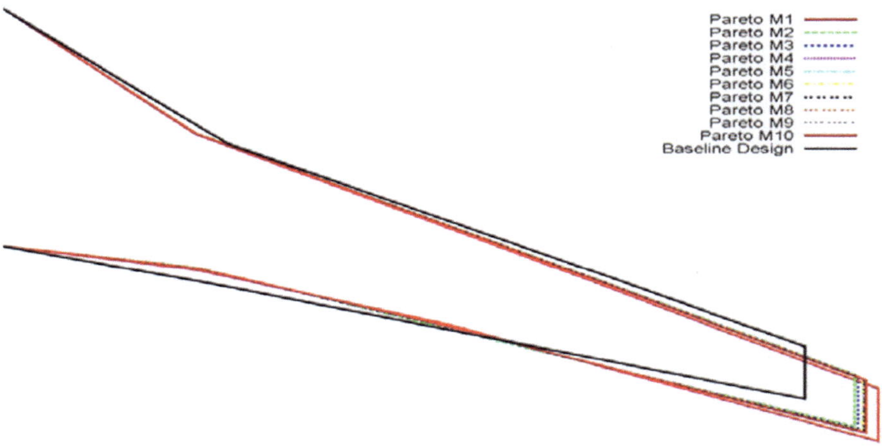

Fig. 9.26 Comparison of wing geometry between Pareto members and baseline design

Figure 9.27 shows the set of airfoils with the Pareto front. It can be seen that classical aerodynamic shapes have been evolved, even considering that the optimisation was started completely from random solutions and the evolution algorithm had no problem specific knowledge of appropriate types of solutions taking certainty into account.

The airfoil sections characteristics for the selected Pareto Members (5 and 6) are compared to the baseline design in Fig. 9.28 and on Table 9.15. The airfoil sections of selected Pareto Members 5 and 6 are 10% thinner than those of the baseline design. Negative cambers are observed at crank 1 and crank 2 airfoil sections while positive cambers are found at root and tip airfoil sections.

Figures 9.29a, b and c compare the pressure coefficient distributions at 0, 20, 40, 60, 80 and 90% of the span between Pareto Members (1, 5, 6 and 10) and the baseline design. The optimal wings have a pressure improvement at the leading edge identified with blue circles from 40~90% of the span when compared to the baseline design.

Figures 9.30a, b and c compare Mach sweeps with the drag coefficient, the lift to drag ratio (L/D) and the wing weight obtained from current optimization (AS-Uncertainty), aerodynamic uncertainty approach (Aero-Uncertainty) (see Sect. 8.5.2) and the baseline design. The two-uncertainty based aero-structure and uncertainty based aerodynamics provide an improvement of the lift to drag ratio

Table 9.14 Comparison of wing specifications between Pareto Members (1, 5, 6 and 10) and baseline design

Specification	b (m)	AR	Λ_{R-C1}	Λ_{C1-T}	λ_{C1}	λ_{C2}	λ_T
Baseline	34.32	11.57	34.03°	21.38°	0.60	0.41	0.22
ParetoM1	37.45	14.03	36.23°	22.55°	0.57	0.40	0.222
ParetoM5	36.84	13.58	36.04°	22.27°	0.56	0.42	0.222
ParetoM6	36.77	13.53	36.05°	22.25°	0.56	0.42	0.222
ParetoM10	36.92	13.64	36.08°	22.28°	0.56	0.418	0.222

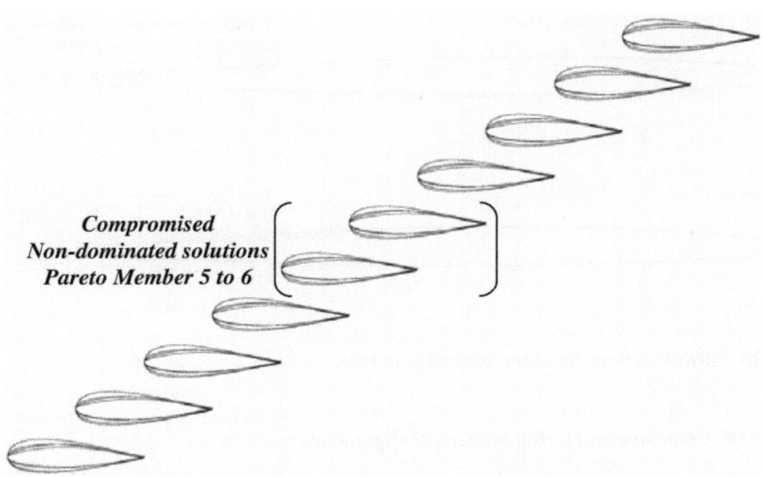

Fig. 9.27 Optimal airfoil sections

when compared to the baseline wing. Both solutions from aero-structure and aerodynamic uncertainty indicating lower drag coefficients are observed from the Mach number 0.7–0.85 as shown in Fig. 9.30a. The lift to drag ratio (Fig. 9.30b) from current optimization using the aero-structure uncertainty concept provides better aerodynamic performance when Mach number is higher than 0.71 with higher lift to drag ratio from aerodynamic uncertainty at all Mach numbers with less sensitivity. However, the wing weights of the Aero-Uncertainty (Pareto member 4 and 5) are higher than the baseline design from Mach number 0.7–0.78 and after 0.835 slightly lower from Mach 0.78–0.835 when compared to the baseline design. The lowest wing weight distributions with less weight sensitivity at all Mach numbers [0.7: 0.85] are achieved by the current solutions (Pareto Members 1, 5, 6 and 10).

Table 9.16 compares the aerodynamic and structural quality at Mach number 0.75–0.85 between current solutions (Aero-Structure Uncertainty), Aero-Uncertainty, and the baseline design. Pareto Members 4 and 5 from aerodynamic uncertainty optimization are higher in terms of aerodynamic quality while their wing weight

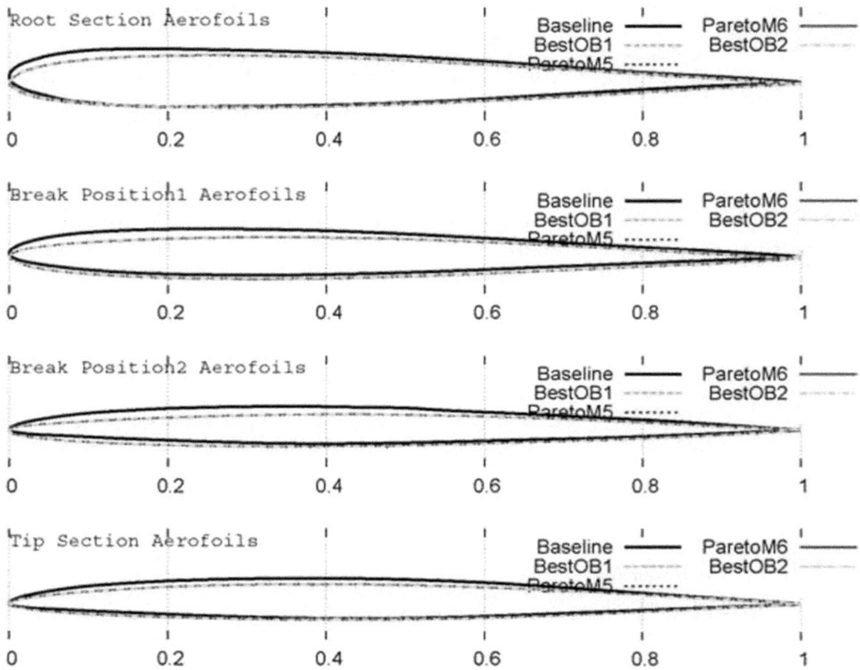

Fig. 9.28 Airfoil sections for compromised solutions

Table 9.15 Comparison of airfoil sections configurations

Design description	Baseline design	Pareto M5	Pareto M6
t/c_{Root}	15.4 @19	13.89 @27.0	13.89 @27.0
$Camber_{Root}$	0.3 @4.6	0.54 @23.0	0.55 @21.0
t/c_{Crank1}	12.5 @28.0	11.4 @32.4	11.4 @32.4
$Camber_{BP1}$	0.7 @10.3	−0.33 @94.0	−0.34 @94.0
t/c_{Crank2}	10.0 @38.0	8.64 @38.6	8.63 @38.4
$Camber_{BP2}$	1.44 @20.4	−0.44 @11.3	−0.43 @11.1
t/c_{Tip}	10.8 @40.3	9.6 @39.7	9.6 @39.5
$Camber_{Tip}$	1.6 @21.2	0.39 @31.9	0.37 @31.8

distributions fluctuate when compared to the baseline design. Pareto Members 5 and 6 from the current optimisation (Aero- Structure Uncertainty) produce not only better aerodynamic performance with stability but also a structurally light weight distribution with stable weight sensitivity at variability of flight conditions when compared to the baseline design and aerodynamic uncertainty solutions.

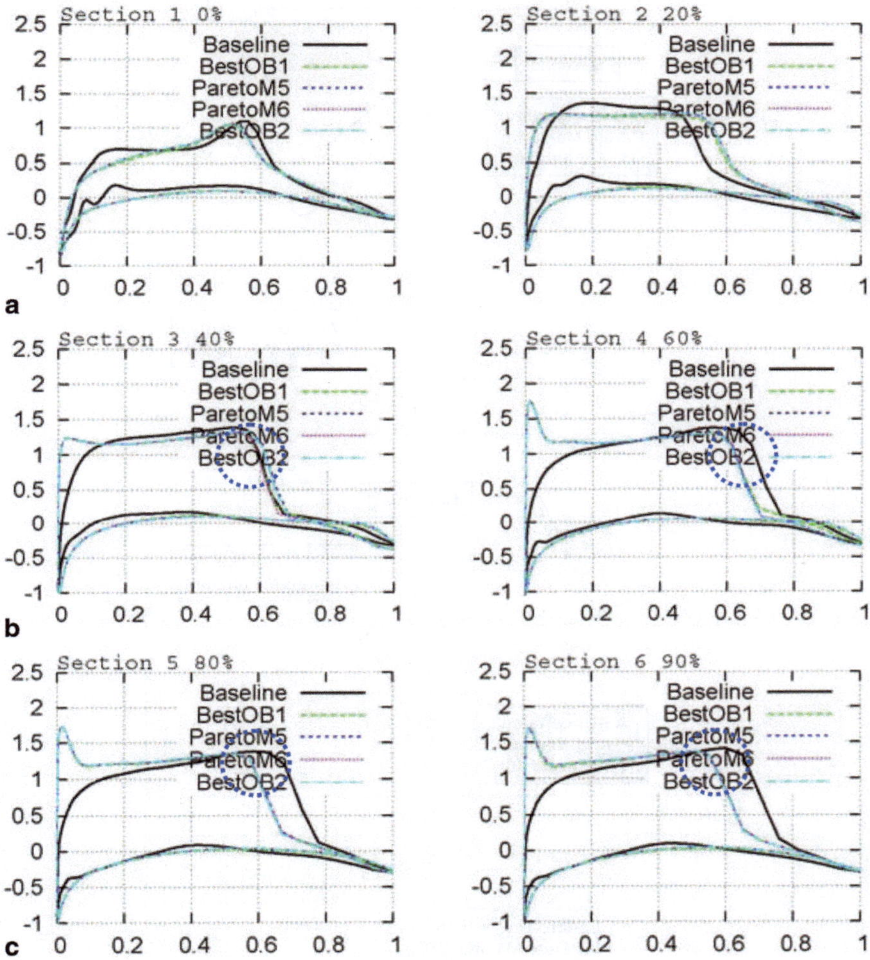

Fig. 9.29 **a** Comparison of pressure coefficients (Cp) distribution at 0 and 20 % of span. **b** Comparison of pressure coefficients (Cp) distribution at 40 and 60 % of span. **c** Comparison of pressure coefficients (Cp) distribution at 80 and 90 % of span

9.5 Robust Aero-Electromagnetic Design Optimization of UAS

Aero-electromagnetics design optimization is also an important problem for manned and unmanned (UAV and UCAV systems) aviation. This section considers two MDO test cases.

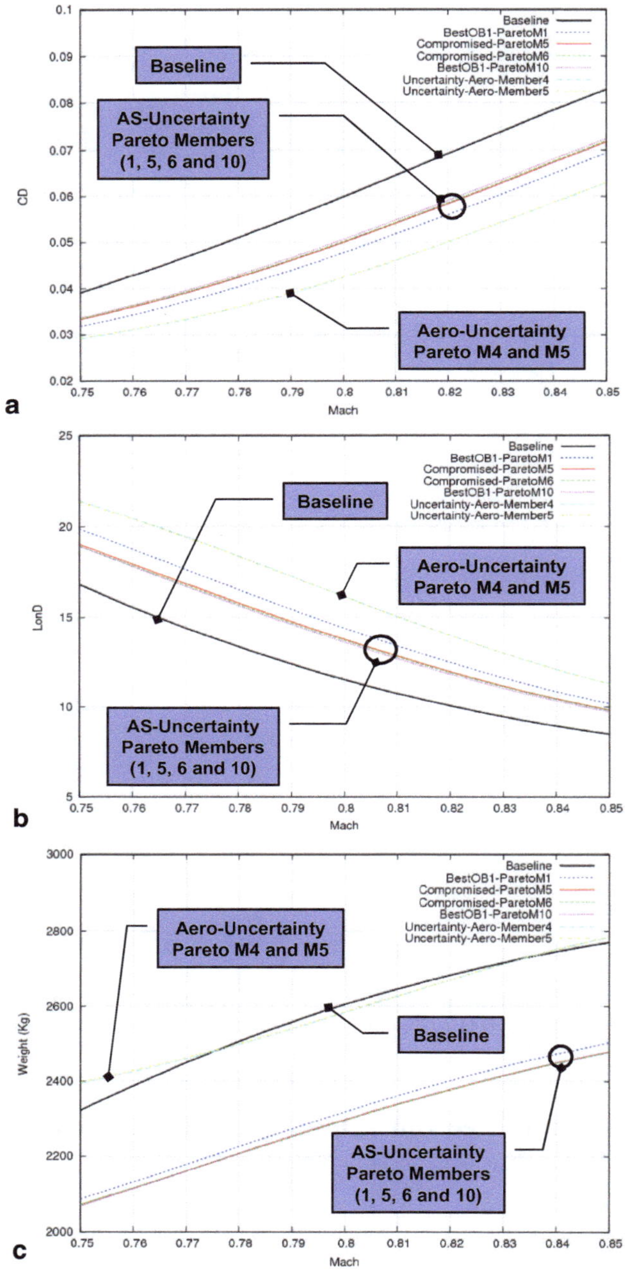

Fig. 9.30 **a** Comparison of total drag coefficients $(C_{D_{Total}})$ vs. Mach numbers. **b** Comparison of lift to drag ratios (*L/D*) vs. Mach numbers. **c** Comparison of wing structure weights (W_{Wing}) vs. Mach numbers

Table 9.16 Comparison of the aerodynamic/structural quality

Description	Baseline	Aero-uncertainty (Sect. 9.2)		Aero-structure uncertainty	
		Pareto M4	Pareto M5	Pareto M5	Pareto M6
AQ	0.5213	0.4702 (−9.8%)	0.4705 (−9.7%)	0.4934 (−5.3%)	0.4941 (−5.2%)
SQ	2.8768	2.8082 (−2.4%)	2.8081 (−2.4%)	2.5111 (−13.0%)	2.5109 (−13.0%)

AQ and *SQ* represent aerodynamic and structural quality respectively

9.5.1 Robust Multi-Disciplinary Design Optimisation of UCAS Using HAPMOEA Software

Problem Definition This test case considers the multi-disciplinary (aero-electromagnetics) design optimization of Joint-Unmanned Air Combat Systems (J-UCAS) as an extended work of Sect. 8.6 Chap. 8. Two (2) disciplines aerodynamics and electro-magnetics are considered to maximize the survivability of the UCAS when operating in a target strike mission. The first objective is to maximize the *mean* of lift to drag ratio in the aircraft Sector 3 while minimizing the second fitness function *variance* of lift to drag ratio (L/D) to reduce sensitivity/fluctuation as shown in Fig. 8.55c. A third fitness function considers *Mono* and *Bi-static* radar analysis at variable radar frequency to produce a stealth UCAS against enemy aircraft when operating at cruise (Sector 2) and the ground level radars when operating at ingress (Sector 4). The RCS quality (performance and stability) is expressed as one combined statistical formula in terms of the *mean* and *variance*. This optimization uses the combination of Algorithms 5 and 6 as shown in Sect. 6.5 of Chap. 6.

The variability of flight conditions (end of Mission 2 to beginning of Sector 4) are:

$$M_{\infty i} \in [0.75, 0.775, M_S = 0.80, 0.825, 0.85] \text{ or } \overline{M_\infty} = 0.8 \text{ and } \sigma M_\infty = 0.03952$$

$$\alpha_{\infty i} \in \left[4.662, 3.968, \alpha_S = 3.275^\circ, 2.581, 1.887\right] \text{or } \overline{\alpha_\infty} = 3.275 \text{ and } \sigma\alpha_\infty = 1.0968$$

$$ATI_{\infty i} \in [30062, 25093, ATI_S = 20125\,ft, 15156, 10187] \text{or}$$
$$\overline{ATI_\infty} = 20{,}124.6 \text{ and } \sigma ATI_\infty = 7{,}856.2$$

where M_S, α_S and ATI_S represents the standard design flight conditions: Mach number, angle of attack and altitude respectively.

The variability of RCS frequency for Mono and Bi-static analysis are:

$$F_{\infty i} \in [1.0, 1.25, F_S = 1.5GHz, 1.75, 2.0] \text{ or } \overline{F_\infty} = 1.5 \text{ and } \sigma F_\infty = 0.3952$$

where F_S represents the standard RCS design frequency.

The fitness functions—*mean and variance*—for this problem are defined as:

$$f_1 = \min\left(\frac{1}{\overline{L/D}}\right) \tag{9.13}$$

$$f_2 = \min\left(\delta\frac{L}{D}\right) \tag{9.14}$$

where $\overline{\frac{L}{D}} = {}^1\!/_K \sum_{i=1}^{K} (L/D_i)\frac{M_{\infty i}^2}{M_S^2}$, $\delta\frac{L}{D} = \frac{1}{(K-1)}\sum_{i=1}^{K}\left(L/D_i \frac{M_{\infty i}^2}{M_S^2} - \overline{L/D}\right)^2$

$$\begin{aligned} f_3 &= \min\left(RCS_{Quality}\right) \\ &= \frac{1}{2}\left[\left(\overline{RCS}_{mono} + \delta RCS_{mono}\right) + \left(\overline{RCS}_{bi} + \delta RCS_{bi}\right)\right] \end{aligned} \tag{9.15}$$

where $\theta = \left[0^\circ : 3^\circ : 360^\circ\right]$ and $\phi = \left[0^\circ : 0^\circ : 0^\circ\right]$ (Mono-static case),

where incident angles are $\theta = 135^\circ$, $\phi = 90^\circ$ at $\theta = \left[0^\circ : 3^\circ : 360^\circ\right]$, $\phi = \left[0^\circ : 0^\circ : 0^\circ\right]$ (Bi-static case)

Design Variables Lower design bounds of airfoil thickness distributions are set as the same as the baseline design. Eight design variables that define wing configurations are considered (λ_{C1}, λ_{C2}, Λ_{R-C1}, Λ_{C1-C2}, Λ_{C2-T}, S_1, S_2 and S_3). The taper ratio (λ_T) at tip is fixed as 0.2 % of the root chord. Two crank positions (C_1 and C_2) and wing span (b) are recalculated without washout or dihedral effects. Upper and lower design bounds are illustrated on Table 8.11 of Chap. 8. Seventy six (76) design variables are considered in total.

Implementation FLO22 and FRICTION CFD solvers are utilised and the following specific parameters are considered for the HAPMOEA optimizer.

1st Layer	Population size of 15 and intermediate recombination with a computational grid of 96 × 12 × 16 cells used by FLO22 solver
2nd Layer	Population size of 20 and intermediate recombination with a computational grid of 68 × 12 × 16 cells used by FLO22 solver
3rd Layer	Population size of 30 and intermediate recombination with a computational grid of 48 × 12 × 16 cells used by FLO22 solver

Numerical Results The algorithm was allowed to run approximately 539 function evaluations and took 200 CPU hours on two 2.4 GHz processors. The resulting Pareto set is shown in Fig. 9.31 where the black inverse triangle (Pareto Member 1) represents the best solution for fitness function1. The red square (Pareto Member

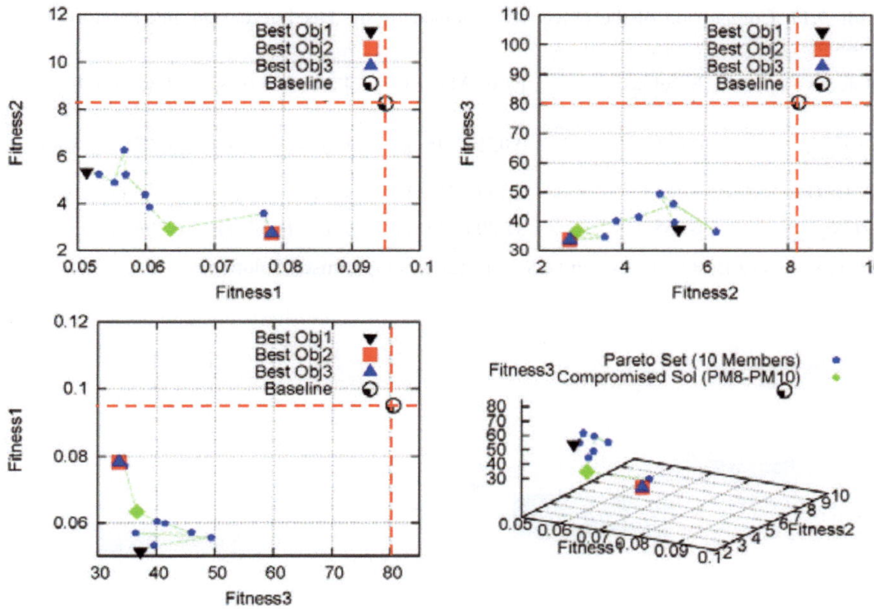

Fig. 9.31 Pareto optimal front

10) represents the best solution for fitness function 2. The blue triangle (Pareto Member 10) indicates the best solution for the third fitness. The light blue square (Pareto member 8) indicates the compromised solution. It can be seen that all Pareto Members produce higher lift to drag ratio (L/D) with low sensitivity while their wing plan form shapes have low signature.

Table 9.17 compares the *mean* and *variance* of lift to drag ratio and RCS quality obtained by Pareto Members (1, 8, 10) and the baseline design. Pareto Member 1(best objective1) produces lower inverse lift to drag ratio $(1/\overline{L/D})$ by 46 % with 35 % reduction of sensitivity $(\delta(L/D))$ when compared to the baseline design. The sensitivity of Pareto Member 10 (best objective 2 and 3) is (67 %) for $\delta(L/D)$ with 18 % for $1/\overline{L/D}$. These results indicate that all Pareto Members produce higher aerodynamic performances with less sensitivity at the start of Sector 3 to end of Sector 3 where is fluctuation as shown in Fig. 8.56c. In addition all Pareto Members have lower RCS quality (more than 50 %) when compared to the baseline design. Pareto Member 8 is selected as a compromised solution for further evaluation.

The *mean* and *standard deviations* obtained by the baseline design and robust Pareto members can be compared using Cumulative Distribution Function (CDF) and Probability Density Function (PDF). Figure 9.32a shows the CDF obtained by the baseline design and the robust compromise Pareto Members (1, 8, 10). It can be seen that all solutions obtained by the robust design method have lower mean inverse lift to drag ratio when compared to the baseline design. Pareto member 1

Table 9.17 Comparison of the objectives obtained by the baseline design and Pareto optimal solutions

Objective	Baseline	ParetoM1 (BO1)	ParetoM8 (CS)	ParetoM10 (BO2-BO3)
$1/\overline{L/D}$	0.095	0.051 (−46 %)	0.063 (−34 %)	0.078 (−18 %)
$\delta(L/D)$	8.25	5.35 (−35 %)	2.91 (−65 %)	2.73 (−67 %)
$RCS_{Quality}$	80.58	37.29 (−53 %)	36.67 (−54 %)	33.62 (−58 %)

BO represents a Best Objective and *CS* indicates a Compromised Solution

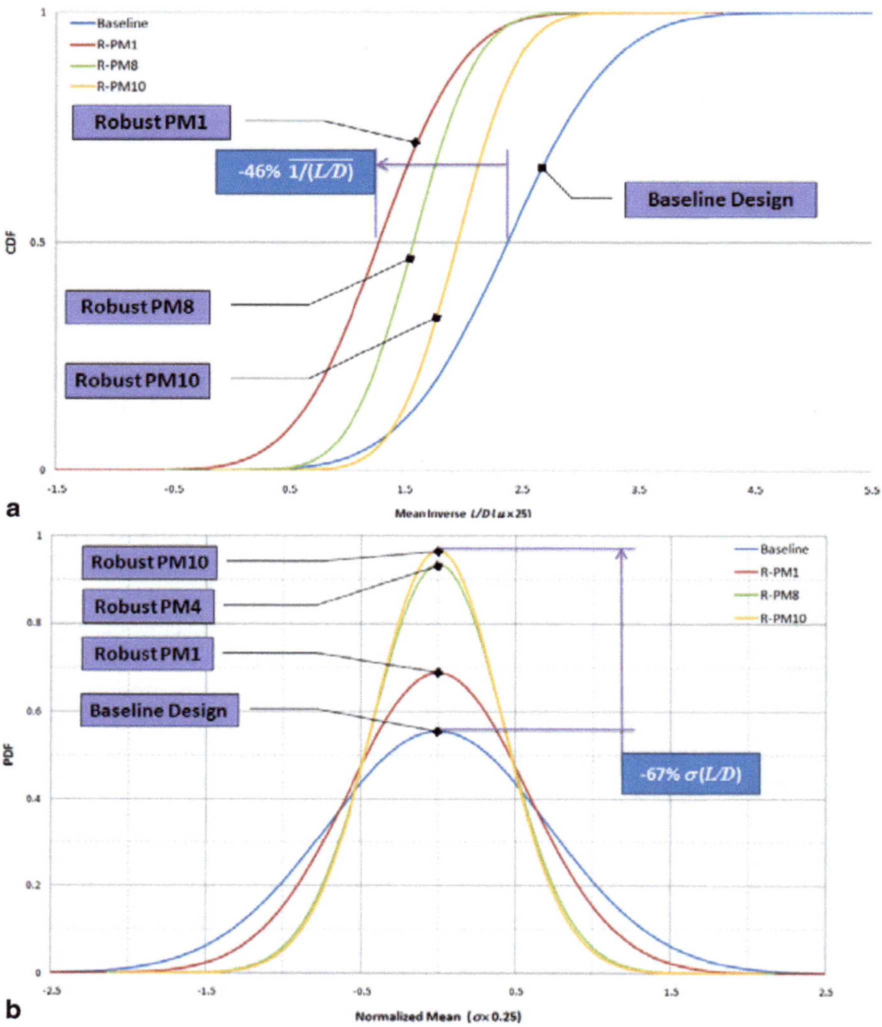

Fig. 9.32 **a** Mean inverse lift to drag ratio comparison using CDF. **b** Inverse lift to drag ratio sensitivity comparison using PDF

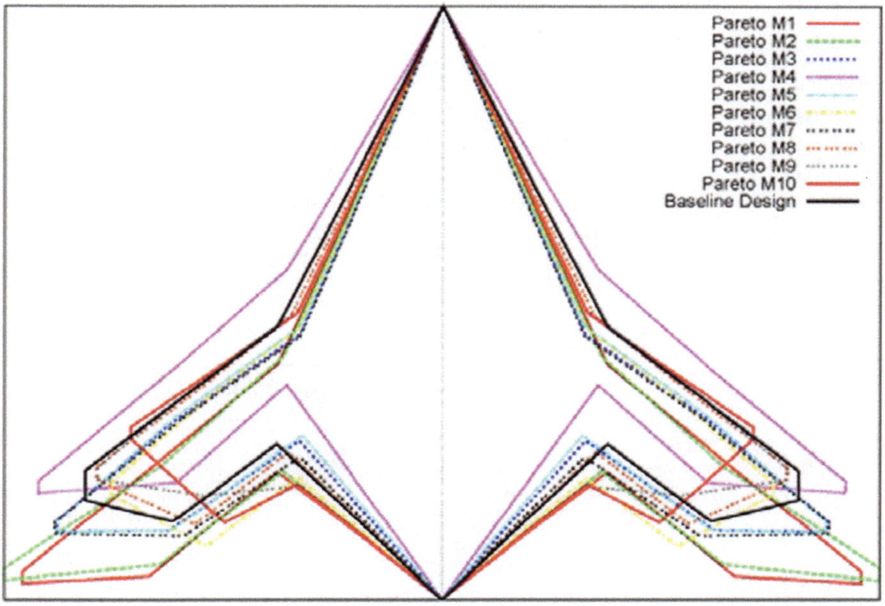

Fig. 9.33 Pareto optimal plan form shapes

improves the mean *L/D* 46 % when compared to the baseline. The standard deviation (sensitivity) can be represented by evaluating the gradient of the lines to the CDF value of 0.5 or 1 (steep gradient signifying low sensitivity). Herein PDF is plotted in Fig. 9.32b to have a clear sensitivity comparison between the baseline design, and robust design solutions. It can be seen that all solutions obtained by the robust design method have lower sensitivity (narrow bell curve). Pareto Member 10 obtained by the robust design method has 67 % *L/D* sensitivity reduction when compared to the baseline design. In other words, the robust design method can produce a set of non-dominated solutions with better performance and sensitivity when compared to the baseline design.

Figure 9.33 and Table 9.18 compare the plan form shapes obtained by Pareto optimal front and the baseline design. It can be seen that the wing span of Pareto member 1 (best objective 1) is 17 % longer with a 35 % increment in aspect ratio when compared to the baseline design. Pareto Member 10 (best objective 2 & 3) has 22 % decrement in aspect ratio and 13 % shorter wing span when compared to the baseline design. However, the taper ratios at crank 1 and crank 2 for Pareto member 10 are higher by 47 and 37 % when compared to the baseline design. This is due to the fitness function 3 (RCS quality) which tried to produce an aero-diamond plan form shape.

Figures 9.34, 9.35 and 9.36 compare the Sector sweep and the lift, drag coefficient and lift to drag ratio obtained from current Pareto members (1 ~ 10), Pareto member 5 (denoted as AeroPM5) from previous test (Sect. 5.1) and the baseline design. The range of Sector sweep (Sector 2 to Sector 4) is $M_\infty = [0.7:0.9]$, $\alpha = [6.05:0.5]$ and at altitude $= [40{,}000:250]$.

Table 9.18 Comparison of wing configurations

Models	AR	b	λ_{C1}	λ_{C2}	Λ_{R-C1}	Λ_{C1-C2}	Λ_{C2-T}
Baseline	4.45	18.9	19.7	19.7	55	29	29
ParetoM1 (BO1)	6.02 (+35 %)	22.20 (+17 %)	18.1 (−8 %)	18.1 (−8 %)	58.0	31.2	30.9
ParetoM8 (CS)	4.30 (−3 %)	18.32 (−3 %)	22.4 (−14 %)	22.4 (−14 %)	56.3	30.0	29.2
ParetoM10 (BO2–BO3)	3.46 (−22 %)	16.47 (−13 %)	29.0 (+47 %)	27.0 (+37 %)	57.26	27.2	26.7

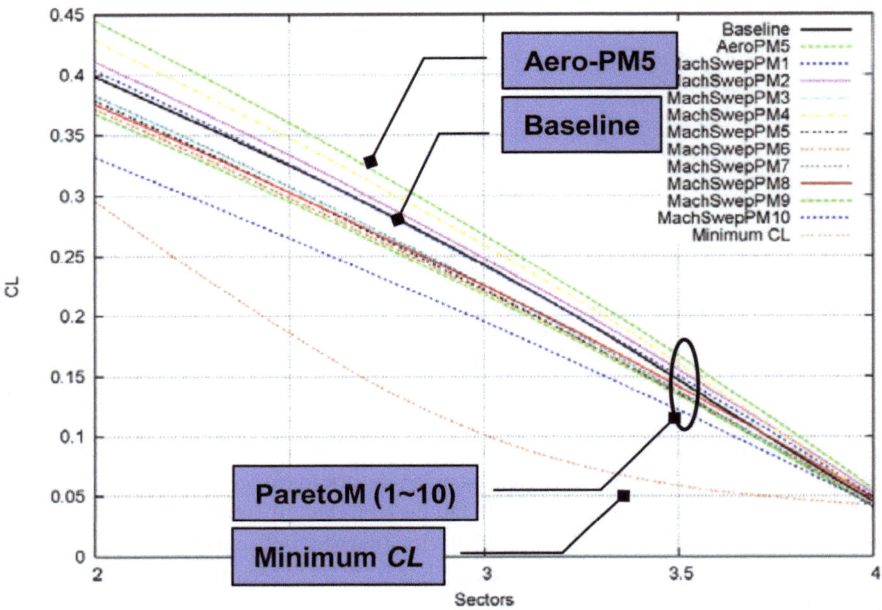

Fig. 9.34 Lift coefficient (*CL*) vs. Sector sweep

Figure 9.34 shows that the current Pareto Members (1 and 2) and Aero PM5 produce higher *Cl* along the Sector sweep when compared to the baseline design. Pareto Members (3–10) are slightly lower when compared to the baseline design while keeping a higher *Cl* instead of minimum *Cl* requirement.

Figure 9.35a and Table 9.19 compare the drag coefficient obtained by Pareto members (1–10), Aero PM5 and the baseline design. Aero PM5 produce lower drag when compared to the current Pareto Members and the baseline, whereas the current Pareto Members have much lower drag sensitivity along the Sector sweep.

Mean and *standard deviations* obtained by the baseline design, multi-objective (Sect. 8.7.5 Chap. 8) and robust Pareto Members can be compared using Cumulative Distribution Function (CDF) and Probability Density Function (PDF). Figure 9.35b shows the CDF obtained by the baseline design, the optimal from multi-objective

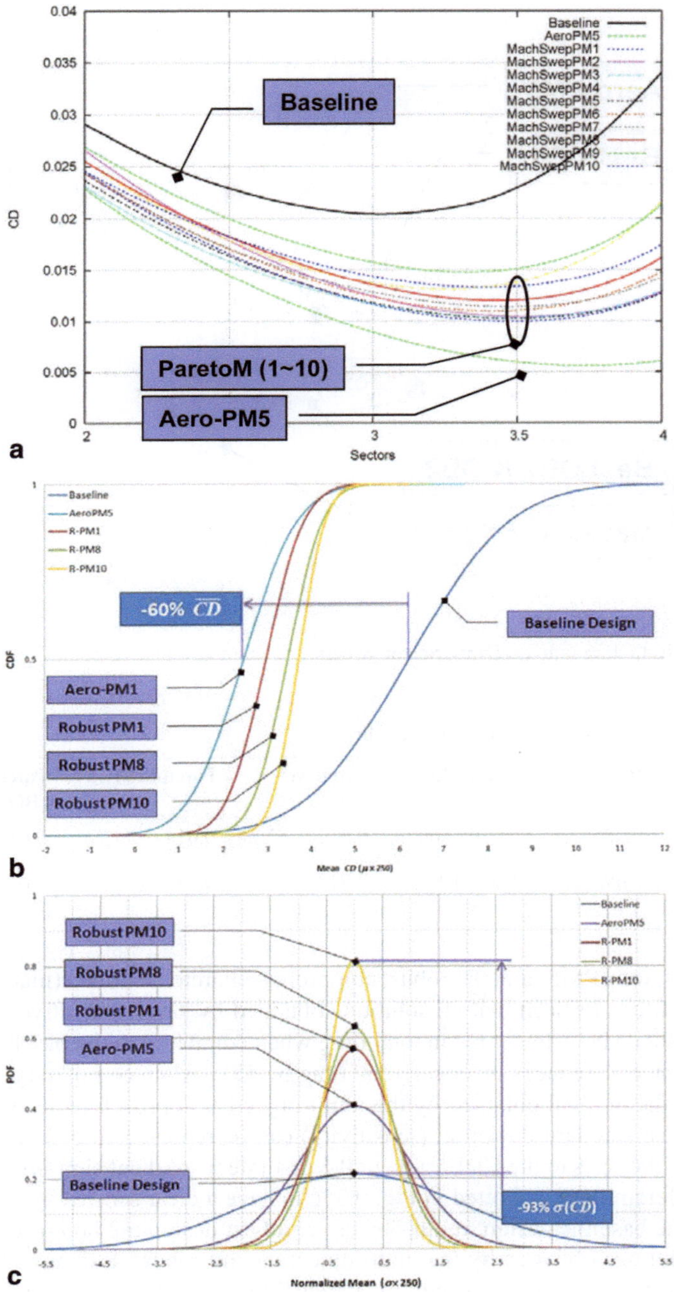

Fig. 9.35 **a** Drag coefficient (*CD*) vs. Sector sweep. **b** Mean drag coefficient comparison using CDF. **c** Drag coefficient sensitivity comparison using PDF

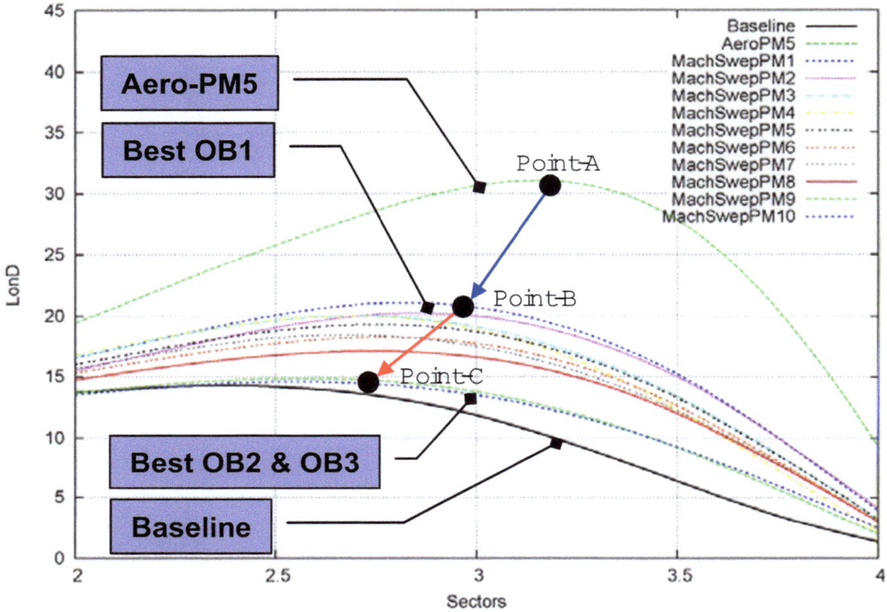

Fig. 9.36 Lift to drag ratio (*L/D*) vs. Sector sweep

Table 9.19 Comparison of aerodynamic quality

Description	Baseline	AeroPM5	ParetoM1 (BO1)	ParetoM8 (CS1)	ParetoM10 (BO2 & 3)
$\overline{C_D}$	0.025	0.010 (−60 %)	0.012 (−52 %)	0.014 (−44 %)	0.015 (−40 %)
δC_D	5.49×10^{-5}	1.54×10^{-5}	7.917×10^{-6}	6.48×10^{-6}	3.83×10^{-6}

(marked as Aero-PM5) and the robust compromise Pareto solutions (marked as Robust PM). It can be seen that all solutions obtained by single-objective and robust design methods have lower mean total drag when compared to the baseline design. Pareto member 1 reduces the mean total drag by 52 % when compared to the baseline while the optimal obtained by the multi-objective approach reduces total drag by 60 %. The standard deviation (sensitivity) can be represented by evaluating the gradient of the lines to the CDF value of 0.5 or 1 (steep gradient signifying low sensitivity). Herein PDF is plotted in Fig. 9.35c to have a clear sensitivity comparison between the baseline design, multi-objective, robust design method. It can be seen that all solutions obtained by the multi-objective and robust design methods have lower sensitivity (narrow bell curve). Pareto Member 10 obtained by the robust design method has 93 % total drag sensitivity reduction when compared to the baseline design while the optimal obtained by a multi-objective approach reduces total drag by only 72 %. In other words, the robust design method can produce a set of solutions which have a better performance and sensitivity when compared to the single-objective optimization method.

Figure 9.36 compares the lift to drag ratio obtained by Pareto Members (1–10), Pareto Member 5 (AeroPM5) from the previous test and the baseline design. It can be seen that Pareto member 1 (best objective1) and AeroPM5 produce higher lift to drag ratios than the baseline design along the Sector sweep while Pareto Member 10 produces lower sensitivity. It can be seen that one of the benefits of uncertainty design technique is that the maximum *L/D* point (Point-A) of AeroPM5 moves to the maximum *L/D* point (Point-B) of Pareto Member 1 (Best objective 1) and then the designs and solutions move to Point-C (maximum *L/D*) of Pareto Member 10 (best objective 2 and 3) which corresponds to and reflects the variance.

Figure 9.37a compares the mono-static RCS analysis obtained by Pareto Members (8 and 10), Pareto Member 5 (Aero PM5) from the previous test and the baseline design at the standard design frequency (1.5 GHz). It can be seen that Pareto Members 8 and 10 produce 9 and 20 % lower RCS when compared to the baseline design while Aero PM5 produces almost the same RCS as the baseline design. Figure 9.37b illustrates a frequency sweep corresponding to mono-static RCS analysis. The variance value for Pareto Member 8 is lower while the baseline design value highly fluctuates at the standard design radar frequency (1.5 GHz).

Figure 9.38a compares the bi-static RCS analysis obtained by Pareto Members (8 and 10), Pareto Member 5 (Aero PM5) from the previous test and the baseline design at the standard design frequency (1.5 GHz). Even though Aero PM5 produces higher aerodynamic performance, its bi-static RCS indicates that it has 17 % more chance to be detected by an enemy radar system when compared to the baseline design. Pareto Members 8 and 10 have 9 and 12 % lower signature when compared to the baseline design. Figure 9.38b compares the bi-static RCS at variable frequencies and shows the lowest RCS variance is achieved by Pareto Member 10 ($\delta\, RCS_{BI}$=0.09).

To conclude with this test case, the method showed that by introducing another discipline and an uncertainty analysis it is possible to compute a set of useful Pareto non-dominated solutions that produces stable design and higher aerodynamic performance with a low signature at variable flight conditions and radar frequencies. Pareto Members 8 and 10 can be selected as compromised solutions for further evaluation and are suitable for stealth strikes. Even though Pareto Member 10 produces similar aerodynamic performance at Sector2 condition when compared to the baseline design, it produces higher aerodynamic performance with less sensitivity at Sector 3 and Sector 4 conditions. Pareto Members 8 and 10 not only have a low signature at mono and bi-static radar system when compared to both baseline design and Pareto Member 5 from aerodynamic design optimization but also have low sensitivity at a set of variable radar frequencies.

9.5.2 Robust Multi-Disciplinary Design Optimization of UCAS Using Hybridized Games

Problem Definition This test case considers the multi-disciplinary design optimisation of UCAS when there is uncertainty in the operating conditions and robust design technique is required. This problem is selected to show the benefits of using

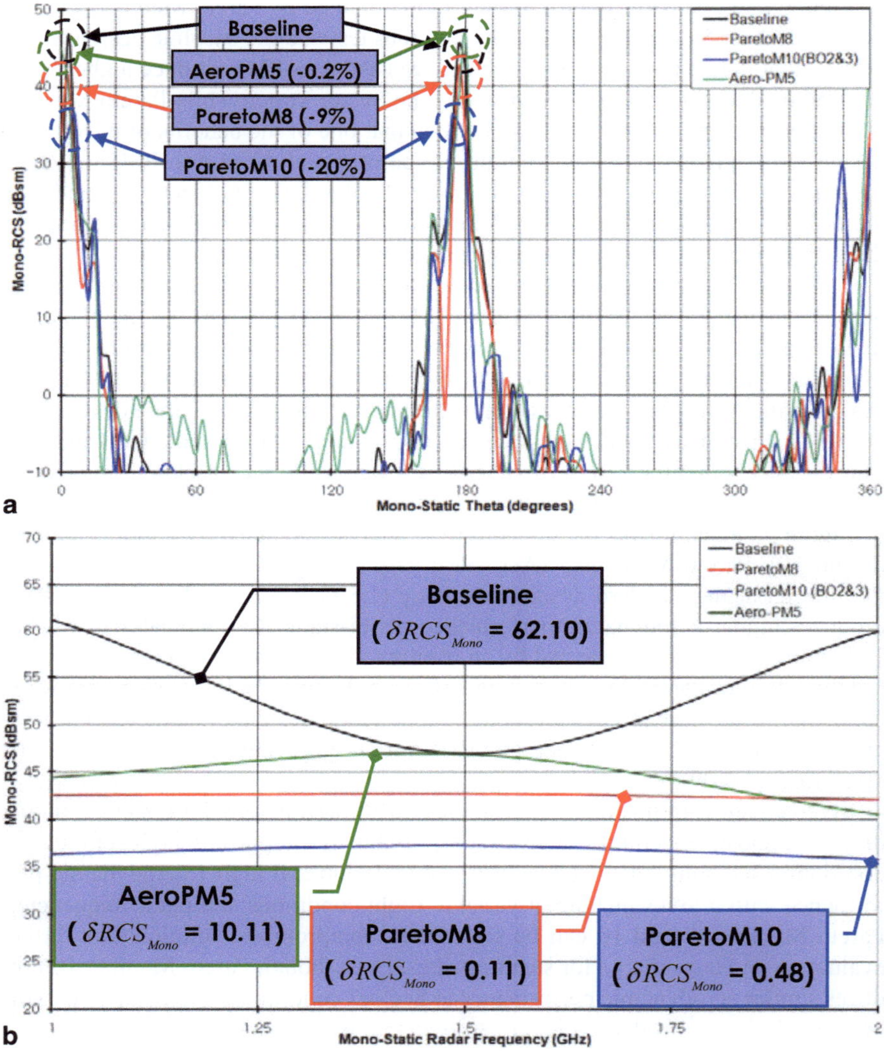

Fig. 9.37 **a** Mono-static RCS vs. Theta at standard design frequency. **b** Mono-static RCS vs. Mono-static radar frequency

the Hybridized Games (Nash and Pareto) method due to the addition of uncertainty which increases the computational cost considerably. The objectives are to maximize Aerodynamic Quality (AQ) while minimizing Electro-magnetic Quality (EQ) in order to maximize the survivability of the UCAS. AQ is defined by fitness functions 1 (mean) and 2 (variance) that represent an aerodynamic performance and sensitivity corresponding to five (5) variability of flight conditions including Mach,

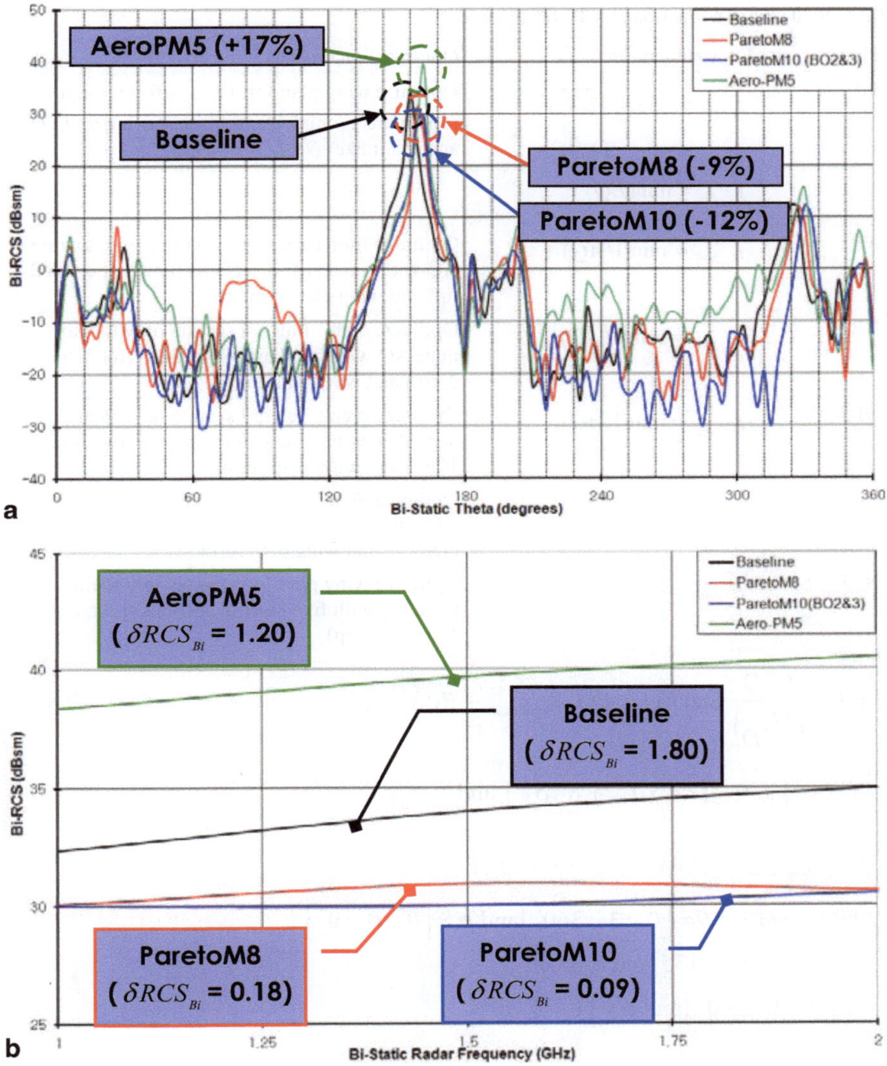

Fig. 9.38 **a** Bi-static RCS vs. Theta at standard design frequency. **b** Bi-static RCS vs. Bi-static radar frequency

angle of attacks and altitude. EQ is expressed using one normalized equation; fitness functions 3 which represents the magnitude and sensitivity of Radar Cross Section (RCS) for a given UAS shape at five variability radar frequencies. UAS will have less chance to be detected by enemy radar systems if the value of EQ is low. In other words, UAS will be stealthier. The fitness functions for Pareto-Player and Nash-Players are indicated on Table 9.20.

Table 9.20 Fitness functions for Hybrid-Game

Player	Fitness function	Optimisation criteria
PP	$f_{1\_PP} = \min\left(1/\left(\overline{L/D}\right)\right)$ $f_{2\_PP} = \min\left(\delta(L/D)\right)$ $f_{3\_PP} = \min(EQ)$	Optimise wing plan form and airfoil sections at root, crank1, crank 2 and tip to maximize $\overline{L/D}$, and minimize $\delta(L/D)$ and EQ
NP1	$f_{1\_NP1} = \min(1/AQ)$	Optimize root aerofoil section only to maximize AQ with fixed airfoil sections (crank 1, crank 2, tip) and wing plan form
NP2	$f_{1\_NP2} = \min(1/AQ)$	Optimize crank 1 airfoil section only to maximize AQ with fixed airfoil sections (root, crank 2, tip) and wing plan form
NP3	$f_{1\_NP3} = \min(1/AQ)$	Optimize crank 2 airfoil section only to maximize AQ with fixed airfoil sections (root, crank 1, tip) and wing plan form
NP4	$f_{1\_NP4} = \min(1/AQ)$	Optimize tip airfoil section only to maximize AQ with fixed airfoil sections (root, crank 1, crank 2) and wing plan form
NP5	$f_{1\_NP5} = \min(EQ)$	Optimize wing plan form shape only to minimize EQ with fixed airfoil sections (root, crank 1, crank 2, tip)

$$AQ = \frac{\overline{L/D}}{\delta\left(L/D\right)} \text{ and } EQ = \delta RCS_{Mono\&Bi} + \overline{RCS_{Mono\&Bi}}$$

$$\overline{RCS}_{Mono\&Bi} = \frac{1}{2}\left(\overline{RCS}_{Mono} + \overline{RCS}_{Bi}\right) \text{ and}$$

$$\delta RCS_{Mono\&Bi} = \frac{1}{2}\left(\delta RCS_{Mono} + \delta RCS_{Bi}\right)$$

$$Mono-static: \theta = \left[0^\circ : 3^\circ : 360^\circ\right] \text{and } \phi = \left[0^\circ : 0^\circ : 0^\circ\right]$$

$$Bi-static: \theta_0 = 135^\circ \text{ and } \phi_0 = 0^\circ$$

$$\theta = \left[0^\circ : 3^\circ : 360^\circ\right] \text{and } \phi = \left[0^\circ : 0^\circ : 0^\circ\right].$$

The possible uncertainty flight conditions (five (5) Mach numbers, angle of attacks and altitudes) and five (5) radar frequencies are:

$$M_{\infty i} \in \left[0.75, 0.775, M_S = 0.80, 0.825, 0.85\right] \text{ or } \overline{M_\infty} = 0.8 \text{ and } \sigma M_\infty = 0.03952$$

$$\alpha_{\infty i} \in \left[4.662, 3.968, \alpha_S = 3.275^\circ, 2.581, 1.887\right] \text{ or } \overline{\alpha_\infty} = 3.275 \text{ and } \sigma\alpha_\infty = 1.0968$$

$$ALT_{\infty i} \in \left[30062, 25093, ALT_S = 20125 ft, 15156, 10187\right] \text{ or}$$

$$\overline{ATI_\infty} = 20{,}124.6 \text{ and } \sigma ATI_\infty = 7{,}856.2$$

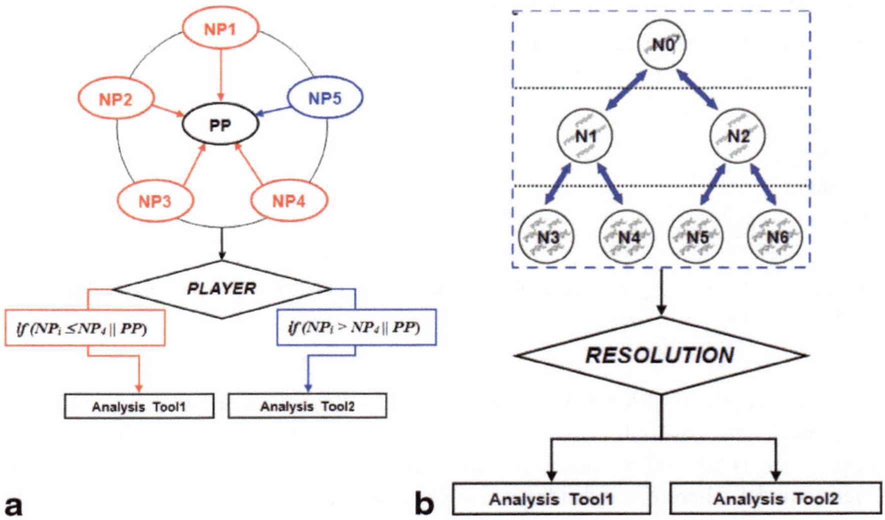

Fig. 9.39 **a** Evaluation mechanism of Hybridized Games. **b** Evaluation mechanism of the HAP-MOEA optimizer

$$F_{\infty i} \in [1.0, 1.25, F_S = 1.5GHz, 1.75, 2.0] \text{ or } \overline{F_\infty} = 1.5 \text{ and } \sigma F_\infty = 0.3952.$$

Uncertainty flight conditions are taken from Sector 2.5 (middle of cruise) to Sector 3.5 (right after high transition dash and before target acquisition) as shown in Fig. 8.49, Chap. 8. The reader will observe critical changes between Sector 2 and Sector 4 where the Mach and altitude number changes from 0.7 to 0.9 and from 41,000 to 250 *ft* respectively. In other words, the changes of flight conditions leads to critical change (fluctuation) in aerodynamic performances which may cause structural or flight control failure. In addition, the altitude change means that the enemy radar systems are also changed from Mono-static to Bi-static with higher radar frequencies. This is the reason that the range from Sector 2.5 to Sector 3.5 is chosen to prevent both aerodynamic and electromagnetic fluctuations.

The objective functions of Nash-Players (1–5) are defined using Variance to Mean Ratio (VMR) to minimize the number of Nash-Players. Otherwise, Hybridized Game will use ten (10) Nash-Players if the objective functions of AQ and EQ are defined by separated *mean* and *variance* formulas. VMR is a statistical formula to minimize variance value while maximizing mean value of objective.

The aerodynamic or electro-magnetic analysis tools used in this multi-disciplinary design optimisation will be determined by the objective of each player for Hybridized Games (Fig. 9.39a). It can be seen that Nash-Players (1–4) in the Hybridized Games approach use aerodynamic analysis tools only to maximise AQ while Nash-Player 5 in Hybridized Games uses electromagnetic analysis tool only to minimize EQ. Pareto-Player in Hybridized Games uses both aerodynamic and electromagnetic analysis tools for both AQ and EQ.

Table 9.21 Distributions of design variables

Design variables	Hybrid-game on HAPMOEA						HAPMOEA-L3
	NP1	NP2	NP3	NP4	NP5	PP	
$Aerofoil_{Root}$	√					√	√
$Aerofoil_{Crank1}$		√				√	√
$Aerofoil_{Crank2}$			√			√	√
$Aerofoil_{Tip}$				√		√	√
Wing					√	√	√

Design variable $Airfoil_{Root}$ indicates airfoil at root section and NP_i represents i^{th} Nash-Player and *PP* indicates the Pareto-Player

In contrast, the HAPMOEA optimizer uses simultaneously aerodynamic and electromagnetic analysis tools shown in Fig. 9.39b. This is explained by the fact that each node in the HAPMOEA optimizer considers both AQ and EQ through the multi-fidelity model. Therefore Hybridized Games have more chance to evaluate candidates.

Design Variables The Hybridized Game employs five (5) Nash-Players and one (1) Pareto-Player as shown on Table 9.21. The Pareto-Player of the Hybridized Game solely considers all seventy six (76) design variables for the shape of airfoil sections and wing plan form. Airfoil sections at root, crank 1, crank 2 and tip are optimized by Nash-Players 1–4 while Nash-Player 5 optimizes wing plan form only. In other words, each Nash-Player from 1 to 4 will optimize seventeen (17) airfoil design variables while Nash-Player 5 will consider eight (8) wing plan form design variables. In contrast, each node (Node 0–6) of the HAPMOEA optimizer will consider all seventy six (76) design variables including airfoil sections and wing plan form.

Numerical Results Both HAPMOEA optimizer and Hybridized Games use two 2.4 GHz processors. The HAPMOEA software was allowed to run approximately 540 function evaluations and took two hundred (200) CPU hours while Hybridized Game algorithm ran approximately only 400 function evaluations and took sixty (60) CPU hours which is 30 % of the computation cost of HAPMOEA optimizer. Pareto fronts obtained by HAPMOEA and Hybridized Games are compared to the baseline design in Fig. 9.40.

It can be seen that Hybridized Game produces similar solutions when compared to HAPMOEA solutions. The black inverse triangle (Pareto Member 1) represents the best solution for fitness function1. The red square (Pareto Member 9) represents the best solution for fitness function 2. The blue triangle (Pareto Member 10) indicates the best solution for the third fitness. The green square (Pareto Member 4 and 5) indicates the compromised solution for Hybridized Games. It can be seen all Pareto Members produce higher lift to drag ratio (fitness 1) with low sensitivity (fitness 2) and also their wing plan form shapes have lower EQ (fitness 3).

Table 9.22 compares the mean and variance of lift to drag ratio and RCS quality of Pareto Members (1, 8, 10) obtained by the HAPMOEA optimizer, Pareto Members (1, 5, 9, 10) from Hybridized Games and the baseline design. Even though the Hybridized Games ran only 30 % of HAPMOEA computational time, it produces

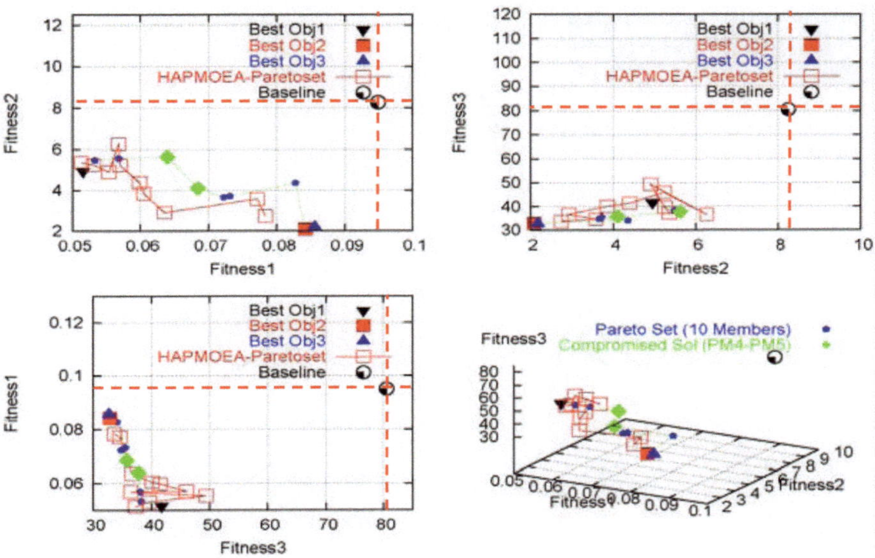

Fig. 9.40 Pareto non-dominated solutions obtained by HAPMOEA optimizer and Hybridized Games

Table 9.22 Comparison of fitness values obtained by HAPMOEA and Hybrid-Game

Objective	HAPMOEA (200 h)			Hybrid-game (60 h)			
	PM1 (BO1)	PM8 (CS)	PM10 (BO2&3)	PM1 (BO1)	PM5 (CS)	PM9 (BO2)	PM10 (BO3)
Fitness1 $1/\left(\overline{L/D}\right)$	0.051 (−46%)	0.063 (−34%)	0.078 (−18%)	0.051 (−46%)	0.068 (−28%)	0.084 (−12%)	0.085 (−11%)
Fitness2 $\delta(L/D)$	5.35 (−35%)	2.91 (−65%)	2.73 (−67%)	4.94 (−40%)	4.10 (−50%)	2.07 (−75%)	2.17 (−74%)
Fitness3 *EQ*	37.29 (−53%)	36.67 (−54%)	33.62 (−58%)	41.74 (−48%)	35.48 (−56%)	32.89 (−59%)	32.69 (−60%)

The fitness values 1, 2 and 3 of Baseline model are 0.095, 8.25 and 80.58 respectively. BO_i represents the best objective solution for i^{th} fitness function. *CS* indicates the compromized solution

similar non-dominated solutions when compared to Pareto non-dominated solutions obtained by HAPMOEA software.

Figures 9.41a and b compare the wing plan form shape corresponding to Pareto non-dominated solutions obtained by HAPMOEA and Hybridized Games, and the baseline design. It can be seen that Hybridized Games software provides more variety on wing plan form shapes. This may be due to the evaluation mechanism of Hybridized Game that allows Pareto-Player and Nash-Player 5 to have a detailed design after running more function evaluations.

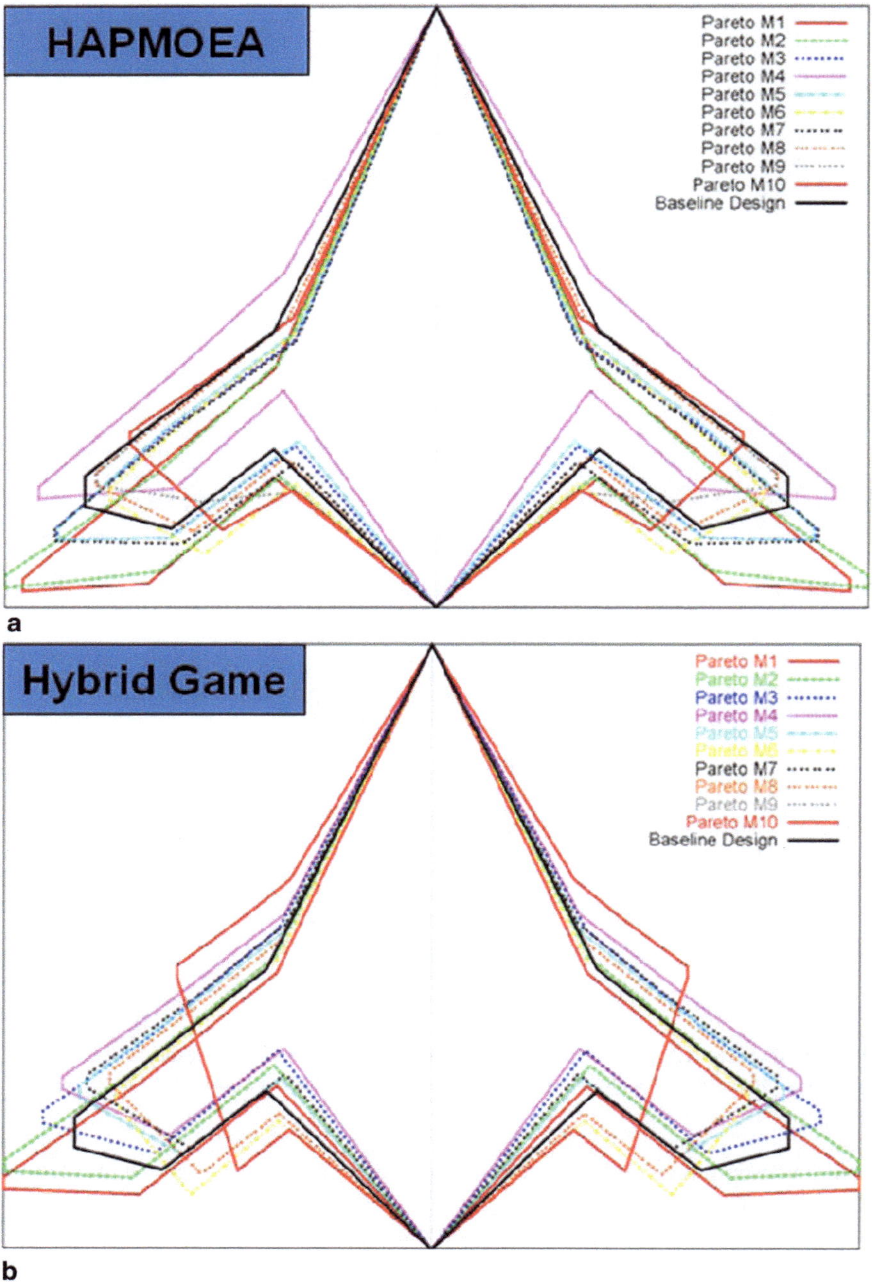

Fig. 9.41 **a** Wing plan form shapes obtained by HAPMOEA. **b** Wing plan form shapes obtained by Hybridized Games software

The Sector sweep is plotted with the lift and drag coefficient obtained by HAPMOEA, Hybrid-Game and the baseline design as shown in Figs. 9.42a and b. The range of sector sweep is $M_\infty \in [0.7:0.9]$, $\alpha \in [6.05°:0.5°]$ and altitude (*feet*) $\therefore$ [40,000:250]. It can be seen that the Hybridized Games software produces a set of comparable solutions to the HAPMOEA software even though Hybridized Games ran only 30 % of HAPMOEA CPU time. The best solution for objective 1 (BO1) from Hybridized Games has higher Cl values while Pareto Member 1 (BO1) from HAPMOEA software produces a lower drag along the sector sweep.

Table 9.23 compares the quality of drag coefficient obtained by HAPMOEA software, Hybridized Games and the baseline design using the uncertainty mean and variance statistical formulas. It can be seen that Pareto Member 1 from HAPMOEA optimizer produces lower drag at [Sector2: Sector4] when compared to Hybridized Games while Pareto Member 9 from Hybridized Games produces a robust design with lower sensitivity in drag.

Figure 9.43 compares the lift to drag ratio (L/D) distribution obtained by Pareto Members (1, 8 and 10) from HAPMOEA, Pareto Members (1, 4, 5 and 10) from Hybrid Game and the baseline design. The Pareto Member 1 (BO1) from both HAPMOEA software and Hybridized Games is not only similarly distributed but also produces higher lift to drag (L/D) ratios than others along the Sector sweep while Pareto Member 9 (BO2) from Hybridized Game produces lower sensitivity in Mach, angle of attack and altitude. It can be seen that all Pareto solutions from HAPMOEA software and Hybridized Games have less fluctuation (stable motions) from Sector 2.5 to Sector 3.5 due to the consideration of uncertainty design during the optimization procedure.

Figure 9.44a compares the mono-static RCS obtained by Pareto Members 8 (CS) and 10 (BO3) from HAPMOEA, Pareto Members 5 (CS) and 10 (BO3) from Hybridized Game and the baseline design at the standard design frequency 1.5 GHz. It can be observed that Pareto Members 8 and 10 from HAPMOEA produce 9 and 20 % lower RCS while Pareto Members 5 and 10 from Hybridized Games produce 13 and 26 % lower RCS when compared to the baseline design. Figure 9.44b illustrates a frequency sweep $F_{\infty i} \in$ [1.0, 1.25, FS = 1.5 GHz, 1.75, 2.0] corresponding to mono-static RCS analysis. The *variance* value (0.024) for Pareto Member 5 (CS) from Hybridized Games is lower than others while the baseline design value highly fluctuates along the frequency sweep.

Figure 9.45a compares the bi-static RCS obtained by Pareto Members (8 (CS) and 10 (BO3)), Pareto Members (5 (CS) and 10 (BO3)) from Hybridized Games and the baseline design at the standard design frequency (1.5 GHz). It can be seen that Pareto Members 8 (CS) and 10 (BO3) from HAPMOEA produce 9 and 12 % lower RCS while Pareto Members 5 (CS) and 10 (BO3) computed with HAPMOEA produce 11 and 15 % lower RCS when compared to the baseline design. Figure 9.45b illustrates a frequency sweep $F_{\infty i} \in$ [1.0,1.25,FS = 1.5 GHz, 1.75, 2.0] corresponding to bi-static RCS analysis. The variance value (0.09) for Pareto Member 10 (BO3) from HAPMOEA is lower than other Pareto Members.

The top, side, front and 3D views of the compromised model used by HAPMOEA software (Pareto Member 8) and Hybridized Games (Pareto Member 5) are

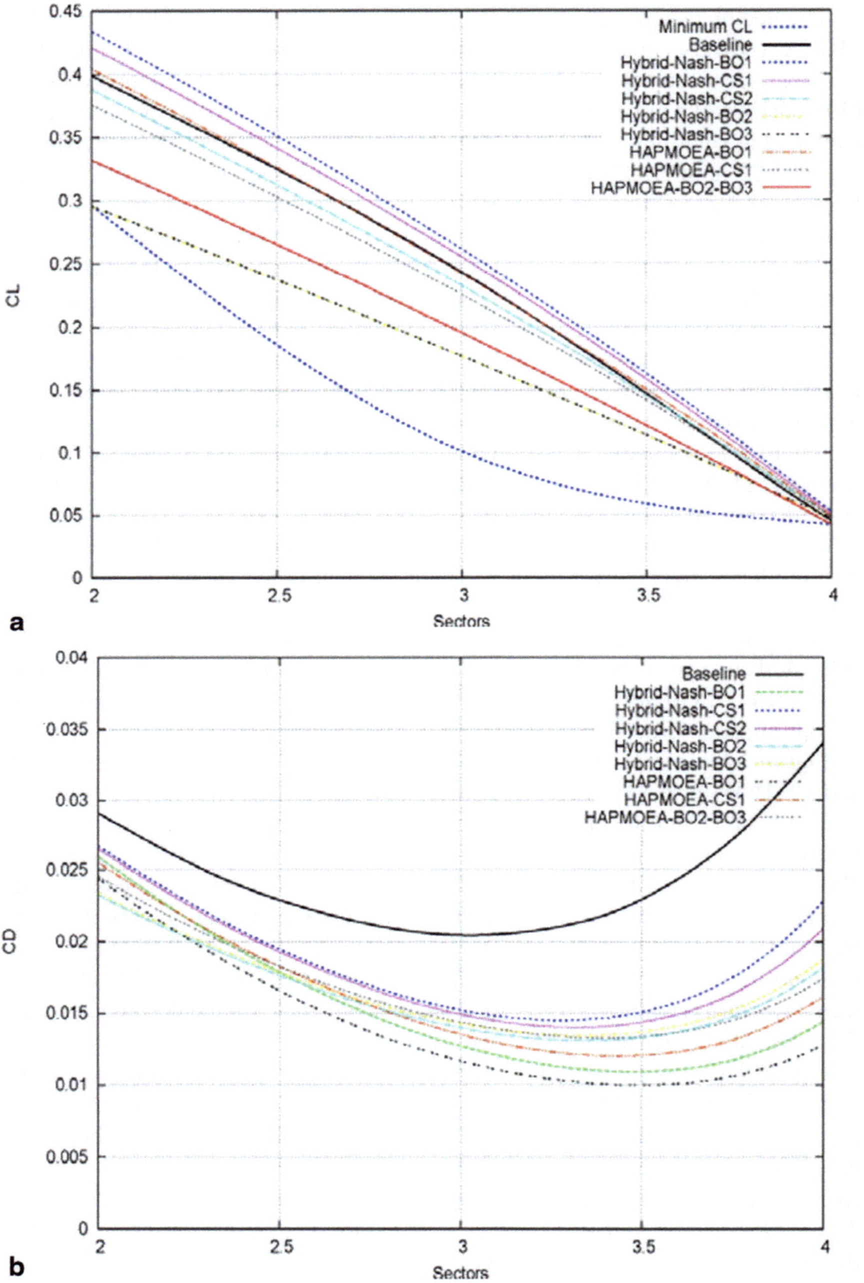

Fig. 9.42 **a** Lift coefficient (*Cl*) vs. Sectors. **b** Drag coefficient (*CD*) vs. Sectors

Table 9.23 Comparison of drag coefficients quality obtained by HAPMOEA and Hybridized Games

Drag Quality	HAPMOEA (200 h)			Hybrid-game (60 h)			
	PM1 (BO1)	PM8 (CS)	PM10 (BO2&3)	PM1 (BO1)	PM5 (CS)	PM9 (BO2)	PM10 (BO3)
$\overline{C_D}$	0.012 (−52 %)	0.014 (−44 %)	0.015 (−40 %)	0.013 (−48 %)	0.015 (−40 %)	0.014 (−44 %)	0.015 (−40 %)
$\delta C_D\,(\times 10^{-6})$	7.92	6.48	3.83	8.50	6.65	3.69	3.74

The $\overline{C_D}$ and δC_D of baseline model are 0.025 and 5.49×10^{-6} respectively. Quality is represented *by mean* (magnitude of performance) and *variance* (sensitivity/stability)

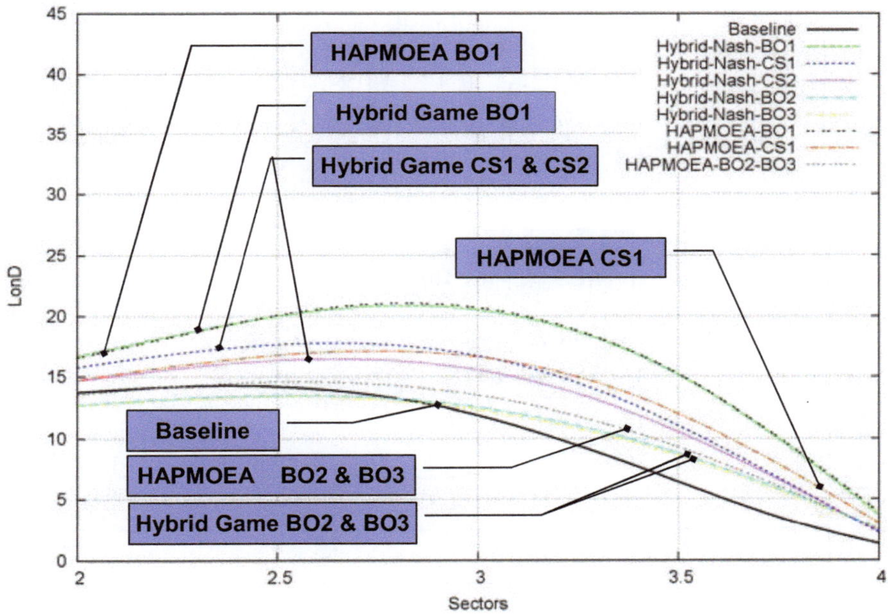

Fig. 9.43 *L/D* vs. Sectors

shown in Figs. 9.46a and b respectively. Even though Hybridized Games spent less CPU time when compared to HAPMOEA, both compromised solutions are geometrically similar.

Pareto Members 8 and 10 from HAPMOEA optimizer and Pareto Member 5 from Hybridized Games can be selected as compromised solutions for further evaluation and are suitable for this RISTA stealth mission: they have not only low observability (stealth property) with mono and bi-static radar system when compared to the baseline design but also have low sensitivity at a set of variable radar frequencies.

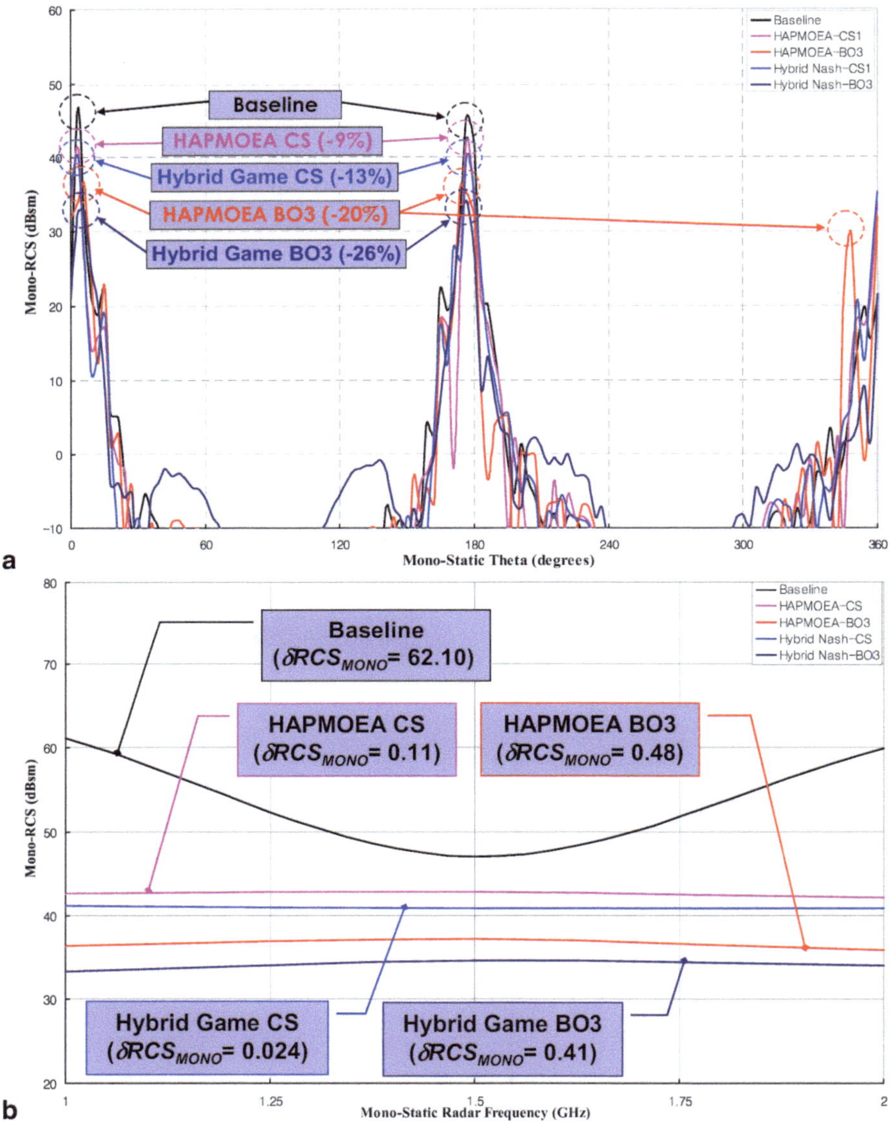

Fig. 9.44 **a** $RCS_{Mono-Static}$ at $F_\infty = 1.5$ GHz. **b** $RCS_{Mono-Static}$ sweep at $F_\infty \in [1.0, 1.25, F_S = 1.5$ GHz, $1.75, 2.0]$

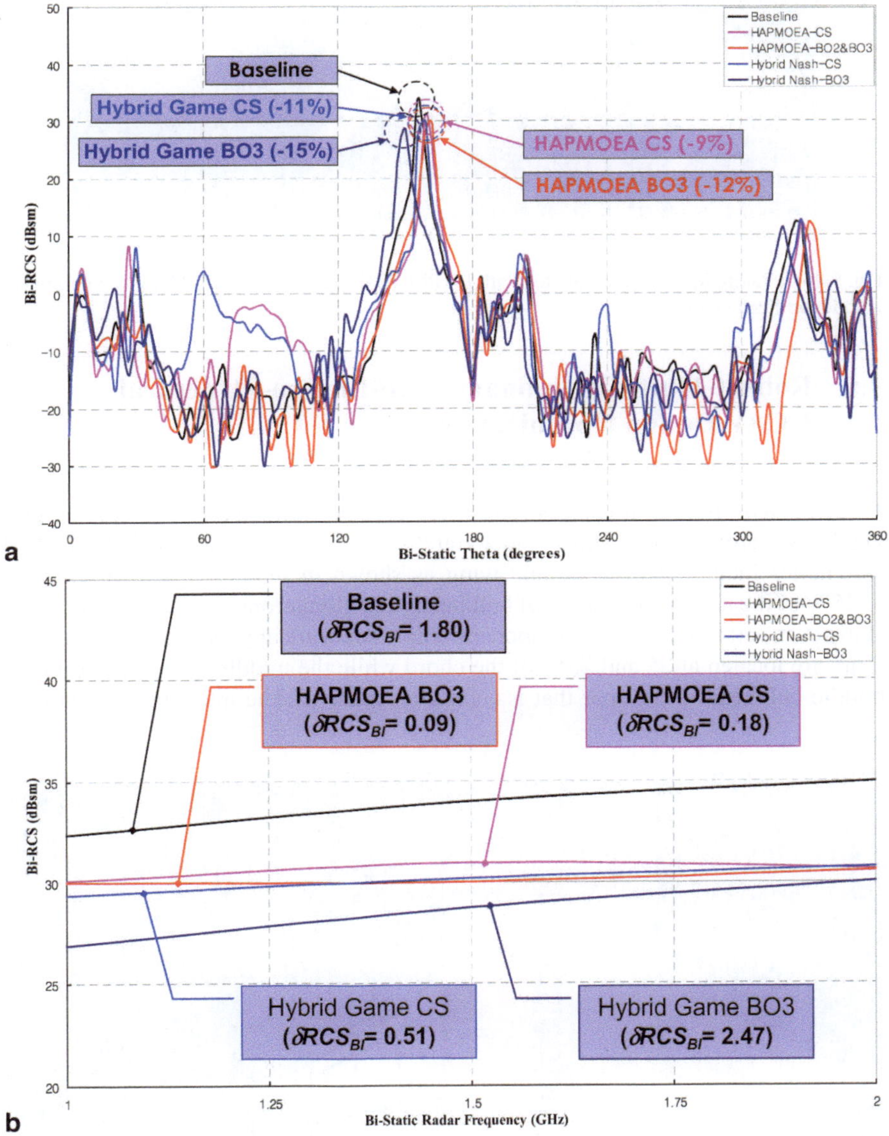

Fig. 9.45 **a** $RCS_{Bi-Static}$ at $F_\infty = 1.5$ GHz. **b** $RCS_{Bi-Static}$ sweep at $F_\infty \in [1.0, 1.25, F_S = 1.5 \text{ GHz}, 1.75, 2.0]$

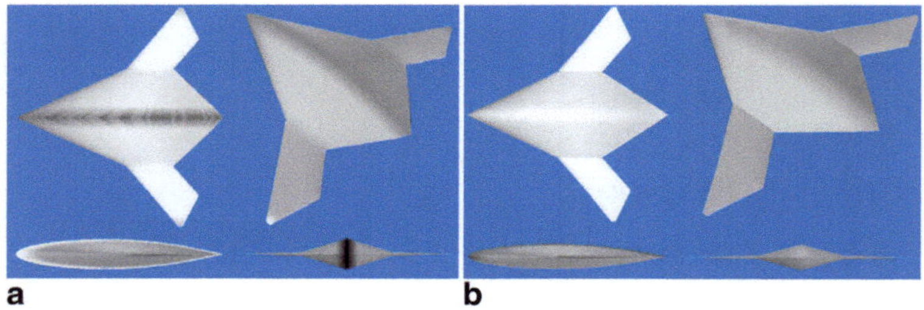

Fig. 9.46 **a** Pareto Member 8 computed with HAPMOEA software. **b** Pareto Member 5 computed with Hybridized Games

9.6 Robust Multi-Disciplinary Aero-Electro-Structural UCAV Design Optimization

Problem Definition This test case considers three performances of J-UCAS including RCS, aerodynamics and manoeuvrability. The J-UCAS is divided into two assemblies including fuselage and wing as shown in Fig. 9.47. The fuselage at 0–25 % of the half span consists of bulkheads and "longerons" while spars, ribs and skin are structured as a wing component. Three spars are applied: the front and rear spars are located at 25 and 75 % of the chord while the middle spar starts from the middle bulkhead to the spar that is right after crank 1. The inter-spar rib starts to

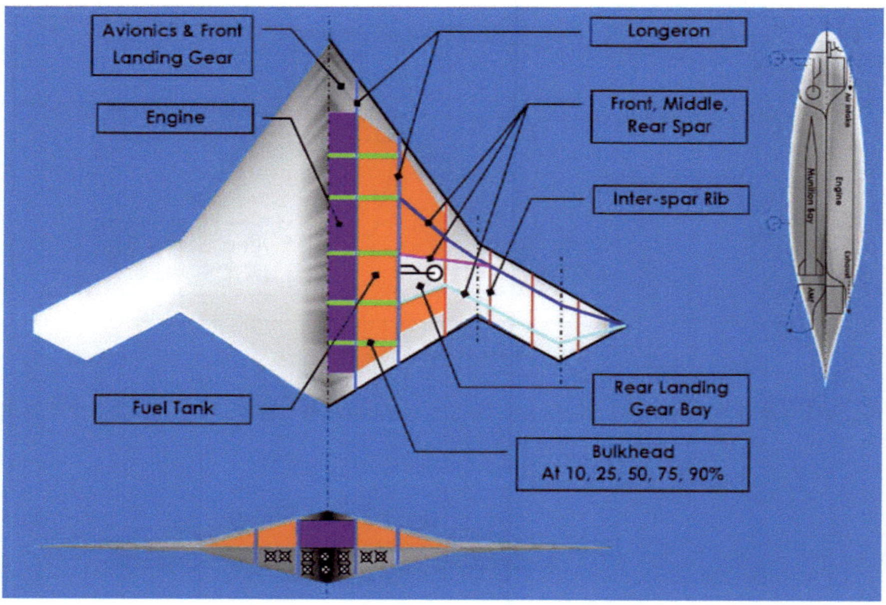

Fig. 9.47 UCAS structural mapping

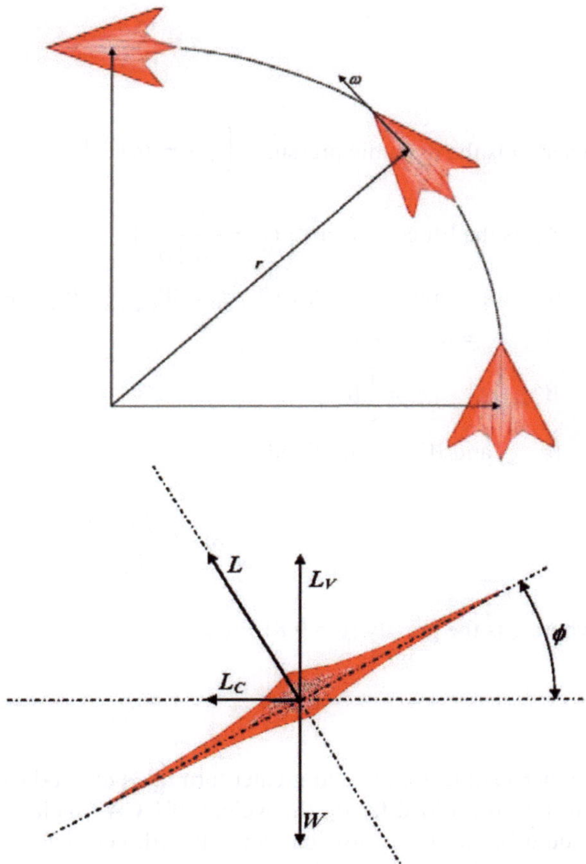

Fig. 9.48 Instantaneous turn

locate after 25 % of the half span. The skin is only applied between front and rear spars.

The wing load-carrying structure (wing box) consists of the wing skin, spar and inter-spar ribs are considered with composites; density (ρ) is 1310 (kg/m^3), ultimate compressive strength of wing ($\sigma_{ultimate}$) is 15.703×10^6 (Kgf/m^2), Young's modulus (E) 15.6×10^9 (Kgf/m^2), safety factor 4. The weight of wing is estimated by using same method in Chap. 8. There are two fixed weights including the engine (23,770 lb or 10,781 kg) and payload (45,000 lb or 2045 kg). It is assumed that the weight of fuselage is half of wing weight.

Manoeuvrability is also considered as one of the fitness functions. A J- UCAS is a design for both air-attack and ground-attack. In air-to-air combat, the fighter desires precision in high rate of turn with a small turn radius to survive. Inspecting manoeuvrability at instantaneous turn, it is necessary to look at the relationship between the load factor or g-loading (n), turn rate (ω or $\dot{\psi}$) and turning radius (r) as shown in Fig. 9.48:

$$n = \frac{qC_L}{W/S} \tag{9.16}$$

where q is the dynamic pressure $\left(q = \frac{1}{2}\rho_\infty V_\infty^2\right)$,

C_L is the lift coefficient $\left(C_L = \frac{C_{L_V}}{\cos(\phi)}\right)$,

W is the weight of UCAS $(W_{UCAV} = W_{fuselage} + W_{wing} + W_{payload})$,

$W_{fuselage} = W_{engine} + W_{structural-fuselage}$

$W_{structural-fuselage} = \frac{1}{2}W_{wing}$

W_{engine} and $W_{payload}$ are fixed.

$$\omega = \frac{g\sqrt{n^2 - 1}}{V_\infty} \tag{9.17}$$

where g is the gravity ($g = 9.81 m/s^2$)

$$\therefore\ r = \frac{V_\infty}{\omega}. \tag{9.18}$$

Higher load factor (n) and greater turn rate (ω) produce smaller turning radius (r). To improve a load factor, the weight of UCAS including $W_{structural-fuselage}$ and W_{wing} should be as low as possible while the lift coefficient (C_{L_V}) should be as high as possible. The structural wing weight is estimated by using the structural estimation code in Sect. 9.3.

Aerodynamic quality at the variability of flight conditions is expressed in terms of *mean and variance* of inverse lift to drag ratio (L/D) while the quality of manoeuvrability is formulated by the *mean* and *variance* of the turning radius at 45 bank angle. RCS quality at the variability of radar frequencies is in terms of mean and variance of mono and bi-static radar systems to produce stealth UCAS against enemy aircraft and ground level radars. In fact, the aerodynamic and manoeuvrability performances at cruise are maximised while their sensitivity is minimized. The values of *mean* and sensitivity of mono/bi-static RCS are minimized with the variability of radar frequencies.

Fitness functions are in conflict by the variability of input parameters: the fitness 1 will look for high and stable lift to drag ratio that requires longer span which leads to heavier structural weight and high radar signature while fitness function 2 will look for high lift to maximize the manoeuvrability (instantaneous turning ratios) which requires UCAS lighter weight. The fitness function 3 will look for the low observable shape “aero-diamond” which can cause low cruise performance and manoeuvrability.

The optimization for this test case is based on the Algorithm described in Fig. 6.6 of Chap. 6.

The variability of flight conditions are:

$$M_{\infty i} \in [0.78, 0.79, M_S = 0.80, 0.81, 0.82] \text{ or } \overline{M_\infty} = 0.8 \text{ and}$$
$$\sigma M_\infty = 0.03952 \text{ and } \alpha_S = 4.3^\circ$$

$$F_{\infty i} \in [1.0, 1.25, F_S = 1.5GHz, 1.75, 2.0] \text{ or } \overline{F_\infty} = 1.5 \text{ and } \sigma F_\infty = 0.3952.$$

The fitness functions for this problem are:

$$\begin{aligned} f_1 &= \min\left(\frac{1}{Aerodyanmic_{Quality}}\right) \\ &= \min\left(\frac{-1}{\ln\left(\overline{1/(L/D)}\right)} + \frac{-1}{\ln\left(\delta(1/(L/D))\right)}\right) + Penalty \end{aligned} \tag{9.19}$$

$$f_2 = \min(\textit{\textbf{Maneuverability}}_{Quality}) = \min(\overline{r} + \delta r) + Penalty \tag{9.20}$$

where r is the instantaneous turning radius at bank angle 45°.

$$f_3 = \min\left(RCS_{Quality}\right) = \frac{1}{2}\left[\left(\overline{RCS}_{mono} + \delta RCS_{mono}\right) + \left(\overline{RCS}_{bi} + \delta RCS_{bi}\right)\right] + Penalty$$

where $\theta = [0^\circ : 3^\circ : 360^\circ]$ and $\phi = [0^\circ : 0^\circ : 0^\circ]$ (Mono-static);

where incident angles are $\theta = 135^\circ$, $\phi = 90^\circ$ at $\theta = [0^\circ : 3^\circ : 360^\circ]$, $\phi = [0^\circ : 0^\circ : 0^\circ]$ (Bi-static);

in which penalties (Chap. 2.7) are computed and added to fitness functions in the following different situations:

- if the aerodynamic quality (AQ) is higher than 0.597 ($AQ \geq 0.597$),
- if the manoeuvrability quality (MQ) is higher than 0.998 ($MQ \geq 0.998km$),
- if the RCS quality (RQ) is higher than 43.63 dBsm ($RQ \geq 43.63dBsm$).

Design Variables The design variables for the aerofoil sections are at four span wise stations as described in Figs. 8.51a–d. Lower design bounds of thickness distributions have 0 %c to provide more space for the payload including avionics and missiles. Eight design variables for wing plan form are λ_{C1}, λ_{C2}, Λ_{R-C1}, Λ_{C1-C2}, Λ_{C2-T}, S_1, S_2 and S_3. The taper ratio (λ_T) at tip is fixed as 0.2 % of the root chord. Two crank positions (C1 and C2) and wing span (b) are recalculated without wash-out and dihedral effects. The upper and lower design bounds were illustrated in Table 8.11. Seventy six (76) design variables are considered in total.

Implementation FLO22 and FRICTION solvers are utilised and the following specific parameters are considered for the HAPMOEA multi layer optimizer:

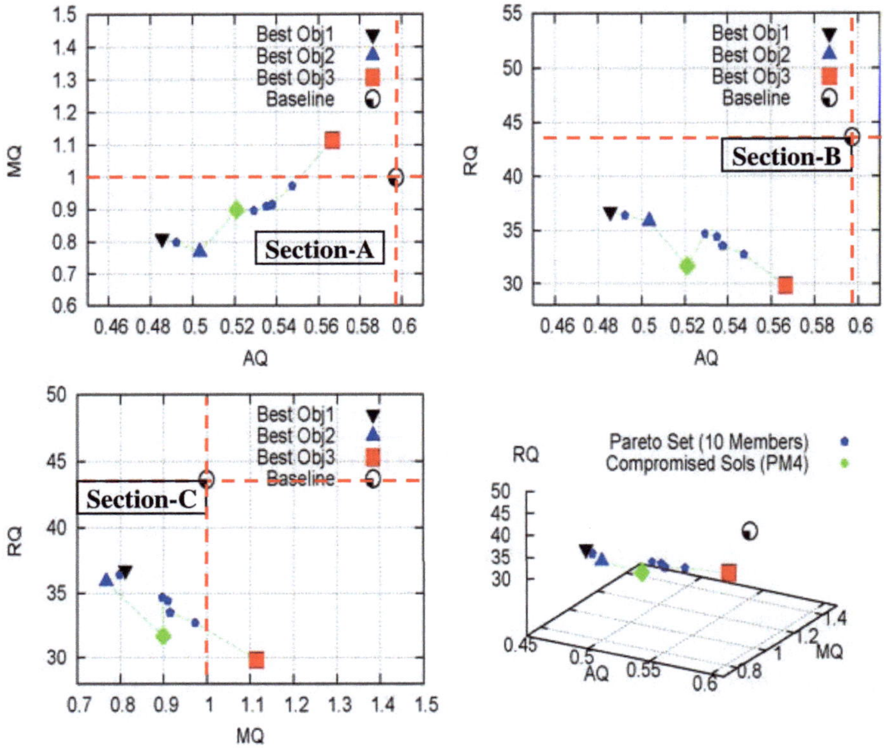

Fig. 9.49 Pareto non-dominated solutions for multi-disciplinary design optimisation of UCAS

1st Layer	Population size of 10 and intermediate recombination with a computational grid of 96 × 12 × 16 cells used by FLO22 solver
2nd Layer	Population size of 20 and intermediate recombination with a computational grid of 68 × 12 × 16 cells used by FLO22 solver
3rd Layer	Population size of 30 and intermediate recombination with a computational grid of 48 × 12 × 16 cells used by FLO22 solver

Numerical Results The algorithm was allowed to run approximately 945 function evaluations and took 150 CPU hours on two 2.4 GHz processors. The resulting Pareto set is shown in Fig. 9.49 where the best solution (Pareto Member 1) for fitness function 1 is marked as a black inverse triangle and the blue triangle is the best solution (Pareto Member 3) for fitness function 2. The red square represents the best solution (Pareto Member 10) for fitness function 3. It can be seen that the baseline UCAS dominates Pareto Member 10 in the aspect of manoeuvrability quality as shown in Section-A and Section-C. However, all Pareto members dominate the baseline UCAS with respect to the quality of cruise aerodynamics and RCS in terms of performance and sensitivity.

The best solutions (Pareto Members 1, 3 and 10) and Pareto Member 4 are selected to compare the aerodynamic, manoeuvrability and RCS quality to the baseline

Table 9.24 Comparison of the objectives

Objective	Baseline	Pareto M1	Pareto M3	Pareto M4	Pareto M10
AQ	0.597	0.485 (−19 %)	0.503 (−16 %)	0.521 (−13 %)	0.566 (−5 %)
MQ	0.998	0.822 (−18 %)	0.768 (−23 %)	0.899 (−10 %)	1.114 (+12 %)
RQ	43.63	36.77 (−8 %)	35.86 (−18 %)	31.68 (−27 %)	29.83 (−32 %)

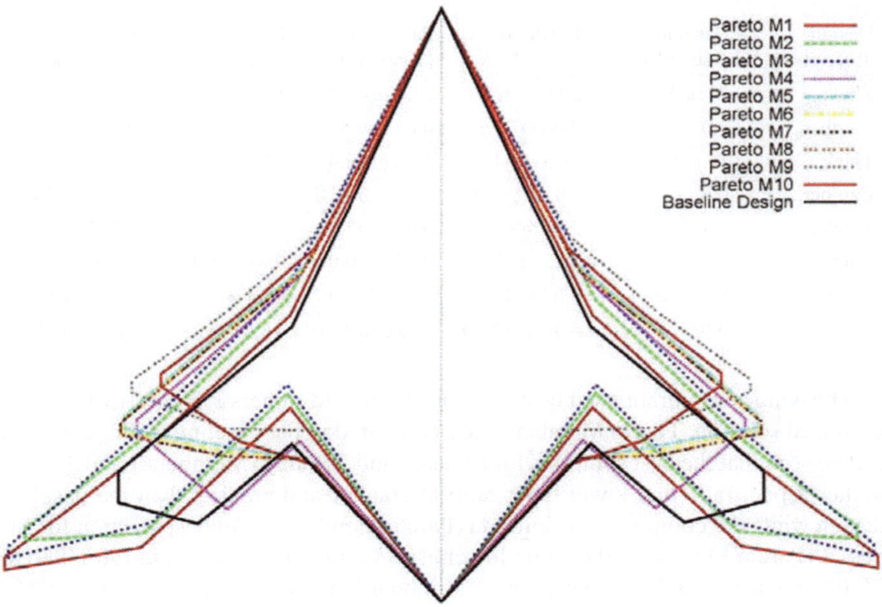

Fig. 9.50 Pareto optimal wing plan form shapes

UCAS on Table 9.24. All Pareto Members exhibited improved quality in aerodynamic parameters. With regards to manoeuvrability quality, Pareto Member 10 is dominated by the baseline UCAS while Pareto Member 10 has 32 % less chance to be detected by an enemy radar system than the baseline UCAS. Pareto Member 4 is selected as a compromised solution for further evaluation since it made an improvement at all aspects of aerodynamic, manoeuvrability and RCS quality.

The wing plan form shapes corresponding to the Pareto front are illustrated in Fig. 9.50. Pareto members 1–3 have longer span length than the baseline due to the fitness function 1 and 2. The Pareto Members 4–10 provide slightly shorter span and the plan form shapes of Pareto Members 5 and 10 are almost aero-diamond due to fitness function 3. The aero-diamond plan form shape is the most efficient shape against radar signature.

Table 9.25 Comparison of wing configurations

Models	AR	b	λ_{C1}	λ_{C2}	Λ_{R-C1}	Λ_{C1-C2}	Λ_{C2-T}
Baseline	4.45	18.9	19.7	19.7	55°	29°	29°
ParetoM1	7.56	25.20	15.1	15.1	59.40°	27.0°	26.60°
ParetoM3	7.48	25.32	16.7	16.7	50.34°	33.80°	29.76°
ParetoM4	3.75	17.81	28.0	28.0	51.70°	31.43°	29.80°
ParetoM10	3.37	16.42	36.5	21.6	51.70°	30.54°	27.11°

Table 9.26 Comparison of airfoil section configurations

Design description	Baseline (%c @%c)	Pareto M1 (%c @%c)	Pareto M3 (%c@%c)	Pareto M4 (%c @%c)	Pareto M10 (%c @%c)
$\text{Thick}_{\text{Root}}$	21 @45	20.0 @42.5	21.0 @41.3	19.6 @44.0	19.0 @41.6
$\text{Camber}_{\text{Root}}$	0.0	0.06 @37.3	0.03 @9.52	0.06 @84.1	0.11 @37.9
Thick_{C1}	15 @50	10.5 @44.7	12.2 @49.4	12.6 @50.2	12.9 @50.5
$\text{Camber}_{\text{C1}}$	0.0	−0.16 @1.45	−0.07 @14.7	−0.09 @4.20	0.14 @2.20
Thick_{C2}	8 @45	7.11 @45.3	6.60 @43.5	6.59 @42.3	6.30 @43.6
$\text{Camber}_{\text{C2}}$	0.0	0.12 @76.1	0.05 @70.0	0.04 @5.41	−0.07 @77.5
$\text{Thick}_{\text{Tip}}$	8 @45	6.02 @46.9	6.91 @46.9	6.33 @44.1	6.92 @44.0
$\text{Camber}_{\text{Tip}}$	0.0	−0.04 @24.4	−0.02 @17.0	−0.01 @82.9	−0.01 @75.4

The wing configurations of best solutions (Pareto Members 1, 3 and 10) and compromised solution (Pareto Member 4) are compared to the baseline in Table 9.25. It can be seen that the best solutions for fitness 1 and 2 (Pareto Members 1 and 3) have higher aspect ratio and lower taper ratio at crank 1 and crank 2 than the baseline design while the compromised solution (Pareto Member 4) and best solution for fitness 3 (Pareto Member 10) contain lower aspect ratio and higher taper ratio. Pareto Member 1 has very high sweep angle at inboard and slightly less sweep angle at outboard when compared to the baseline design.

The airfoil sections are compared in Table 9.26 and Fig. 9.51 where all Pareto Members have less thickness when compared to the baseline design. Pareto Members 1, 3 and 4 have negative camber at crank 1 and tip section while Pareto Member 10 has negative camber at crank 2 and tip.

Figure 9.52a, b, c compare the span wise *Cp* distributions between best solutions (Pareto Members 1, 3 and 10), compromised solution (Pareto Member 4) and the baseline design. Pareto Members 1 and 3 start to have leading edge pressure improvement at 40 % of the span while Pareto Members 4 and 10 improve pressure from 60 % of the span when compared to the baseline design.

Figure 9.53 shows the comparison of mono-static RCS at standard design frequency (1.5 GHz) between Pareto member 4 (compromised solution), Pareto Member 10 (best solution for fitness function 3) and the baseline design. Pareto Members 4 and 10 produce 25 and 35 % lower mono-static radar signature respectively when compared to the baseline design. The RCS peak (black circles) of the baseline was found to be 3 and 177 degrees where the baseline produces higher RCS than Pareto

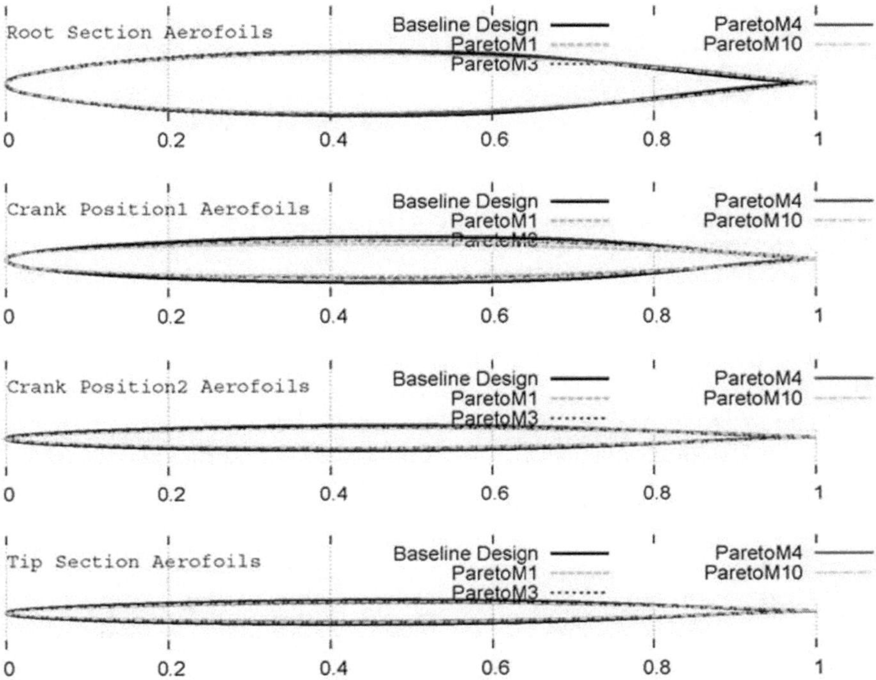

Fig. 9.51 Airfoil section shape comparison

Members 4 and 10 marked with blue and red circles. Figure 9.54a and b compare the bi-static RCS between Pareto Members 4, 10, and the baseline design. The results show that the current non-dominated solutions (Pareto member 4 and 10) produce lower observability by 17 and 20 % than the baseline design.

Figures 9.55a, b, c, d, e compares the aerodynamic performance and manoeuvrability (instantaneous turning radius) at the range of Mach numbers 0.7–0.85. The lift coefficient is compared in Fig. 9.55a. Pareto Member 3 (best solution for fitness 2) indicates the highest lift coefficient along the Mach numbers. The lift distribution of Pareto Member 4 is slightly higher than the baseline lift curve from Mach 0.70 to 0.76 while Pareto Member 3 shows highest lift coefficient. Figure 9.55b showing the drag coefficient distribution exhibits a common trend where all Pareto Members produce lower drag compared to the baseline design. Pareto member 1 shows the lowest drag along the Mach sweep. The comparison of lift to drag ratio is shown in Fig. 9.55c where Pareto Member 10 produces lower aerodynamic performance when Mach number is lower than 0.765 while Pareto Member 1 (best solution for fitness function 1) made the highest aerodynamic performance when compared to the other Pareto Members and baseline design. Figures 9.55d and 9.55e shows the comparison of manoeuvrability (instantaneous turn radius) at 45 bank angle. Pareto

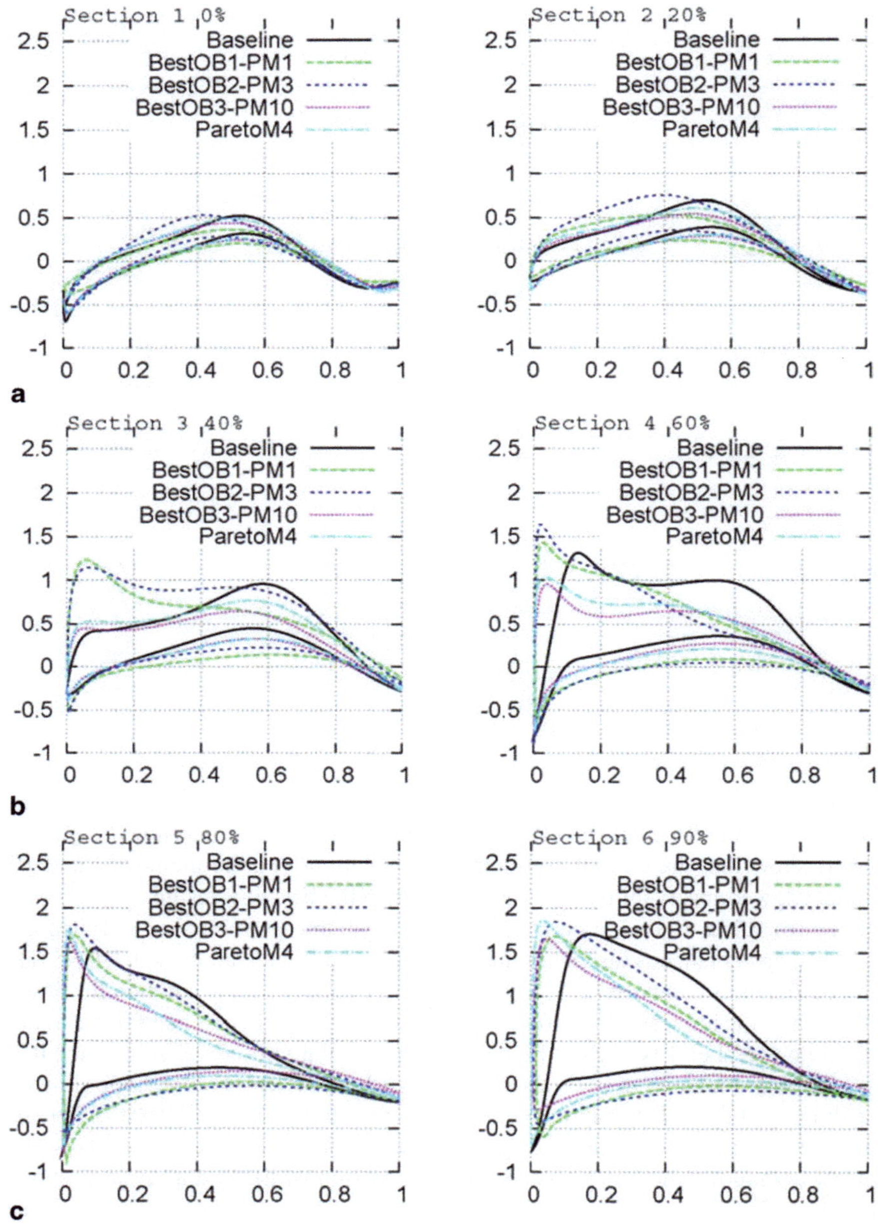

Fig. 9.52 **a** Pressure coefficients (Cp) distribution at 0 and 20 % of span. **b** Pressure coefficients (Cp) distribution at 40 and 60 % of span. **c** Pressure coefficients (Cp) distribution at 80 and 90 % of span

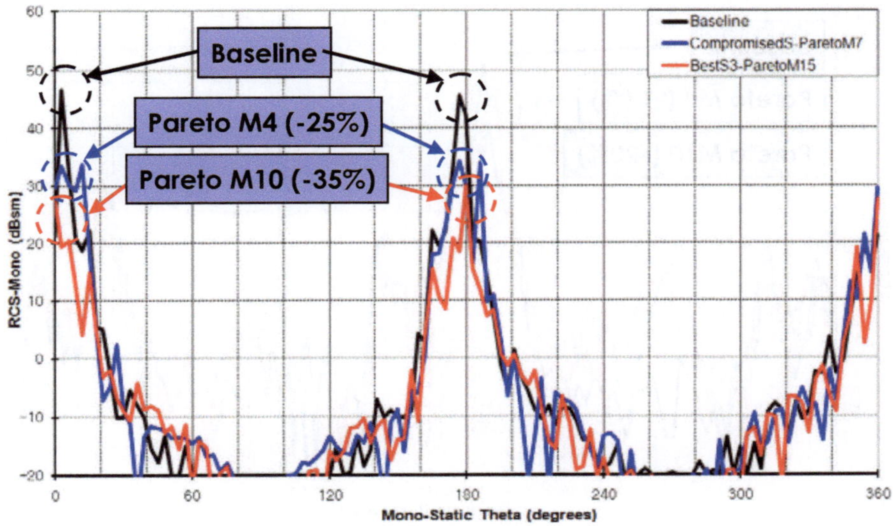

Fig. 9.53 Mono-static radar signature comparison

Member 3 made the smallest turning ratio which means it is superior when it competes with the other solutions.

Top, side, front and three-dimensional view for compromised solution's (Pareto Member 4) wing geometry are respectively shown in Fig. 9.56.

To conclude with this test case, it was demonstrated that all Pareto non-dominated solutions are better solutions with respect to aerodynamic performance, manoeuvrability and RCS signature. From the designer's point of view, Pareto Member 4 can be selected as J-UCAS prototype since it does not only produce higher cruise aerodynamic performance with higher manoeuvrability but also its shape provides low radar observability at mono and bi-static radar systems when compared to the baseline design.

9.7 Summary of Chapter

This chapter presents and analyses numerical experiments obtained from several robust multi-disciplinary design optimisation problems. The reader can easily observe through computation and optimization of these problems that as the complexity of the problem is increasing, the need of advanced EA methods is required.

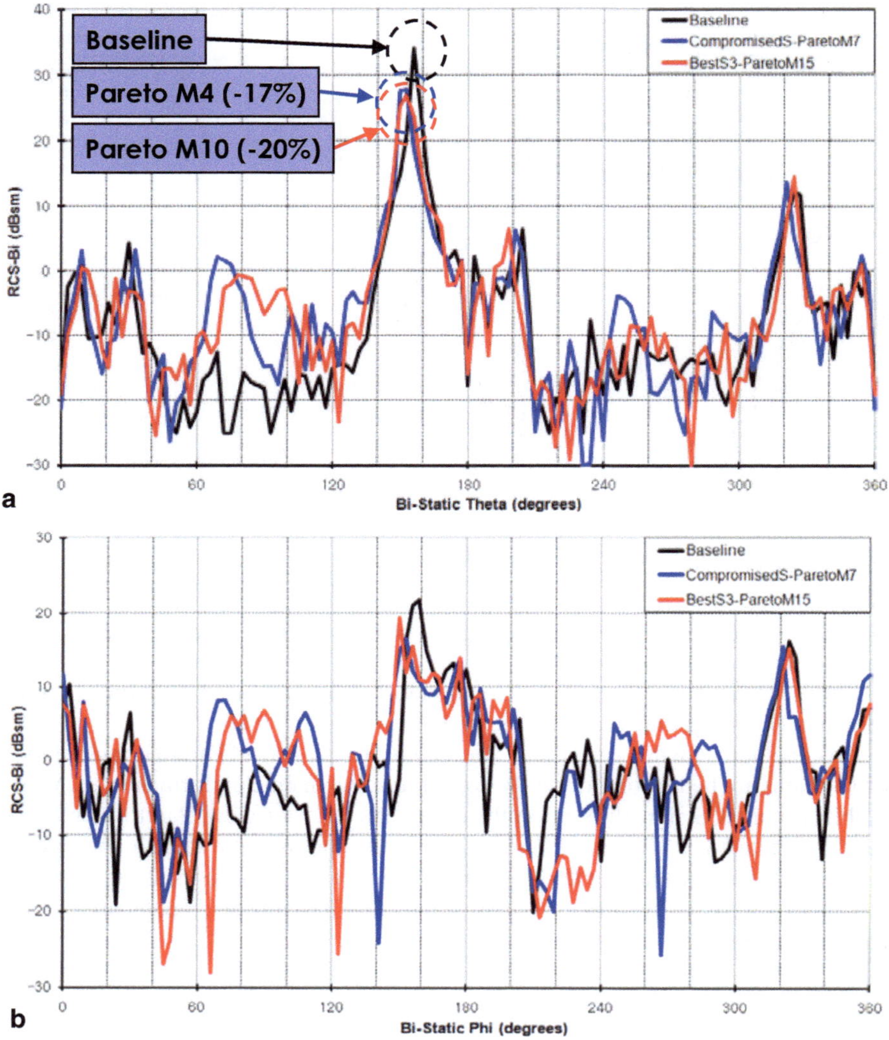

Fig.9.54 **a** Bi-static (θ) radar signature comparison. **b** Bi-static (ϕ) radar signature comparison

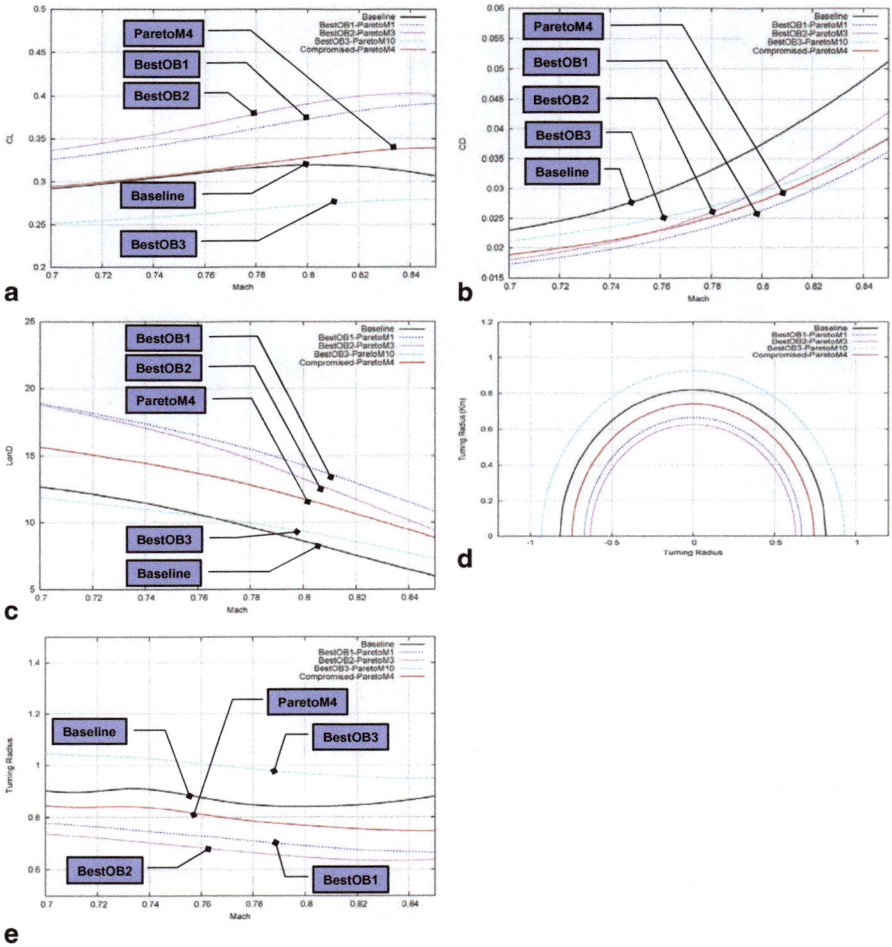

Fig. 9.55 **a** Lift coefficients (*Cl*) vs. Mach numbers comparison between Pareto Members (1, 3, 4 and 10) and baseline design. **b** Drag coefficients (*CD*) vs. Mach numbers comparison between Pareto members (1, 3, 4 and 10) and baseline design. **c** Lift to drag ratio (*L/D*) vs. Mach numbers comparison between Pareto Members (1, 3, 4 and 10) and baseline design. **d** Turning radius obtained by Pareto Members (1, 3, 4 and 10) and baseline design at the standard flight condition. **e** Turning radius vs. Mach numbers comparison between Pareto Members (1, 3, 4 and 10) and baseline design

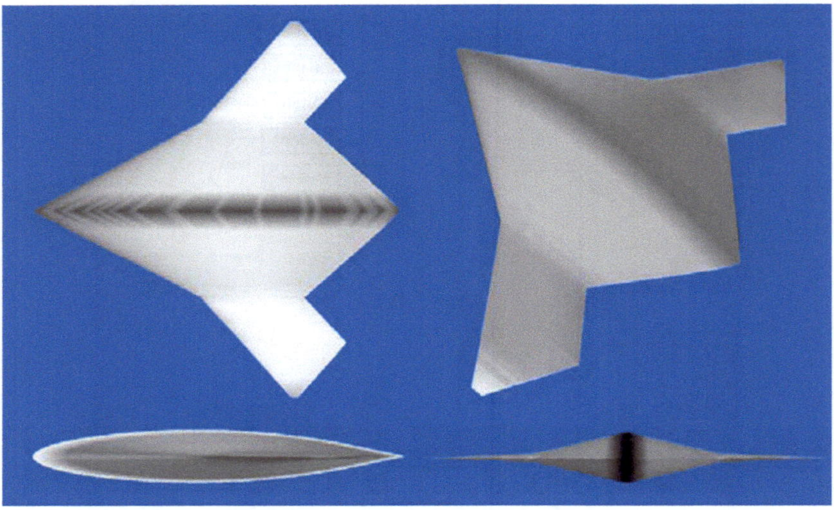

Fig. 9.56 Pareto Member 4 at top, side, front and 3D view

9.8 Appendix

References

1. Drela M. A user's guide to MSES V 2.3. MIT
2. Lee DS, Periaux J, Gonzalez LF, Srinivas K, Onate E (2012) Robust multidisciplinary UAS design optimisation. Struct Multidiscip Optim 45(3):433–450
3. Hague D (January 1978) General aviation synthesis program (GASP), NASA
4. Jameson, Caughey D, Newman P, Davis R A brief description of the Jameson Caughey NYU Transonic Swept-Wing Computer Program FLO22, NASA Technical Memorandum, NASA TM X–73996

Chapter 10
Robust Airfoil Design Optimization with Morphing Techniques

10.1 Overview

It is well known that morphing aircraft can change its shape for mission agility and control of flight. The main advantage of a morphing airfoil is its integrated shape control into an airfoil smart structure [1].

In this chapter, we describe and compare another technique, namely morphing design, with those discussed in previous chapters for single and multi- objective design optimization. This approach is based on the leading and trailing edge (LTED) deformation technique without taking into account a mechanism for actuation.

LTED morphing technology is simple and uses either a piezo or a hydraulic actuator. This approach can change significantly the shape of the wing or the airfoil and requires less modifications to the structure when compared to similar morphing techniques. The LTED technology is still sensitive to uncertain flow conditions especially lift coefficient (Cl) numbers and boundary layer transition positions (x_{tr}) which are directly related to the position/strength of shock on the suction/pressure side of the wing. Therefore it is critical to find suitable and reliable actuator positions and deformation angles using robust design methods.

This Chap. 10 also investigates the shape design optimization for morphing aerofoil/wing to maximize the lift coefficient at take-off and landing conditions using Advanced Multi-Objective Genetic Algorithms (MOGA).

J. Periaux et al., *Evolutionary Optimization and Game Strategies for Advanced Multi-Disciplinary Design,* Intelligent Systems, Control and Automation: Science and Engineering 75, DOI 10.1007/978-94-017-9520-3_10

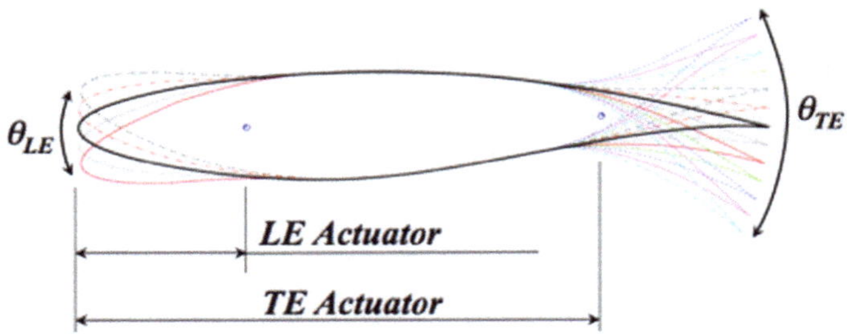

Fig. 10.1 Control parameters for morphing techniques at fixed leading and trailing edge actuator positions

10.2 Morphing Airfoil/Wing Design Mechanism: Leading and Trailing Edge Deformation

10.2.1 Parameterization of Morphing Aerofoil/Wing: Leading and Trailing Edge Deformation

The Leading and Trailing Edge Deformation (LTED) morphing technique can be defined by four control parameters as shown in Fig. 10.1: the leading edge actuator position (x_{LE}), the trailing edge actuator position (x_{TE}), the deformation angle for leading edge (θ_{LE}), and the deformation angle for trailing edge (θ_{TE}).

Deformation angles comply with the right-hand rule. Figure 10.2 shows a morphing technique mechanism consisting of four steps:

- Step 1: Find actuator positions for leading and trailing edges; x_{LE}, y_{LE}, x_{TE}, y_{TE} (marked as green crosses),
- Step 2: Deform leading edge by θ_{LE},
- Step 3: Deform trailing edge by θ_{TE},
- Step 4: Smooth sharp joint (marked as red circles shown in Fig. 10.2) using Bezier Spline Curves denoted by BSC1, BSC2, BSC3, and BSC4.

If the problem considers only trailing edge deformation then the morphing mechanism will consist only of three following steps: Step1, Step 3 and Step 4.

10.2.2 Baseline Design

For the baseline design, a natural laminar flow airfoil RAE 5243 is selected as shown in Fig. 10.3. The baseline design has a maximum thickness ratio (*t/c*) of 0.14 at 41 % of the chord and a maximum camber of 0.018 at 54 % of the chord.

In the following Sects. 10.3 and 10.4, the shape of the baseline design will be adapted to control the transonic flow especially to maximize aerodynamic efficiency including lift to drag ratio (*L/D*) at cruise conditions and fixed lift coefficient (*Cl*).

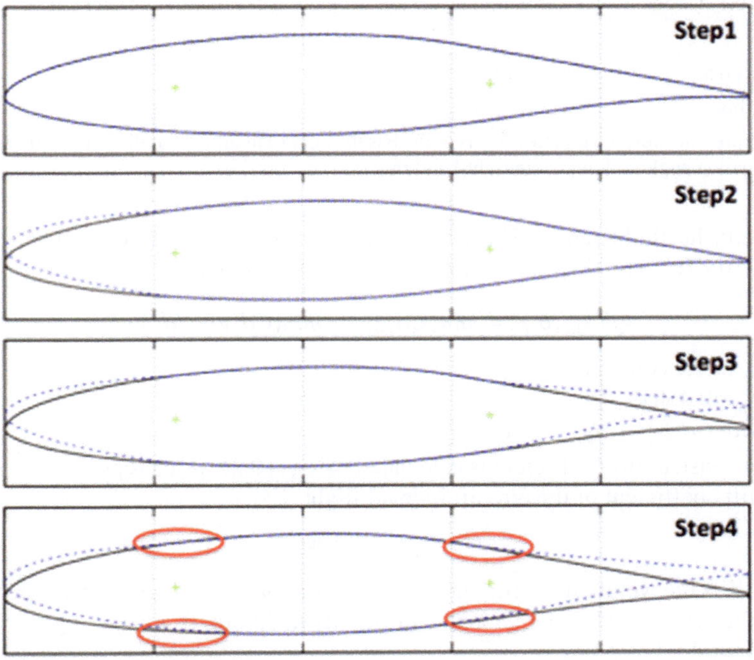

Fig. 10.2 LTED morphing mechanism

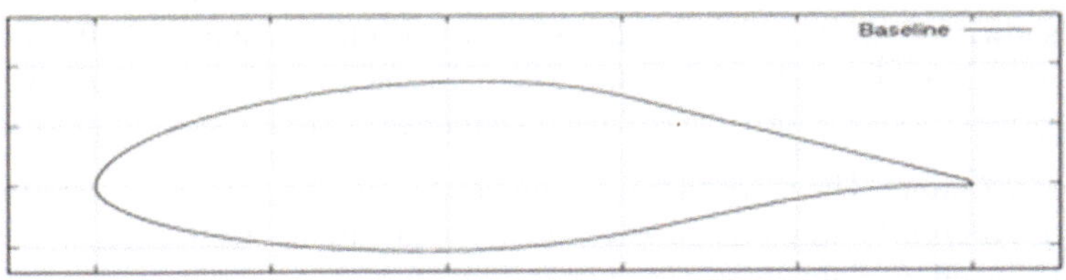

Fig. 10.3 Baseline design (RAE 5243 airfoil) geometry (Note max $t/c = 0.14$ at 41 %c and max camber = 0.018 at 54 % of the chord)

10.3 Morphing Airfoil/Wing Design Optimisation at Cruise Flight Conditions

10.3.1 Trailing Edge Deformation (TED) Design Optimisation

In this section, only TED Design Optimization is considered.

Problem Definition This test case considers a single objective morphing airfoil design optimization to minimize the total drag at a constant lift coefficient. This objective aims to maximize the lift to drag ratio (L/D) allowing the aircraft

Table 10.1 Design bounds for morphing air foil/wing using Trailing Edge Deformation (TED)

DVs	X_{TE}	θ_{TE}
Lower bound	65.0	−5.0°
Upper bound	75.0	+5.0°

DVs represents design variables: x_{TE} is in the baseline chord length (% of the chord) [0:100] and deformation angle following the right-hand rule

to extend its flight range. The flow conditions are $M_\infty=0.75$, $Cl_\infty=0.45819$ and $\mathrm{Re}=19.0\times10^6$. The fitness function for DMOGA is shown in (10.1).

$$f(x_{TE},\theta_{TE}) = min\left(Cd_{Total}\right) = min(Cd\ v + Cd\ w) \tag{10.1}$$

subject to the constraint

$Cl_{min}=0.45819$

where $Cd_{Total}=Cd_{Visous}+Cd_{Wave}$

The constant lift coefficient is calculated using (10.2) which represents the minimum lift coefficient of the aircraft in level flight.

$$C\,l_{min} = 2W / \rho V^2 S \tag{10.2}$$

where W is the weight force (m×g) of the aircraft, mass m = 77,564 kg and acceleration of gravity $g=9.81$ m/s², ρ is the air density at 35,000 ft, $\rho=0.41$ kg/s³, S is the wing area, $S=124.58$ m².

Design Variables Upper and lower bounds values of design variables for trailing edge deformation (TED) are provided on Table 10.1. All Bezier Spline Curves (BSC) shown in Fig. 10.1, BSC3 and BSC4 have a constant length of 20 % of chord length. BSC3 and BSC4 have the same x-axis Bezier control points (starting, peak, finishing points). Therefore two (2) design parameters are considered in total.

Numerical Results The algorithm was allowed to run one hour CPU time (as time stopping criterion) and 6,020 fitness function evaluations. Convergence occurred after 617 function evaluations as shown in Fig. 10.4.

Table 10.2 compares the aerodynamic characteristics obtained by the baseline design (RAE 5243 airfoil) and the optimal solution obtained by Distributed Multi-objective GA (DMOGA) [2]. Even though applying optimal morphing configuration slightly increases the wave drag, the viscous drag is reduced by 27.2 % which results in improving the lift to drag ratio (L/D) by 16.4 %.

Table 10.3 describes the optimal morphing (trailing deformation) configurations obtained by MOGA. Figure 10.5 compares the geometry of the baseline design and the baseline with the optimal morphing configurations. The optimal solution has the same maximum thickness ratio (t/c) value 0.14 as the baseline design at 41 % of the chord, while the maximum camber is lowered by 1.2 % of the chord (max camber = 0.0066 at 29.0 %c).

Figure 10.6 compares the lift coefficient sweep obtained by the baseline design and the optimal morphing configuration at range of [0.1:0.8] with $M_\infty=0.75$ and $\mathrm{Re}=19.0\times10^6$. Even though the optimal morphing solution produces lower total

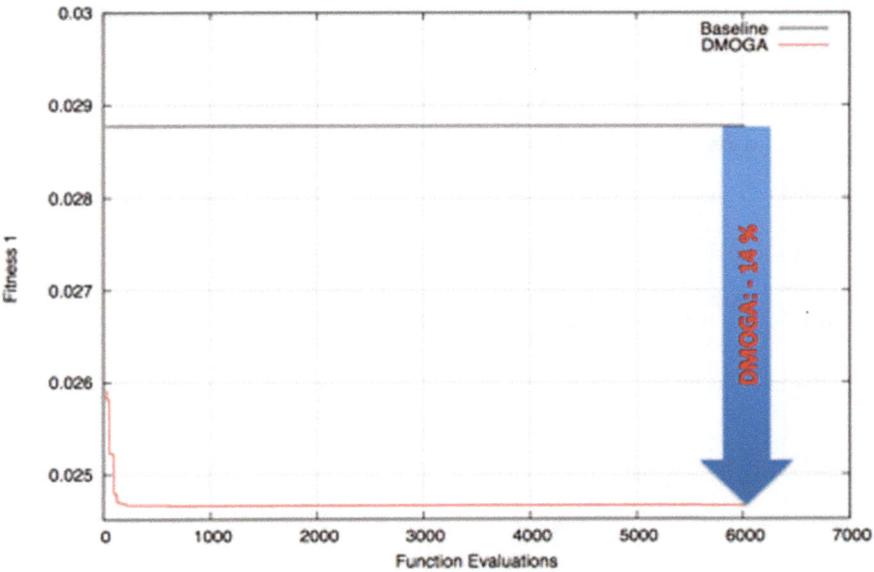

Fig. 10.4 Convergence history for TED design optimisation

Table 10.2 Aerodynamic characteristics obtained by TED design optimization

Aerofoil	$Cd_{Viscous}$	Cd_{Wave}	L/D
Baseline (RAE 5243)	0.0213	0.0074	15.93
With optimal SCB	0.0155 (−27.2 %)	0.0088 (+19.0 %)	18.55 (+16.4 %)

$Cd_{Viscous}$ and Cd_{Wave} represent viscous and wave drag respectively and the sum of them is the total drag of airfoil

Table 10.3 Optimal TED morphing configuration

Variables	X_{TE} (%c)	θ_{TE}
Optimal TED	65.04	−4.5195°

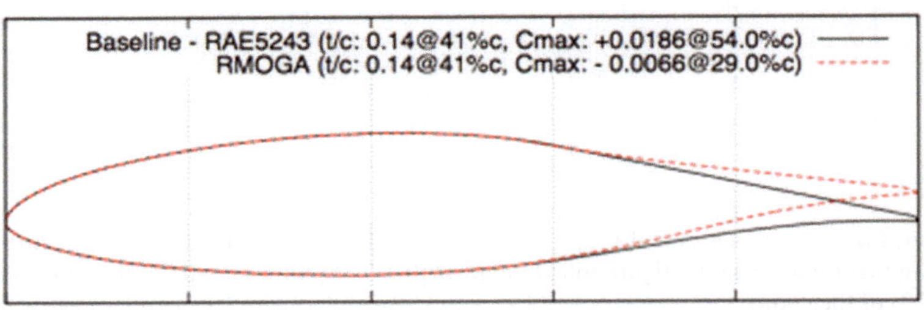

Fig. 10.5 Comparison of geometry obtained by baseline design and the optimal morphing configurations obtained by DMOGA (max $t/c = 0.14$ at 41 %c and max camber $= -0.0066$ at 29.0 %c)

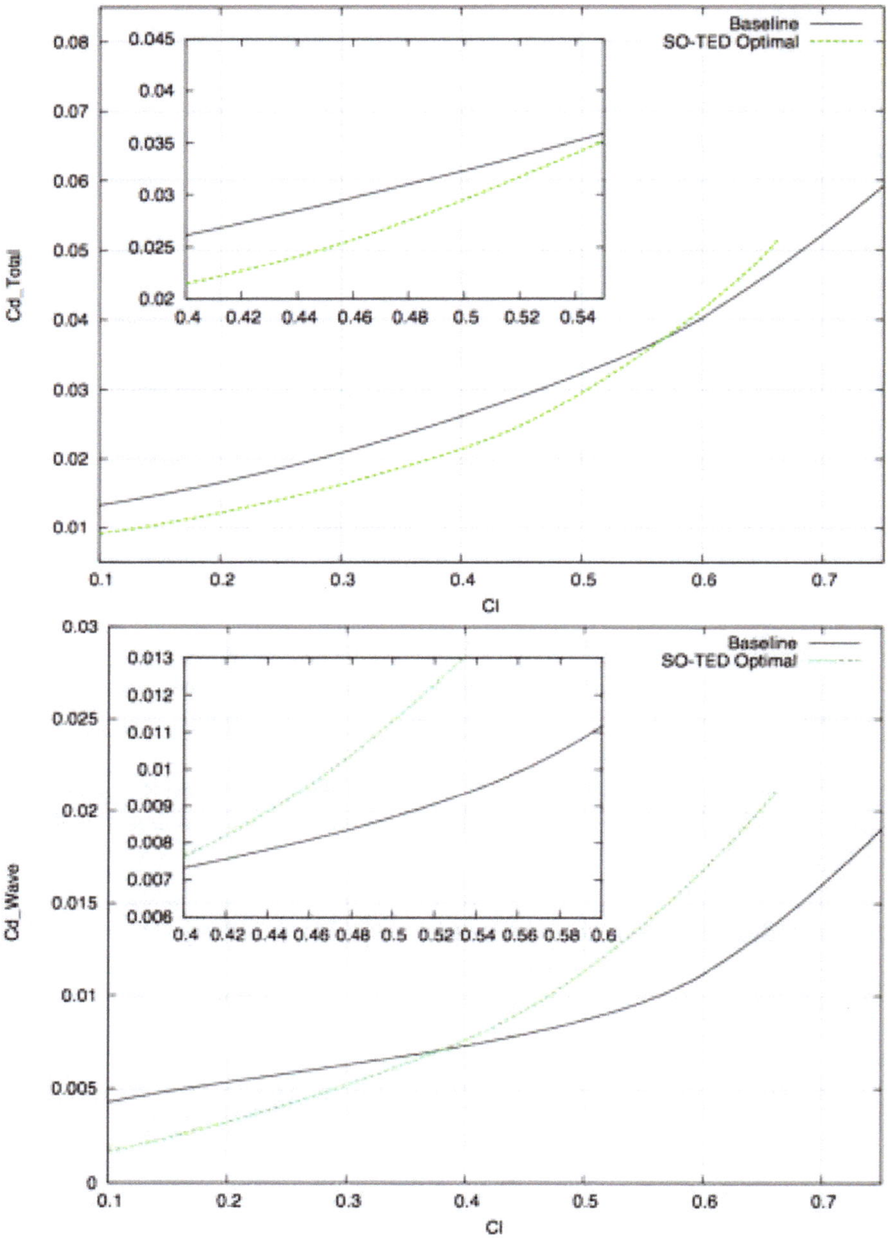

Fig. 10.6 Cd_{Total} vs. *Cl* (*top*) and Cd_{Wave} vs. *Cl* (*bottom*) sweep obtained by the baseline design and the optimal morphing configuration at range of [0.1:0.75] with $M_{\infty}=0.75$, Re $=19.0\times10^{6}$ and $\mu x_{tr}=45\,\%$ of the chord

drag when the lift coefficient is lower than 0.55 (Fig. 10.6 top), it produces higher wave drag when the lift coefficient is higher than 0.38 (Fig. 10.6 bottom).

From the variations of Cd values with respect to Cl values on Fig. 10.6, it is necessary to optimize the morphing configuration with uncertainty parameters in order to find the model which can produce lower and stable total drag. For uncertainty parameters, lift coefficients and boundary transition positions (x_{tr}) are considered since both are directly affecting the strength and shock position.

10.3.2 Robust Leading and Trailing Edge Deformation (LTED) Design Optimization

Problem Definition This test case considers a robust multi-objective morphing airfoil design optimisation on the RAE 5243 airfoil to minimize the average and standard deviation of the total drag ($\mu\, Cd_{Total}$, $\sigma\, Cd_{Total}$) at flow conditions $M_\infty=0.75$ and $Re=19.0\times10^6$.

Introducing uncertainties on operating flow conditions, two sets of ten values of lift coefficient (Cl) and then of boundary layer transition positions (x_{tr}) are considered. Both uncertainty parameters; Cl and x_{tr} can be statistically defined as $\mu Cl=0.5558$, $\sigma Cl=0.0611$ in a range of $Cl_\infty\in$ [0.458, 0.648], and $\mu x_{tr}=0.2565$, $\sigma x_{tr}=0.1604$ in a range of $x_{tr\infty}\in$ [0.0 %, 50.0 % of the chord]. The candidate morphing airfoil/wing models will be evaluated at hundred flight conditions ($10x_{tr}\times 10\, C_l$). The robust fitness functions are shown in (10.3) and (10.4) respectively.

$$f_1=\min\left(\mu Cd_{Total}\right)=\frac{1}{n\times m}\sum_{i=1}^{n}\sum_{j=1}^{m}Cd_{Total_{ij}} \tag{10.3}$$

$$f_2=\min\left(\sigma Cd_{Total}\right)=\sqrt{\frac{1}{n\times m-1}\sum_{i=1}^{n}\sum_{j=1}^{m}\left(Cd_{Total_{ij}}-\mu Cd_{Total}\right)^2} \tag{10.4}$$

where n and m represent the number of boundary layer transition positions and C_l conditions respectively.

This optimisation procedure uses the Algorithm 6 in Sect. 6.5.7 Chap. 6.

Design Variables The lower and upper bounds of design variables for a morphing geometry search space of leading and trailing edge deformation are given on Table 10.4. The length of all Bezier Spline Curves (BSC shown in Fig. 10.2): BSC1, BSC2, BSC3, and BSC4, are selected as additional design parameters to find more sophisticated morphing shapes. Finally eight design variables are considered in total.

Numerical Results The algorithm was allowed to run twenty (20) CPU hours (with a selected time stopping criterion) and 600 function evaluations. Figure 10.7 compares the Pareto optimal front obtained by Robust Optimization (RO-LTED with the baseline design and the optimal solution obtained in Sect. 10.3.1 denoted as Single Objective Trailing Edge Deformation (SO-TED). It can be seen that all Pareto

Table 10.4 Design bounds for morphing airfoil using leading and trailing edge deformation

DVs	X_{LE}	θ_{LE}	L_{BSC1}	L_{BSC2}	X_{TE}	θ_{TE}	L_{BSC3}	L_{BSC4}
Lower	15.0	−25.0°	10.0	10.0	65.0	−5.0°	10.0	10.0
Upper	25.0	+10.0°	25.0	25.0	75.0	+35.0°	30.0	30.0

Note: DVs represents leading and trailing edge design variables. x_{LE}, x_{TE} are expressed with respect to the baseline chord length (%c) [0:100] and deformation angle follows the right-hand rule. L_{BSCi} represents the length of the *i*th BSC in chord percentage (%)

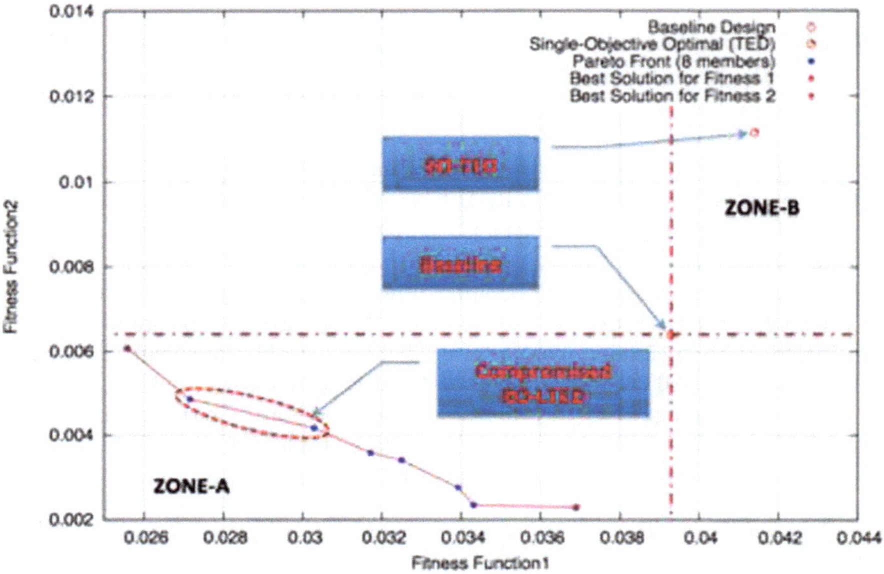

Fig. 10.7 Pareto optimal front for robust LTED design optimisation

optimal solutions in Zone-A dominate the baseline design and the optimal solution: SO-TED (obtained in Sect. 10.3.1) in both the mean total drag and total drag standard deviation. In Zone-B, the baseline design dominates SO-TED. Pareto members 2 and 3 are selected as a compromised solution and denoted as RO-LTED.

Table 10.5 compares the fitness values obtained by the baseline design, the optimal solutions from Sect. 10.3.1 and the Pareto members 1 (best solution for the objective 1), 2 and 3 (compromised solutions), and 8 (best solution for the objective 2). It can be seen that compromised solutions, Pareto Members 2 and 3 have better drag quality in terms of mean and standard deviation when compared to the baseline design, SO-TED and best solutions (Pareto Members 1 and 8).

It can be seen that the optimal solution obtained in Sect. 10.3.1 has higher mean total drag and higher total drag standard deviation by 5.3 and 73 % respectively when compared to the baseline design. Even though the best solutions (Pareto Members 1 and 8) for objectives 1 and 2 have lower mean total drag and lower total drag standard deviation respectively, one of the compromised solutions Pareto Member 2 has

Table 10.5 Comparison of fitness values obtained by the robust design approach

Aerofoil	μCd	σCd	$\mu L/D$
Baseline	3.93×10^{-2}	6.40×10^{-3}	14.14
SO-TED	4.14×10^{-2} (+5.3%)	11.1×10^{-3} (+73.4%)	13.42 (−5.1%)
BO1-PM1	2.55×10^{-2} (−35.1%)	6.07×10^{-3} (−5.1%)	21.79 (+54.1%)
CS-PM2	*2.71×10^{-2} (−31.0%)*	*4.84×10^{-3} (−24.4%)*	*20.51 (+45.0%)*
CS-PM3	*3.02×10^{-2} (−23.1%)*	*4.16×10^{-3} (−35.0%)*	*18.40 (+30.1%)*
BO2-PM8	3.68×10^{-2} (−6.4%)	2.28×10^{-3} (−64.3%)	15.10 (+6.8%)

PM_i represents Pareto optimal member obtained by robust design optimization. BO *i* and CS denote the best solution for the *i*th objective and compromised solution

lower mean total drag and lower standard deviation of total drag when compared to the baseline design. The mean and standard deviations obtained by the baseline design, single-objective and robust Pareto members can also be compared using Cumulative Distribution Function (CDF) and Probability Density Function (PDF). Figure 10.8 shows that the robust optimal solution lowers mean total drag by 31 % when compared to the baseline design. The standard deviation (sensitivity) can be analysed by evaluating gradients of the lines to the CDF value of 0.5 or 1 (steeper gradient means lower sensitivity). The PDF plotted in Fig. 10.8 (bottom) shows sensitivity comparison between the baseline design, single-objective and robust optimal solutions. It can be seen that the solution (Pareto Member 2) obtained by the robust design method has lower sensitivity (narrower and taller bell curve). Pareto Member 2 obtained by the robust design method reduces the total drag sensitivity by 24 % while the optimal solution obtained in Sect. 4.3 (SO-TED) has higher total drag sensitivity by 73 % when compared to the baseline design. In other words, the robust design method has capabilities to produce a set of solutions, which have better performance and sensitivity when compared to the single-objective optimization method.

The optimal morphing configurations (Pareto Member 2) for robust design optimisation are described in Table 10.6. Figure 10.9 compares the geometry of the baseline design and the baseline with the red-dot geometry of optimal morphing configurations. The optimal solution has the same maximum thickness ratio (*t/c*) of 0.14 as the baseline design at 41 % of the chord, while the maximum camber is lowered by 0.6 %c (max camber=—0.006 at 17.4 %c).

Table 10.7 compares the aerodynamic characteristics obtained by the baseline design (RAE 5243) and the robust optimal solution (Pareto member 2) at the flight conditions of $M_\infty=0.75$, $Cl_\infty=0.5558$, $X_{tr}=25.65$ % of the chord and $Re=19.0\times10^6$. The robust optimal solution reduces the viscous and wave drag by 43 and 26 % respectively that results in improving the lift to drag ratio (*L/D*) by 66 %.

Figure 10.10 shows pressure coefficient contours obtained by the baseline design and the robust optimal solution (Pareto member 2) at flight conditions of $M_\infty=0.75$, $Cl_\infty=0.5558$, $X_{tr}=25.65$ %c and $Re=19.0\times10^6$. It can be seen that the position of shock is delayed and its strength is reduced.

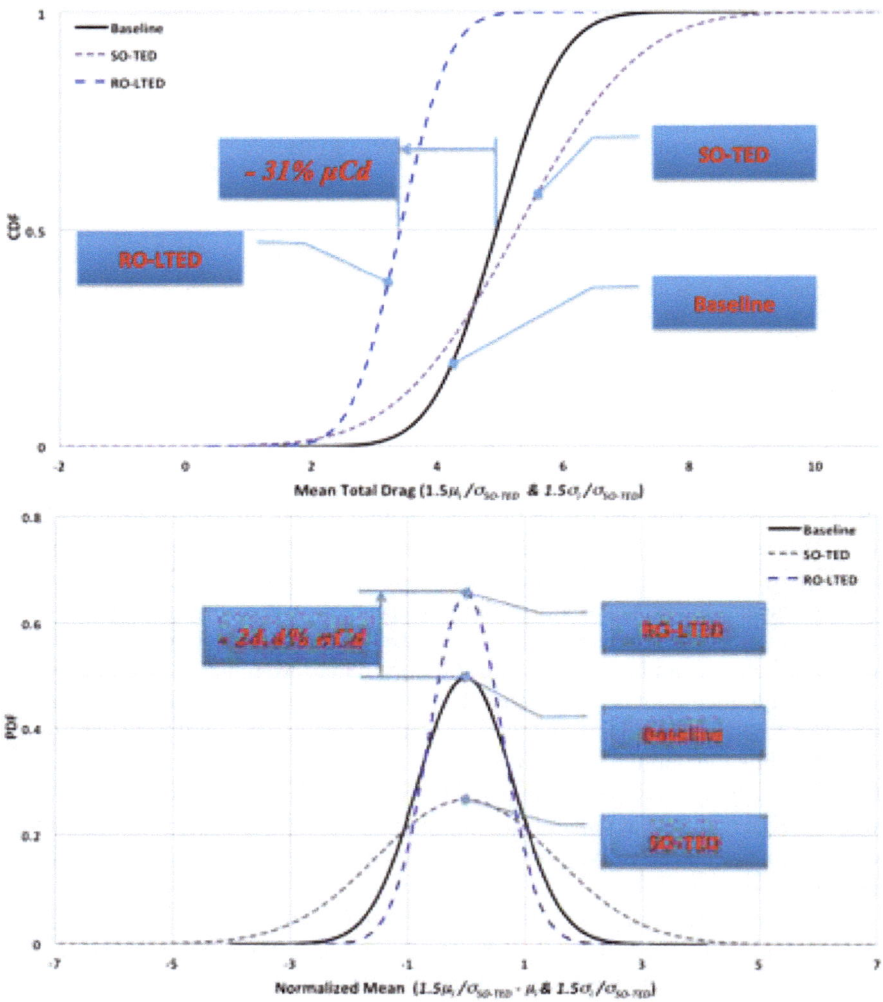

Fig. 10.8 Mean and standard deviation of total drag comparison using CFD (*top*) and PDF (*bottom*)

Table 10.6 Optimal morphing (LTED) configurations for the robust design optimisation

DVs	X_{LE}	θ_{LE}	L_{BSC1}	L_{BSC2}	X_{TE}	θ_{TE}	L_{BSC3}	L_{BSC4}
LTED	17.2	+6.5°	13.4	10.8	73.0	−3.0°	21.5	25.3

Note: DVs represents leading and trailing edge design variables. X_{LE}, X_{TE} are expressed with respect to in the baseline chord length (%c) [0:100] and deformation angle follows the right-hand rule. L_{BSCi} represents the length of the *i*th BSC (%c)

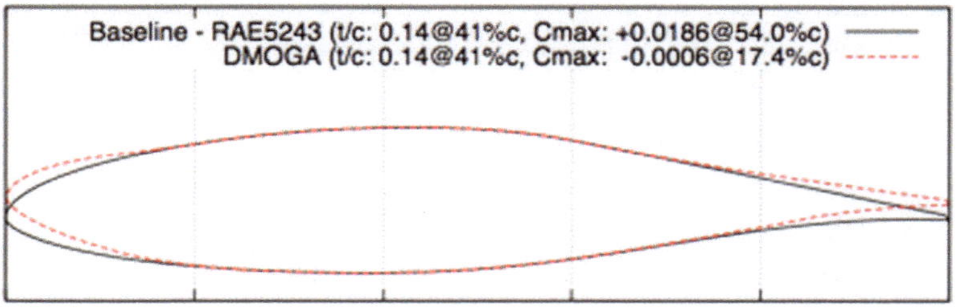

Fig. 10.9 Baseline design with the optimal LTED morphing configurations (max $t/c=0.14$ at 41 %c and max camber=−0.006 at 17.4 %c)

Table 10.7 Aerodynamic characteristics obtained by LTED design optimisation

Aerofoil	$Cd_{Viscous}$	Cd_{Wave}	L/D
Baseline (RAE 5243)	0.0304	0.0078	14.498
With optimal SCB	0.0172 (−43.4 %)	0.0058 (−25.6 %)	24.070 (+66.0 %)

$Cd_{Viscous}$ and Cd_{Wave} represent viscous and wave drag, and the sum of them is the total drag of airfoil

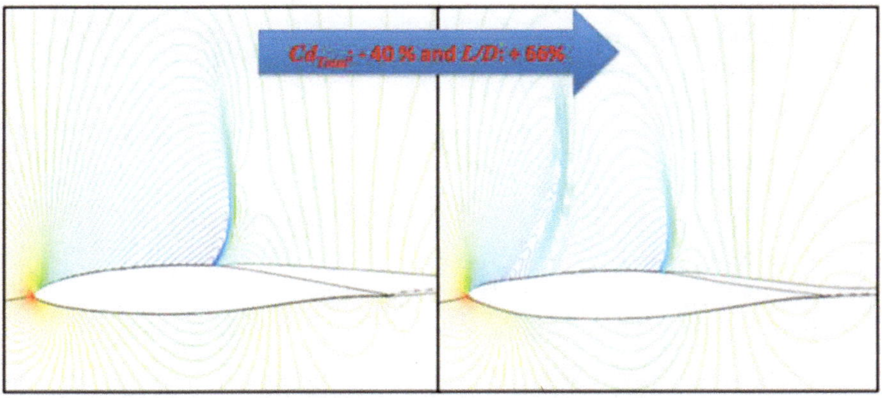

Fig. 10.10 *Pressure* (C_p) contours obtained by the baseline design and the robust optimal solution (Pareto member 2) at the flight conditions of $M_\infty=0.75$, $Cl_\infty=0.5558$, $X_{tr}=25.65$ %c and $Re=19.0\times10^6$

Figure 10.11 compares the lift coefficient sweep obtained by the baseline design, the optimal solution obtained in Sect. 10.3.1 (SO-TED) and the robust optimal morphing configuration (Pareto Member 2: RO-LTED) at range of $Cl_\infty \in [0.1:0.7]$ with $M_\infty=0.75$, $X_{tr}=25.65$ %c and $Re=19.0\times10^6$. It can be seen that SO-TED produces lower total drag when Cl is lower than 5.5 while the robust optimal solution produces lower total drag for all Cl values.

Figure 10.12 compares the lift coefficient sweep obtained by the baseline design, the optimal solution obtained in Sect. 10.3.1 (SO-TED) and the robust optimal

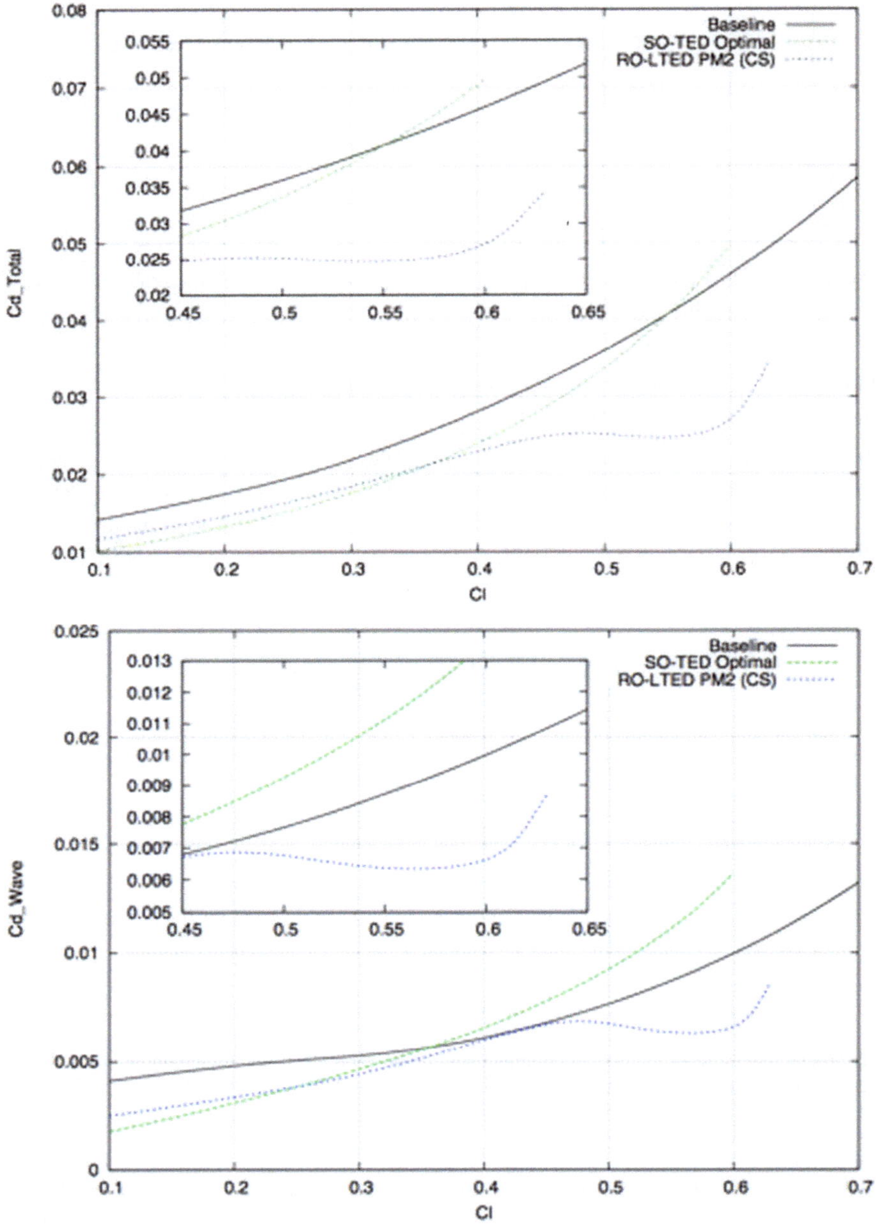

Fig. 10.11 Cd_{Total} vs. Cl (*top*) and Cd_{Wave} vs. Cl (*bottom*) sweep obtained by the baseline design and the optimal morphing configuration at Cl range of [0.1:0.7] with $M_{\infty}=0.75$, $Re=19.0\times10^{6}$ and $\mu x_{tr}=25.65\,\%$ of the chord

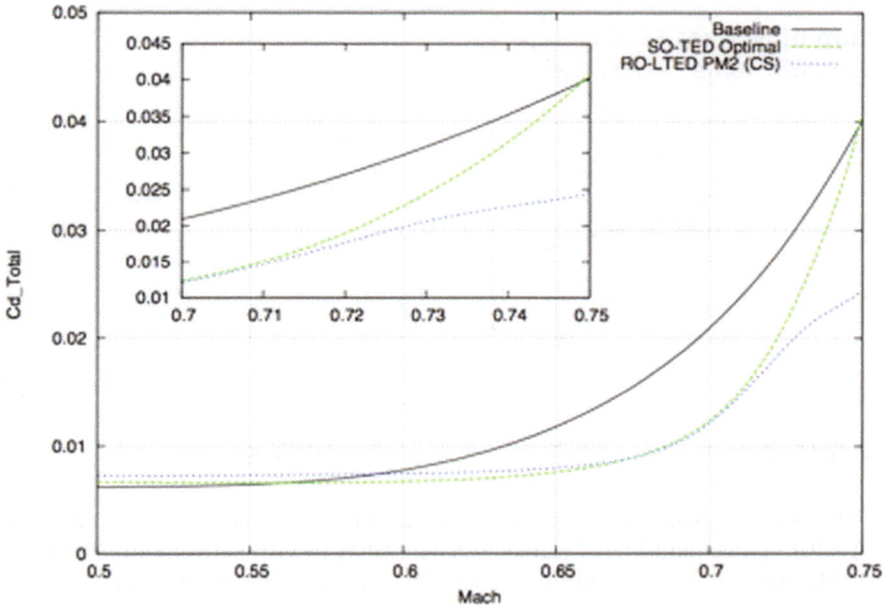

Fig. 10.12 Cd_{Total} vs. Cl (*top*) and Cd_{Wave} vs. Cl (*bottom*) sweep obtained by the baseline design and the optimal morphing configuration at Cl range of [0.1:0.7] with $M_\infty=0.75$, $Re=19.0\times10^6$ and $\mu x_{tr}=25.65\,\%$ of the chord

morphing configuration (Pareto Member 2) at range of $M_\infty \in [0.5{:}0.75]$ with $\mu Cl=0.5558$, $\mu x_{tr}=25.65\,\%$c and $Re=19.0\times10^6$. It can be seen that SO-TED produces higher total drag when Mach number is higher than 0.748 when compared to the baseline design while the robust optimal solution produces lower total drag when Mach number is higher than 0.58.

In this section, a robust evolutionary optimization technique has been applied to the morphing wing/aerofoil design optimisation. An analytical research shows the benefits of coupling an optimization method with robust design techniques to produce stable and high performance solutions. The use of a LTED morphing technique on an existing airfoil, can reduce significantly the transonic drag which will save operating and manufacturing cost as well as emission reduction.

10.4 Morphing Airfoil/Wing Design Optimization at Take-Off and Landing Conditions Using Moga and Hybridized Game with Moga

Problem Definition This test case considers a multi-objective optimization of a LTED morphing method with low/middle design complexity using MOGA and HMOGA. For the baseline design, a natural laminar flow around an airfoil RAE

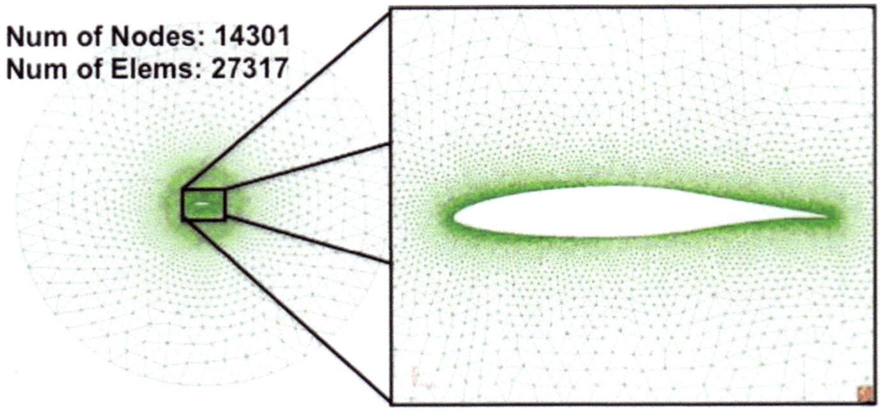

Fig. 10.13 Mesh conditions for the baseline design obtained by GiD software

5243 is selected as shown in Fig. 10.3. Figure 10.13 shows the computational mesh of 14,301 vertices and 27,317 elements generated by by GiD software from CIMNE [3]. The mesh conditions on the suction and pressure sides of the airfoil are set by constant ratio. In other words, there will be the same mesh conditions around the airfoil even though the shape of the airfoil is changed. Figure 10.14 shows the pressure and Mach contour obtained by the baseline design at both the take-off (M_∞=0.2, α=15.0°) and landing (M_∞=0.12, α=17.18°) conditions.

The shape of the baseline design will be adapted to control the subsonic/transonic flow especially to maximise the lift coefficient. For the optimization, the population size of Pareto-Game and Nash-Game is set as twenty (20) and the probabilities of mutation and crossover are set as $1/n$ and 0.9 respectively (where n is the number of design variables). With the hybridized games approach, a Pareto-Game considers the total number (n) of design variables as the sum of design variable numbers for Nash players. In other words, a Nash-Game will have higher mutation probability providing a better exploration when compared to a Pareto-Game. At a frequency of twenty (20) function evaluations, elite designs obtained by Nash-Game are seeded to Pareto-Game only if its fitness value dominates other fitness values in the Pareto's buffer, a well known Darwinian "above average" strategy based on the concept of survival of the fittest.

The objectives of this optimization are to maximize lift coefficients (Cl) at both the take-off (M_∞=0.2, α=15.0°) and landing (M_∞=0.12, α=17.18°) conditions. Hybrid-Game (HRMOGA) consists of three players; one Pareto-Player (PP), two Nash-Players ($NP1$ and $NP2$). The fitness functions for Pareto-Player (f_1 and f_2) and Nash-Game (f_{NP1} and f_{NP2}) are shown in (10.5)—(10.8).

$$f_1\left(x_{LE},\theta_{LE},L_{BSC1},L_{BSC2},x_{TE},\theta_{TE},L_{BSC3},L_{BSC4}\right)=\min\left(1/Cl_{TakeOff}\right) \tag{10.5}$$

$$f_2\left(x_{LE},\theta_{LE},L_{BSC1},L_{BSC2},x_{TE},\theta_{TE},L_{BSC3},L_{BSC4}\right)=\min\left(1/Cl_{Landing}\right) \tag{10.6}$$

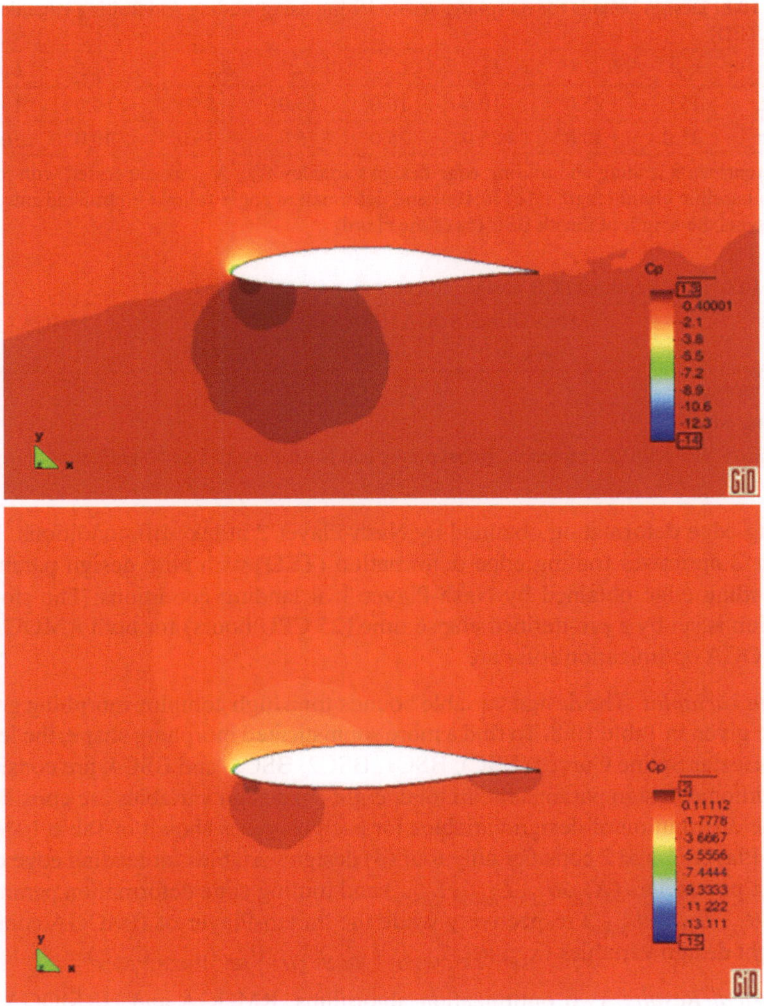

Fig. 10.14 *Cp* contours obtained by the baseline design at the take-off (*top*)—$Cp_{Take\ Off}$ range [−14:1.3] and landing (*bottom*)—$Cp_{Landing}$ range [−15:2] conditions

$$f_{NP1}\left(x_{LE}, \theta_{LE}, L_{BSC1}, L_{BSC2}, x^*_{TE}, \theta^*_{TE}, L^*_{BSC3}, L^*_{BSC4}\right) = \min\left(1/Cl_{Landing}\right) \quad (10.7)$$

$$f_{NP2}\left(x^*_{LE}, \theta^*_{LE}, L^*_{BSC1}, L^*_{BSC2}, x_{TE}, \theta_{TE}, L_{BSC3}, L_{BSC4}\right) = \min\left(1/Cl_{Landing}\right) \quad (10.8)$$

where design variables with * are elite designs obtained by Nash-Players; $x^*_{LE}, \theta^*_{LE}, L^*_{BSC1}, L^*_{BSC2}$ are elite designs obtained by Nash-Player 1, and $x^*_{TE}, \theta^*_{TE}, L^*_{BSC3}, L^*_{BSC4}$ are elite designs from Nash-Player 2. The Pareto-Player will optimise both leading and trailing edge deformation parameters to maximize *Cl* at both the take-off and landing conditions. Simultaneously, Nash-Player 1 will optimize only leading edge deformation (LED) parameters with an elite design of

Table 10.8 Design bounds for morphing aerofoil/wing using leading and trailing edge deformation (LTED)

DVs	x_{LE}	θ_{LE}	L_{BSC1}	L_{BSC2}	x_{TE}	θ_{TE}	L_{BSC3}	L_{BSC4}
Lower	15.0	−25.0°	10.0	10.0	65.0	−5.0°	10.0	10.0
Upper	25.0	+10.0°	25.0	25.0	75.0	+35.0°	30.0	30.0

DVs represents leading and trailing edge design variables. X_{LE}, X_{TE} are expressed with respect to the baseline chord length (%c) [0:100] and deformation angle follows right-hand rule. L_{BSCi} represents the length of the *i*th BSC (% of the chord)

Table 10.9 Pareto optimal SCB design parameters

Types	Hybridized MOGA (HMOGA)			MOGA
	Pareto P	*NashP1*	*NashP2*	
Leading edge	√	√		√
Trailing edge	√		√	√

Note: *GP, NP1 and NP2* represent global player and Nash-Players 1 and 2 respectively

trailing edge deformation obtained by Nash-Player 2 at take-off conditions. Nash-Player 2 optimises trailing edge deformation (TED) with elite design parameters for leading edge obtained by Nash-Player 1 at landing conditions. The stopping criterion is set by a pre-defined elapse time (25 CPU hours) for both RMOGA and HRMOGA optimization software.

Design Variables The design variable bounds for a high complex morphing geometry are given in Table 10.8. To find a more sophisticated morphing shape, the lengths of all Bezier Spline Curves (BSC); BSC1, BSC2, BSC3, and BSC4, are considered as additional design parameters. In total, eight (8) design variables are considered.

The distribution of design variables for each player is shown in Table 10.9. The Nash-Players 1 and 2 consider only four (4) design variables for leading edge deformation parameters (X_{LE}, θ_{LE}, L_{BSC1}, L_{BSC2}) and trailing edge deformation parameters (X_{TE}, θ_{TE}, L_{BSC3}, L_{BSC4}) respectively while the Pareto-Player of HMOGA considers all eight design variables (x_{LE}, θ_{LE}, L_{BSC1}, L_{BSC2}, X_{TE}, θ_{TE}, L_{BSC3}, L_{BSC4}).

Numerical Results Two optimization algorithms, RMOGA and HRMOGA were allowed to run 25 hours of CPU time and 640 and 285 function evaluations respectively using a single 4×2.8 GHz processor. Figure 10.15 compares the Pareto optimal front obtained by RMOGA and HRMOGA software with the baseline design. It can be seen that both RMOGA and HRMOGA produce better solutions when compared to the baseline design. The Pareto optimal solutions obtained by the HRMOGA optimization software dominate the optimal solutions computed with RMOGA (without hybridization).

Figure 10.16 compares the convergence history obtained by RMOGA and HR-MOGA software in terms of the normalised function evaluations and the best fitness value for Objective 1. Even though RMOGA software needs more computational time or more function evaluations to capture Pareto optimal solutions obtained by HRMOGA, two similar solutions (<3.5 % fitness value difference) are selected to compare the computational efficiency: HRMOGA (67 function evaluations (5.8

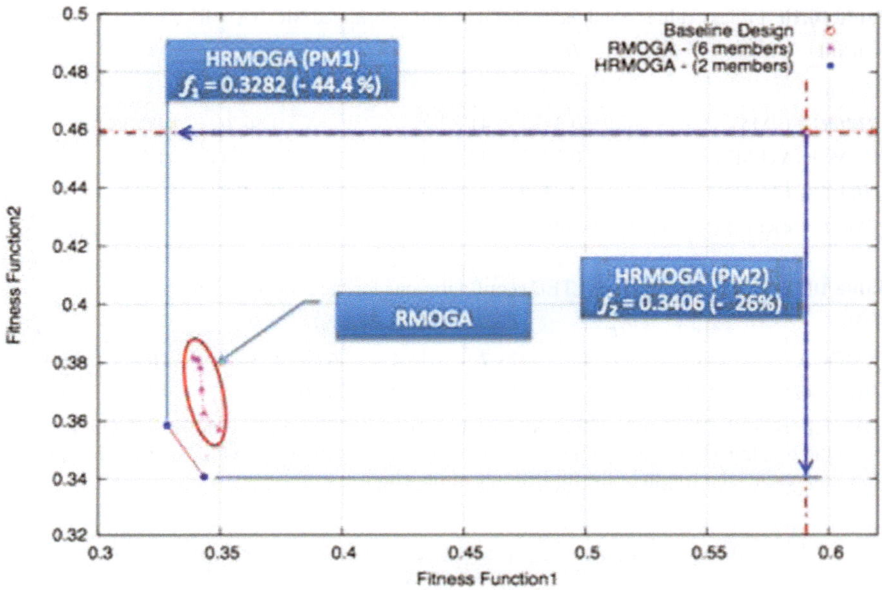

Fig. 10.15 Optimal Pareto fronts obtained by MOGA and HMOGA software

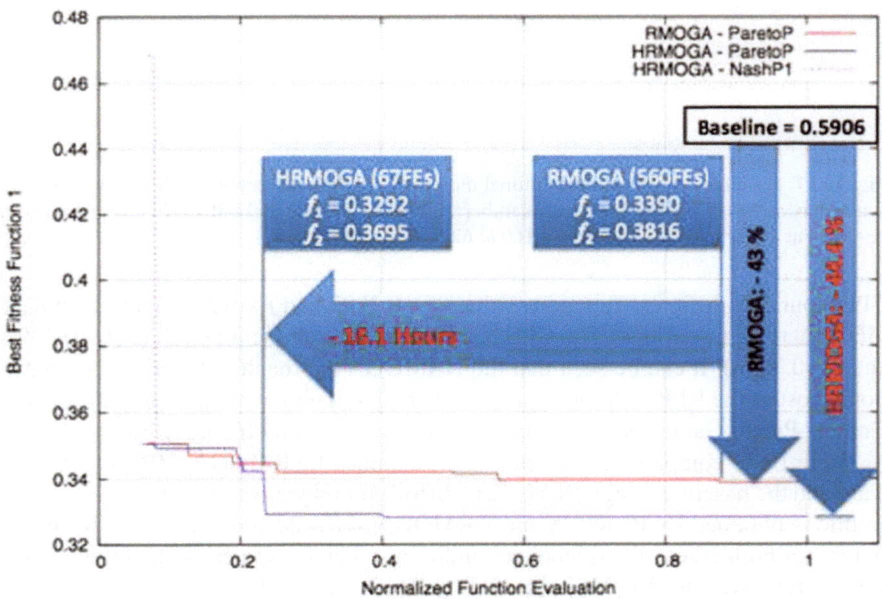

Fig. 10.16 Convergence history (f_1 vs. normalised function evaluation) obtained by RMOGA and HRMOGA software

Table 10.10 Fitness values obtained by MOGA and hybridized MOGA software

Airfoil	$1/Cl_{Take\ Off}$	$1/Cl_{Landing}$
Baseline	0.5906	0.4589
RMOGA (PM5)	0.3433 (−41.8 %)	0.3626 (−21.0 %)
HRMOGA (PM1)	0.3282 (−44.4 %)	0.3584 (−22.0 %)
HRMOGA (PM2)	*0.3434 (−41.8 %)*	*0.3406 (−25.7 %)*

PM *i* represents the *i*th Pareto member

Table 10.11 Optimal morphing (LTED) configurations for the robust design optimization

DVs	x_{LE}	θ_{LE}	L_{BSC1}	L_{BSC2}	x_{TE}	θ_{TE}	L_{BSC3}	L_{BSC4}
RMOGA	23.7	−22.4°	23.7	11.4	69.3	34.3°	26.2	15.4
HRMOGA	21.9	−20.6°	12.3	21.4	66.0	34.4°	20.2	13.6

DVs represents leading and trailing edge design variables. X_{LE}, X_{TE} are expressed with respect to in the baseline chord length (%) [0 %:100 %] and deformation angle follows right-hand rule. L_{BSCi} represents the length of the *i*th BSC (% of the chord)

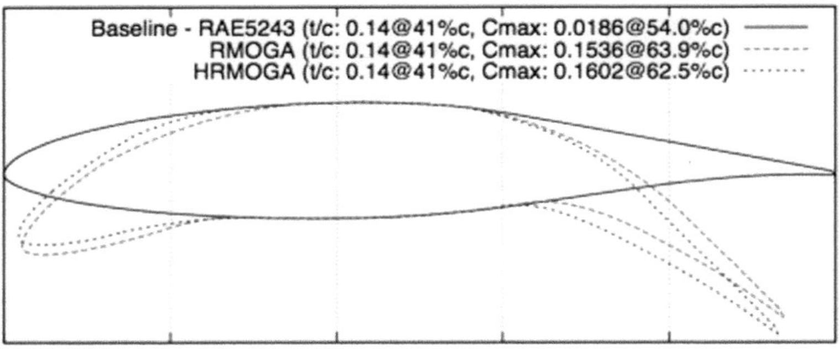

Fig. 10.17 Baseline design with the optimal morphing configurations obtained by RMOGA software (max $t/c = 0.14$ at 41 %c and max camber = 0.153 at 63.9 %c) and HRMOGA software (max $t/c = 0.14$ at 41 %c and max camber = 0.160 at 62.5 % of the chord)

CPU hours) with fitness functions values $f_1 = 0.3292$ and $f_2 = 0.3695$) and RMOGA (560 function evaluations (22.0 CPU hours) with fitness functions values: $f_1 = 0.3390$ and $f_2 = 0.3816$). It can be seen that the HRMOGA approach reduces the computational cost of the RMOGA run by 16.1 CPU hours. In other words, applying a Nash Game—Pareto Game coalition improves the optimization efficiency by 73 %!

Table 10.10 compares the fitness values obtained by RMOGA, HRMOGA software and the baseline design (RAE 5243 airfoil) for Objectives 1 and 2. The optimal solutions obtained by RMOGA and HRMOGA software provide higher lift coefficient for both take-off and landing conditions when compared to the baseline design. Pareto Member 5 (PM5) obtained by RMOGA, and Pareto Member 2 (PM2) obtained by HRMOGA are selected for further investigation.

The optimal morphing configurations obtained by RMOGA and HRMOGA optimization procedures are described in Table 10.11. Figure 10.17 compares the

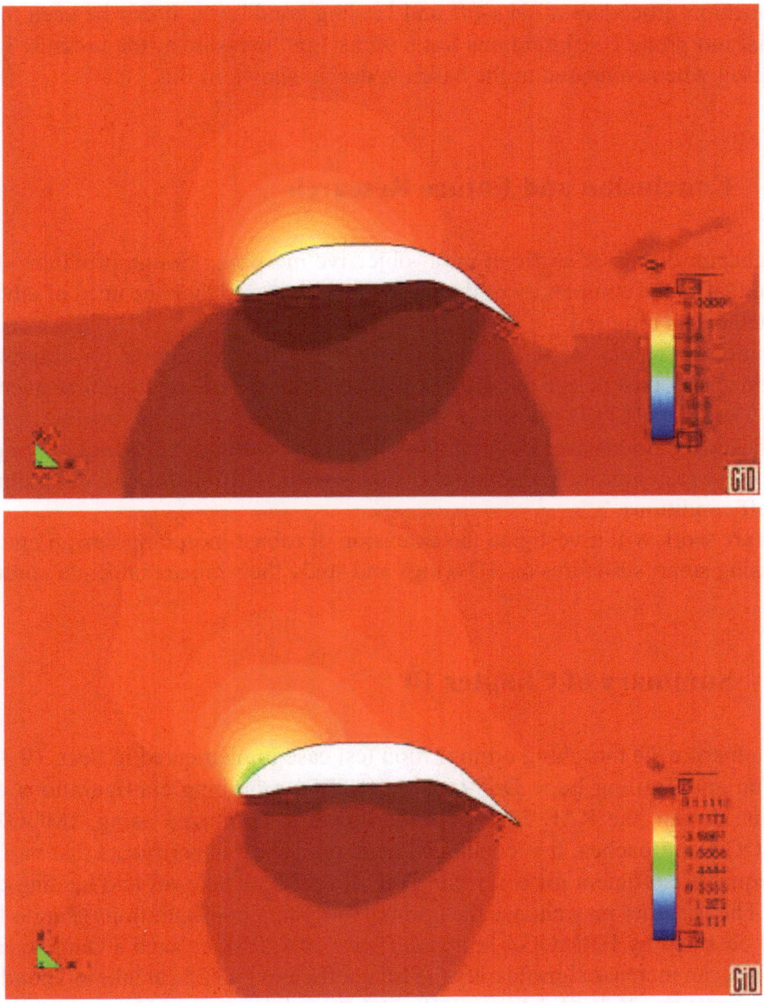

Fig. 10.18 *Cp* contours obtained by Pareto member 2 (PM2) at the take-off (*top*)—$Cp_{Take\ Off}$ range [−14:1.3] and landing (*bottom*)—$Cp_{Landing}$ range [−15:2] conditions

geometry of the baseline design and the baseline coupled with the optimal morphing configurations obtained by RMOGA and HRMOGA software. Optimal solutions have the same maximum thickness ratio (*t/c*) of 0.14 at 41 % of the chord as the baseline design while increasing the maximum camber; with RMOGA approach (max camber = 0.154 at 64.0 %c and c = 0.919) and with HRMOGA approach (max camber = 0. 160 at 62.5 %c and c = 0.920). Both optimal solutions have higher camber by 14 % of the chord and lower chord length by 8.0 %.

Figure 10.18 compares pressure contours obtained by the baseline design with the optimal morphing configurations (Pareto Member 2) obtained by the HRMOGA optimization procedure at take-off and landing conditions. It can be seen that the optimal morphing configurations has a wider high pressure at the pressure side of the airfoil when compared to the baseline design shown in Fig. 10.14.

10.5 Conclusion and Future Research

This chapter discussed difficult multi-objective morphing design problems. It was shown that as the complexity of the problem is increased then the need of advanced EA methods is required.

Numerical results of this section show benefits of coupling an optimization method with robust design techniques to produce stable and high performance solutions. The use of LTED-morphing techniques on an existing 2D airfoil can reduce significantly the transonic drag which will save operating and manufacturing cost as well as CO2 emission reduction and also maximize lift in the case of landing and take off conditions.

Future work will investigate the extension of robust morphing design optimization using smart structures on 3D wings and study their impact on flight control.

10.6 Summary of Chapter 10

To summarize the morphing optimization test case implemented in Sect. 10.4.2 via Leading and Trailing Edge Deformation (LTED) morphing configurations on the baseline design, the RAE 5243 baseline airfoil is optimized using RMOGA and HRMOGA approaches. It is concluded from numerical experiments that morphing techniques are efficient for controlling the flow at both take-off and landing conditions. The designer may choose the optimal morphing configurations (Pareto Member 2) obtained by HRMOGA software (Sect. 10.4.2) since such a Game coalition approach can increase significantly the lift coefficient by 42 % at take-off conditions while improving 26 % of the lift coefficient at landing conditions.

References

1. Friswell MI, Inman D (2006) Morphing concepts for UAVs. 21st Bristol UAV systems conference
2. Lee DS, Bugeda G, Periaux J, Onate E (2011) Robust active shock control bump design optimisation using parallel hybrid-game. In: 23rd international conference on parallel computational fluid dynamics, Barcelona Spain
3. GID Manual CIMNE. http://gid.cimne.upc.edu/support/manuals)

Appendix: Two "Hands-On" Examples of Optimization Problems

A.1 Overview

In this Appendix the numerical implementation of multi-objective problems of increasing complexity using a step-by-step optimization procedure is described in detail. However it is highly suggested that the reader consult Chaps. 6–10 before attempting to implement the hands-on examples. More detailed results of these "Hands-on" test cases can be obtained by contacting the authors (felipe.gonzalez@qut.edu.au, jperiaux@gmail.com, ds.chris.lee@gmail.com).

A.2 Reconstruction of BI-NACA Using HAPEA and NASH-EA

A.2.1 Introduction

In this hands-on example, two different evolutionary optimisation methods described in previous chapters are investigated; the first method is an EA with multi-population/multi-fidelity Pareto optimality; Hierarchical Asynchronous Parallel Evolutionary Algorithm (HAPEA mentioned in Chap. 2). The second method is Nash strategy, named Nash-EA and described in Chap. 2. These optimisation methods are coupled to a pre/post processor GiD and aerodynamic analysis software PUMI for the reconstruction/inverse aerodynamic design problem.

A.2.2 Definition of the Test Case

Computational Domain The baseline design is shown in Fig. A.1 where two NACA0012 aerofoils are placed at a distance of 50 % of chord length of NACA0012.

J. Periaux et al., *Evolutionary Optimization and Game Strategies for Advanced Multi-Disciplinary Design,* Intelligent Systems, Control and Automation: Science and Engineering 75, DOI 10.1007/978-94-017-9520-3

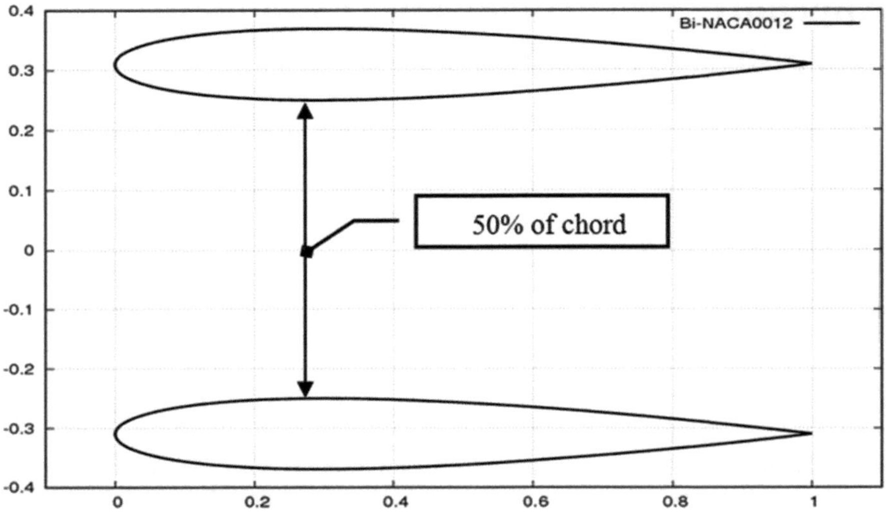

Fig. A.1 Baseline design (Bi-NACA0012)

This reconstruction design is a single-objective problem to capture the shape of the baseline design (BI-NACA0012).

Figure A.2 shows the unstructured mesh around BI-NACA0012 with 7785 nodes and 14,884 triangles. The mesh generation takes only 9 s.

Mathematical Modelling The chord length of each NACA0012 aerofoil is one i.e. $c=1$ m. The outer boundary of the computational domain size is 10 m.

Boundary Conditions Upstream entrance: $v_x = v_\infty \cos(\alpha)$, $v_y = v_\infty \sin(\alpha)$, angle of attack $\alpha = 6.0°$

Downstream exit: free boundary conditions

Aerofoil surface: no-slip condition

Flow Conditions The baseline design is tested at the following flow conditions; Mach (M_∞)=0.3, angle of attack (α)=6.0° and Reynolds number (Re)=1.0E+6. The flow field is solved by PUMI software (see Sect. A.2.4) using an unstructured finite Euler analyser.

Figure A.3 shows the coefficient pressure (Cp) distribution of the baseline design and the Mach contour is shown in Fig. A.4. They are automatically plotted by GiD.

A.2.3 Optimization

The objective is to minimize the pressure deviation between the baseline configuration (BI-NACA0012) and the geometries of candidates. Both the computational efficiency and quality design obtained by HAPEA and Nash-EA are compared.

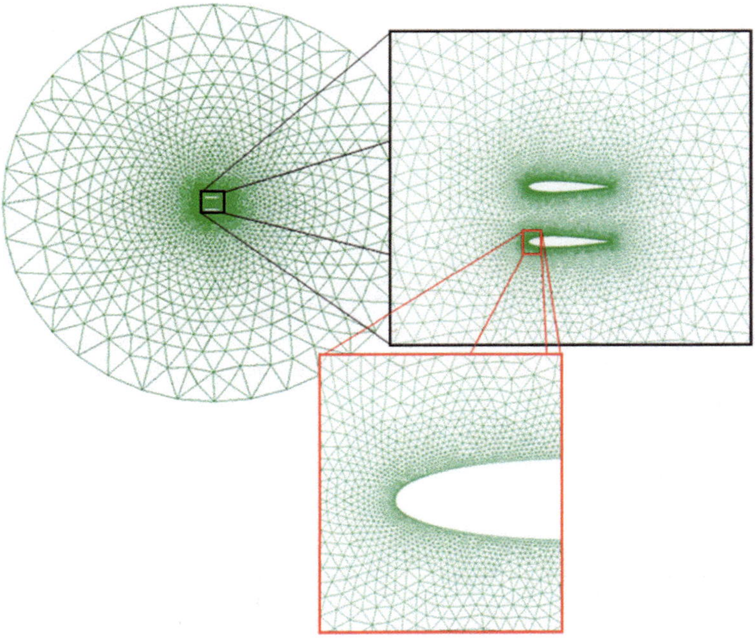

Fig. A.2 Computational domain and unstructured mesh around a BINACA 0012 obtained by GiD

Objective Functions The optimization considers the minimization of pressure deviation between target (BI-NACA0012) and candidate geometry using HAPEA software and Nash-EA software (see A.2.4). Two different types of fitness functions are used:

HAPEA:

When using HAPEA software (see A.2.4), the fitness function is:

$$f = \min\left(P_{Deviation}\right),\ P_{Deviation} = \sum_{i=1}^{n}\left(P_{T_i} - P_{C_i}\right)^2 + \sum_{j=1}^{m}\left(P_{T_j} - P_{C_j}\right)^2 \quad \text{(A.1)}$$

where n (and respectively—is the number of pressure positions for upper and lower aerofoils respectively. P_T is the target pressure coefficient distribution while P_C is the candidate pressure coefficient distribution.

NASH:

The optimisation problem consists of minimizing the pressure deviation between BI-NACA0012 and candidate using Nash-EA with two Nash-Players (NP1, NP2). The fitness function defined in (A.1) is now split into (A.2) for Nash Player 1) and (A.3) for Nash Player 2:

$$f_{NP1} = \min\left(P_{Deviation}\right),\ P_{Deviation} = \sum_{i=1}^{n}\left(P_{T_i} - P_{C_i}\right)^2 + \sum_{j=1}^{m}\left(P_{T_j} - P_{C_j}^{*}\right)^2 \quad \text{(A.2)}$$

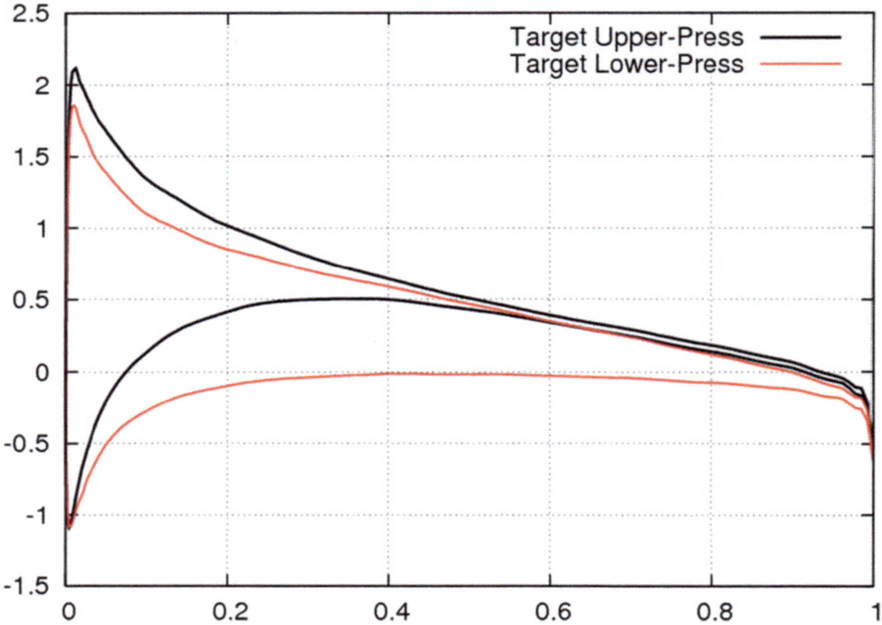

Fig. A.3 The target *Cp* distribution obtained by PUMI software

Fig. A.4 Target mach contours computed with PUMI software

$$f_{NP2} = \min\left(P_{Deviation}\right),\ P_{Deviation} = \sum_{i=1}^{n}\left(P_{T_i} - P^{*}_{C_i}\right)^2 + \sum_{j=1}^{m}\left(P_{T_j} - P_{C_j}\right)^2 \quad \text{(A.3)}$$

where n and m represent the number of points considered on the upper and lower aerofoils respectively. P_T is the target Cp distribution while P^{*}_{Ci} and P^{*}_{Cj} are the Cp distributions obtained on elite airfoils from Nash-Player 1 and Nash-Player 2 respectively.

Design Variables/Encoding Design bounds for the reconstruction inverse problem are shown in Fig. A.5 where the normal triangles represent the upper thickness bounds (30 % of the baseline) while the inverse triangles are for lower thickness bounds (5 % of the baseline). The mean camber line design bounds are marked by upper half-filled and lower half-filled circles. The airfoil geometry is represented using Bezier curves with a combination of mean line and thickness distribution. Each airfoil contains eighteen (18) mean and thickness control points therefore, a total of thirty six (36) design variables is considered for the reconstruction of this BI-NACA0012 baseline airfoil.

Constraints Geometry constraints are shown on Table A.1 and Table A.2.

Definition of the EA Strategy Two optimization approaches are implemented: the first approach uses the HAPEA software using three (3) layers with seven (7) different populations. Each node (Node 0 ~ Node 6) in HAPEA considers all of the airfoil

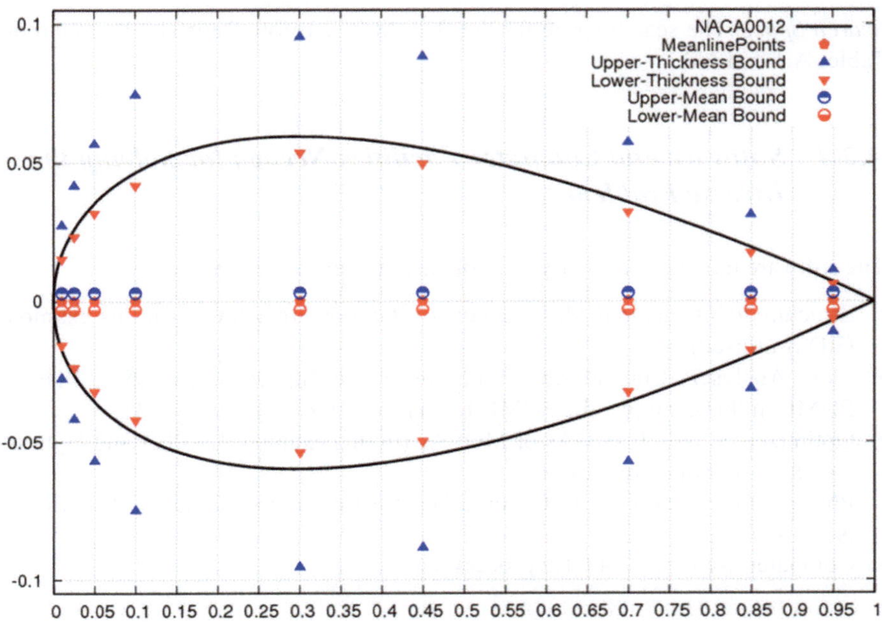

Fig. A.5 Design bounds for NACA0012

Table A.1 Mean camber line control points for upper and lower NACA0012 aerofoil (Remark: Mx represents a set of fixed mean camber *x*-coordinates while LMy and UMy are lower and upper bounds of mean camber *y*-coordinates)

Mx	0.01	0.025	0.05	0.10	0.30	0.45	0.70	0.85	0.95
LMy	−0.01	−0.01	−0.01	−0.01	−0.01	−0.01	−0.01	−0.01	−0.01
UMy	0.01	0.01	0.01	0.01	0.01	0.01	0.01	0.01	0.01

Table A.2 Thickness control points for upper and lower NACA0012 aerofoil (Note: Tx represents a set of fixed thickness *x*-coordinates. LTy and UTy are lower and upper bounds of thickness *y*-coordinates.)

Tx	0.01	0.025	0.05	0.10	0.30	0.45	0.70	0.85	0.95
LTy	0.0238	0.021	0.0396	0.0554	0.0420	0.0490	0.0256	0.0143	0.0056
UTy	0.0472	0.055	0.0569	0.0748	0.0960	0.0892	0.0586	0.0328	0.0129

design variables made of upper and lower airfoils with different time step sizes of the aerodynamic PUMI analysis tool (see Sect. A.2.4). The second approach uses the Hybridized EA using one Global-Player and two Nash-Players. The Global-Player (GP) optimises upper and lower airfoil shapes (with thirty six (36) design variables) to minimise the pressure deviation between the target (shown in (A.1) and Fig. A.1) while Nash-Player 1 (NP1) and Nash-Player 2 (NP2) minimize the pressure deviation by optimising both the upper airfoil geometry (NP1: with eighteen (18) design variables) and the lower airfoil shape (NP2: with eighteen (18) design variables).

Search Space The search space for the Bi-NACA aerofoil shape can be found on Tables A.1 and A.2.

A.2.4 Software and Computer Facilities Needed for Solving the Inverse Problem

The software modules necessary to implement the above test case are:

- Structured/Unstructured Mesh Generator: GiD or FreeFem++ (in this example GiD is utilised) [1].
- Flow Analyser: Compressible Euler or Navier-Stokes solver—NSC2KE or PUMI (in this example, the PUMI (Euler) software is utilised) [2, 3].
- Optimizer: Simple EAs/GAs or MOEAs (in this example HAPEA and Hybridized EAs software are utilised).
- Pre/Post-Processor: GNUPLOT or GiD software (in this example GiD software is utilised) [4, 1].
- Computer used: two 2.4 GHz processors.

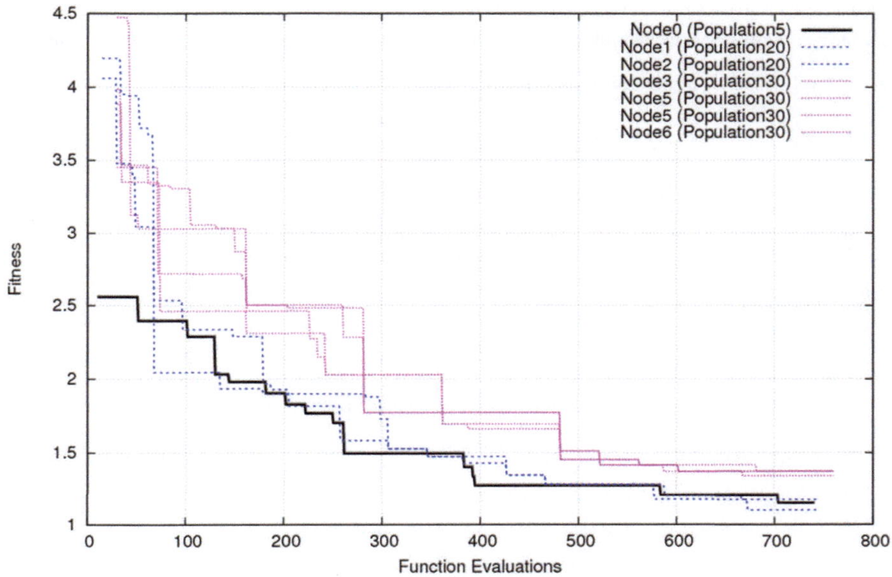

Fig. A.6 Sequential procedure of the HAPEA algorithm

A.2.5 A Step by Step Design Optimisation Procedure

With HAPEA The PUMI solver is utilised and the following specific parameters are considered for the evolutionary optimiser using a hierarchical topology (HAPEA) as shown in Fig. A.6.

1st Layer	Population size of 5 and intermediate recombination with tolerance residual 1×10^{-6} of the PUMI solver
2nd Layer	Population size of 20 and intermediate recombination with tolerance residual 1×10^{-5} of the PUMI solver
3rd Layer	Population size of 30 and intermediate recombination with tolerance residual 1×10^{-4} of the PUMI solver

With NASH This reconstruction design is a single-objective problem to capture the shape of the baseline design (BI-NACA0012) using Hybridized Games as shown in Algorithm 3 introduced in Sect. 6.5.4 and Fig. A.7.

The PUMI solver is utilized and the following specific parameters are considered for the evolutionary optimizer with two Nash-Players.

Nash Player 1: Population size of 20 (Upper airfoil) with residual 1×10^{-6}.

Nash Player 2: Population size of 20 (Lower airfoil) with residual 1×10^{-6}.

Define design variables *x*, parameters *p*, and constraints *g*: $\mu(x,p,g)$

Define number of subpopulations (nodes) of hierarchical levels *i* and integrated analysis *k*: μi with *Analysisk*

for all levels initialise subpopulations $\left(\mu_0^0, \mu_1^0, \mu_2^0, \ldots, \mu_6^0\right)$

Layer 1: Uses Type 1 integrated analysis: $init\left(\mu_0^0\right)$: $Analysis_1$

Layer 2: Uses Type 2 integrated analysis: $init\left(\mu_1^0, \mu_2^0\right)$: $Analysis_2$

Layer 3: Uses Type 3 integrated analysis: $init\left(\mu_3^0, \mu_4^0, \mu_5^0, \mu_6^0\right)$: $Analysis_3$

loop

while stopping condition not met,

Update the asynchronous solver.

if solved individuals are available for $\left(\mu_0, \mu_1, \mu_2, \ldots, \mu_6\right)$: incorporate them.

A new individual is received: *Inew*

Age population: For each μi in μ: $\mu' i\_age = \mu i\_age + 1$

Age selection buffer: For each *bi* in B: $b'i\_age = bi\_age + 1$

Perform tournament selection: *tournament* (*Inew* , *b*)

if tournament accept = true

Perform replacement in μ: *replace* (μ, *Inew*)

Delete dominated and oldest *bi* in selection buffer, insert *Inew*

Sort μ

endif

endif

Fig. A.7 Sequential procedure of the Nash-Game algorithm

A.2.6 Analysis and Synthesis of Results Obtained by HAPEA and Nash-EAs Software

The computational CPU cost in terms of function evaluations run with HAPEA and Nash-EAs is compared on two 2.4 GHz processors.

Results obtained with HAPEA The HAPEA algorithm was allowed to run approximately 740 function evaluations and was stopped after 50 h of CPU as shown in Fig. A.8. The converged fitness value 1.1556 obtained by Node 0 (f_{Node0}=1.1556) in HAPEA indicates a difference of 0.77 % between the optimal solution and the target.

Figure A.9a compares the geometry of the baseline design to the optimal airfoils obtained by HAPEA and Mach contours are illustrated in Fig. A.9b; *Cp* distribution comparisons for upper and lower airfoils are shown in Fig. A.10. A good agreement is observed in the shapes and Mach numbers distribution between target and final geometries.

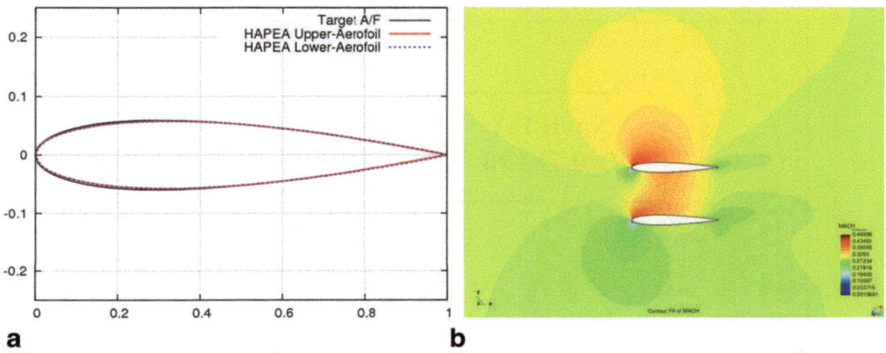

Fig. A.8 Convergence history obtained by the HAPEA software

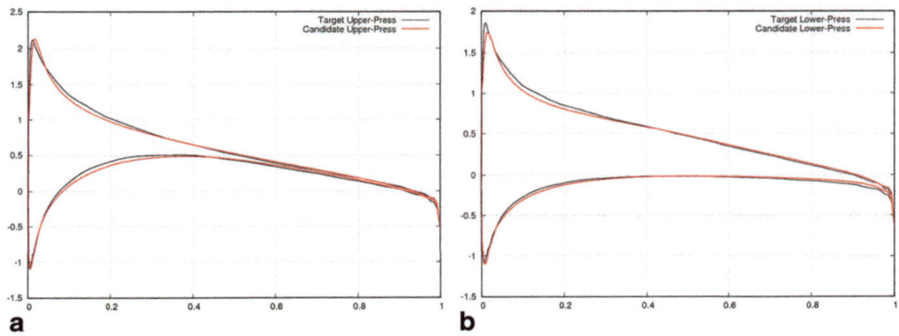

Fig. A.9 Comparison of airfoil shape (**a**) and Mach contours (**b**) obtained by HAPEA software

Results obtained with NASH-EAs The Nash-EA algorithm was allowed to run approximately 2400 function evaluations and was stopped after 50 h CPU. Figure A.11 compares the optimization progress achieved in generations with HAPEAs (Node 0) and Nash-EA. Even though the initial guess (f=3.32172) of Nash-EAs marked as a red circle is 23 % further when compared to the initial guess (f=2.55915) of HAPEA, Nash-EA reached the HAPEA converged fitness value (f=*1*.1556) after only 21.25 h and a better solution after 50 h. In addition, the converged fitness value (f=0.78196) of Nash-EA indicates 0.43 % difference between the optimal airfoils and BI-NACA0012 configuration. The solution obtained by Nash-EAs is 32 % accurately closer to the baseline design when compared to the HAPEAs optimal solution.

Figure A.12a compares the airfoil geometries between the baseline design and an optimal solution from Nash-EAs; Mach contours are shown in Fig. A.12b. A good *Cp* agreement can be observed with the baseline design as shown in Fig. A.13.

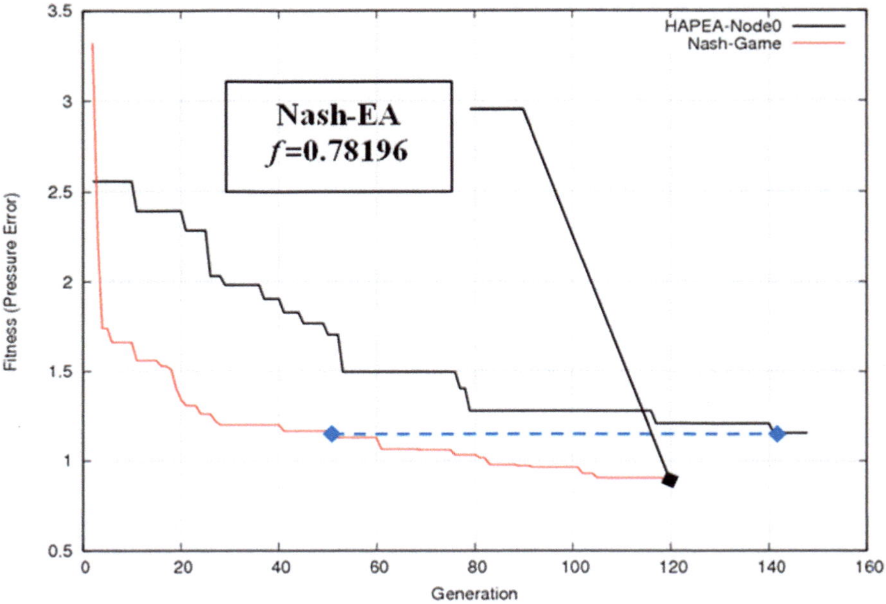

Fig. A.10 Comparison of pressure distributions (*upper airfoil: left*) and (*lower airfoil: right*)

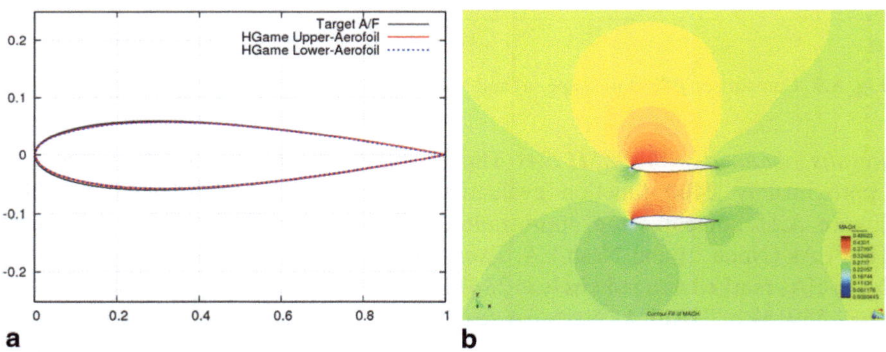

Fig. A.11 Optimization convergence speed-up obtained by Nash-EAs

A.2.7 Conclusion

The numerical results for the shape reconstruction of a BI-NACA0012 configuration clearly shows the benefits of coupling an EAs with Nash Games and illustrate the potential of the above evolutionary optimization methodology making this approach promising for solving the second complex "hands-on" test case optimization problems.

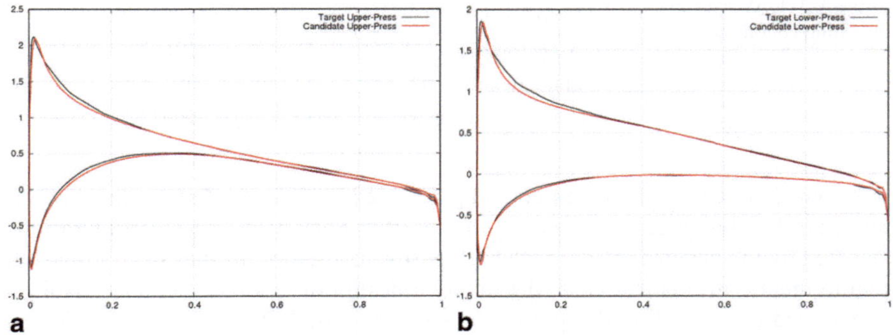

Fig. A.12 Comparison of airfoil shape (**a**) and Mach contours (**b**)

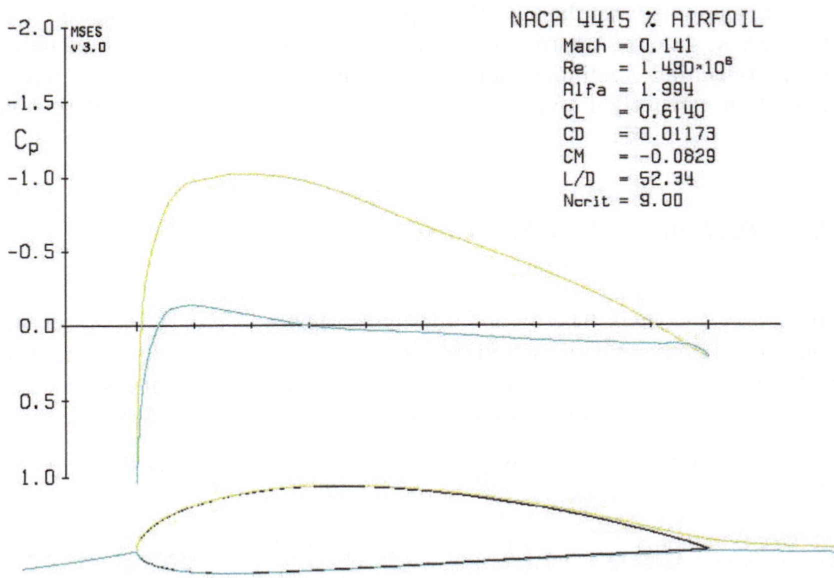

Fig. A.13 Comparison of pressure distributions (*upper: left*) and (*lower: right*)

A.3 Unmanned Aerial Vehicle Design: Multi-Objective Optimization

A.3.1 Introduction

The use and development of Unmanned Aerial Vehicles for military and civilian applications are rapidly increasing despite many difficulties in the design of these vehicles because of the varied and non-intuitive nature of the configurations and

Table A.3 UAV data and operating conditions

Airfoil section	NACA4415
Wingspan b, ft	12.8
Chord (approx), ft	2.0
Cruising altitude, m	3000

missions that can be implemented and performed. Similar to their manned counterparts, the challenge is to develop trade-off studies of optimal configurations selected by the designer to produce high performance aircraft that satisfy mission requirements. In this test case, the detailed design of a single element airfoil is considered for a small UAV application similar to the RQ-7A Shadow 200 Tactical UAV and use the airfoil design module for this task. The aircraft maximum gross weight is approximately 320 lbs, it has a wingspan of approximately 12.8 ft, a mean chord of approximately 2 ft, length of 11 ft, and a plan form shape without sweep. It is assumed that the aircraft is operating between a slow cruise 33.3 m/s and fast cruise 46.6 m/s approximately. This results in the airframe, flight parameters and operating conditions indicated on Table A.3. These conditions assume an aircraft at mid-weight cruise during an extended cruise phase at intermediate altitude.

A.3.2 Definition of the Test Case

Computational Domain The baseline design is shown in Fig. A.14. An existing airfoil geometry operating at two suggested design points is initially assumed, and then a design and airfoil that preserves the original thickness while reducing the drag coefficient. The assumed baseline geometry is the NACA4415 airfoil.

Mathematical Modelling The chord length of the NACA4415 airfoil is one i.e. $c=1$ m. Its airfoil thickness is 15 %.

Boundary Conditions Upstream entrance: $v_x = v_\infty \cos(\alpha)$, $v_y = v_\infty \sin(\alpha)$,

Downstream exit: free boundary conditions

Airfoil surface: no-slip condition

Flow Conditions The baseline design was tested at the following two flow conditions:

- Flight Condition One—*Slow*: Cruise Mach (M∞)=0.1025, Cl=1.18 and Reynolds number (Re)=1.085 × 10**6;
- Flight Condition Two—*Fast*: Cruise Mach (M∞)=0.141, Cl=0.6140 and Reynolds number (Re)=1.490 × 10**6. The flow field is solved by MSES software (using structured finite Euler analyser.

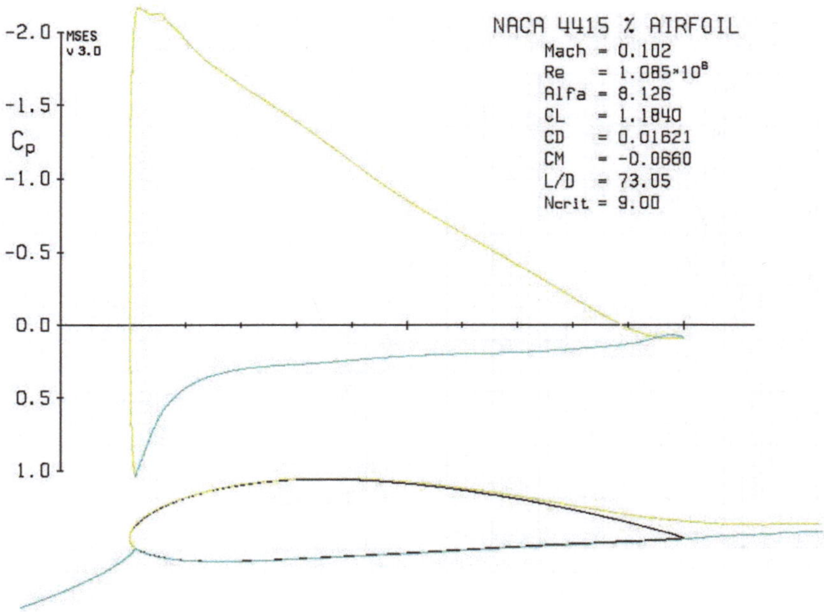

Fig. A.14 *Cp* distribution obtained by MSES for the NACA 4415 airfoil—Flight Condition Two—*Fast* Cruise

Figures A.15 and A.16 show the pressure coefficient (Cp) distribution and some aerodynamic data for the two flight conditions considered. The combined polars for the NACA4415 aerofoil are shown in Fig. A.10. It is noticed that both cruise points operate inside the invariant drag region of the airfoil; the low speed cruise condition giving approximately Cd=0.016 and the high speed giving approximately Cd=0.012.

A.3.3 Optimization

The objective is to minimize the drag at two flight conditions. Both the computational efficiency and quality design obtained by HAPMOEAs software are discussed and compared.

Objective Functions The two fitness functions to be optimised are defined as minimisation of drag (cd) at the two flight conditions.

$$\min(f_1) = c_d \quad \mathrm{Re} = 1.085\times10^6,\ \mathrm{C_l} = 1.184 \tag{A.3}$$

$$\min(f_2) = c_d \quad \mathrm{Re} = 1.490\times10^6,\ \mathrm{C_l} = 0.6140 \tag{A.4}$$

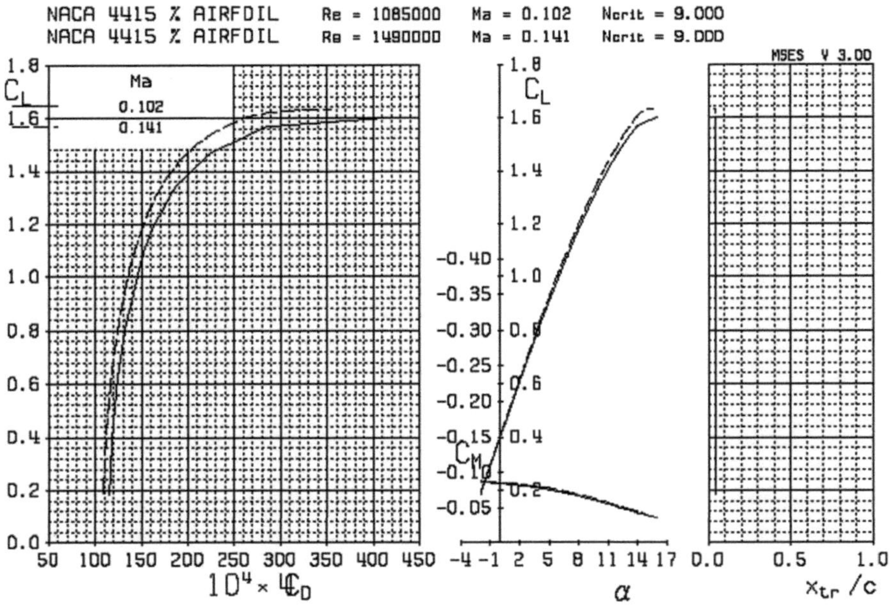

Fig. A.15 *Cp* distribution obtained by MSES for the NACA 4415 airfoil—Flight Condition One—*Slow* Cruise

Design Variables/Encoding The airfoil geometry is represented by two Bezier curves, one for the mean line and one for the thickness distribution. The mean line-thickness distribution is a standard method for representing airfoil geometries, as it closely couples the representation with the results; the mean line has a powerful effect on cruise lift coefficient and pitching moment, while the thickness distribution has a powerful effect on the cruise drag. In this case six (6) free control points are considered on the mean line and ten (10) free control points on the thickness distribution. An individual is defined as an object containing a vector of sixteen (16) problem variables.

Constraints There are three types of constraints: maximum thickness, maximum thickness location and pitching moment (cm). The thickness and maximum thickness location of each aerofoil must exceed 15 % ($t/c \geq 0.15$) and be between 20 and 40 % chord, respectively. If a constraint on pitching moment is applied this must not be more severe than −0.0660 (cm ≥ −0.0660). When all constraints are considered they are added up and applied by equally penalizing both fitness values via a linear penalty method.

In designing a replacement airfoil for this UAV platform, the following design factors are considered:

- Maintain approximately the same Cl in order not to impinge upon the assisted launch and landing length;

Define design variables x, parameters p, and constraints g: μ(x,p,g)

Define number of subpopulations (nodes) of hierarchical levels i and integrated analysis k: μ_i with $Analysis_k$

for all levels initialise subpopulations $\left(\mu_0^0, \mu_1^0, \mu_2^0, \ldots, \mu_6^0\right)$

Layer 1: Uses Type 1 integrated analysis: $init\left(\mu_0^0\right): Analysis_1$

Layer 2: Uses Type 2 integrated analysis: $init\left(\mu_1^0, \mu_2^0\right): Analysis_2$

Layer 3: Uses Type 3 integrated analysis: $init\left(\mu_3^0, \mu_4^0, \mu_5^0, \mu_6^0\right): Analysis_3$

loop

while stopping condition not met,

Update the asynchronous solver.

if solved individuals are available for $\left(\mu_0, \mu_1, \mu_2, \ldots, \mu_6\right)$: incorporate them.

A new individual is received: I_{new}

Age population: For each μ_i in μ: $\mu'_{i\_age} = \mu_{i\_age} + 1$

Age selection buffer: For each b_i in B: $b'_{i\_age} = b_{i\_age} + 1$

Perform tournament selection: *tournament* (I_{new}, b)

if tournament accept = true

Perform replacement in μ: *replace* (μ, I_{new})

Delete dominated and oldest b_i in selection buffer, insert I_{new}

Sort μ

endif

endif

Update CMA parameters (ref alg 5)

if new individuals can be provided: generate them.

$\boldsymbol{x}_{new}$, $dist(\rho)$)=$recombine(\mu)$

Mutate offspring via CMA:

for $i = 1, \ldots N$,

$z_i = N(0,1)$

loop

$\boldsymbol{x}' = \boldsymbol{x} + BD\boldsymbol{z}$

Store $\boldsymbol{z}$ in $\boldsymbol{I}_{new}$

endif

Evaluate candidate using specific integrated analysis type:

$$\lambda^{g+1} = f\left(\lambda_M^{g+1}\right): Analysis_i$$

Get output analysis a_i, parameters p, check constraints g and add penalty:

$$\lambda^{g+1} = f\left(\lambda_M^{g+1}\right): Analysis_i + Penalty$$

if epoch completed:

Start migration: $\mu_i^{g_k} = mig\left(\mu_i^{g_k} \rightarrow \mu_{i\pm1}^{g_k}\right): Analysis_i$

Fig. A.16 NACA 4415—Polar

Layer 1: Receive best solutions from layer 2 re-evaluate using Type 1 - Analysis:

$$\left(\mu_1^{g_k}, \mu_2^{g_k} \rightarrow \mu_0^{g_k}\right), \left(\mu_o^{g_k} = f\left(x_{1,2}^{*}\right)\right): Analysis_1$$

Layer 2: Receive random solutions from layer 1 and best from layer 3 re-evaluate them using Type 2 - Analysis:

$$\left(\mu_0^{g_k} \rightarrow \mu_{1,2}^{g_k}, \mu_{3,45,6}^{g_k} \rightarrow \mu_{1,2}^{g_k}\right), \left(\mu_{1,2}^{g_k} = f\left(x_0^{*}\right), \mu_{1,2}^{g_k} = f\left(x_{3,4,5,6}^{*}\right)\right): Analysis_2$$

Layer 3: Receive random solutions from layer 2 re-evaluates them using Type 3 - Analysis:

$$\left(\mu_{1,2}^{g_k} \rightarrow \mu_{3,45,6}^{g_k}\right), \left(\mu_{3,4,5,6}^{g_k} = f\left(x_{1,2}^{*}\right)\right): Analysis_3$$

loop

Fig. A.16 (continued)

- Maintain at least the current thickness, in order not to increase the weight of the wing;
- Lower the drag at both cruise points, in a multi-objective approach.

Definition of the EA Strategy The complexity, non-linearity and multi-objective characteristics of this problem make it well suited to be solved with an EAs optimizer. The computational CPU cost for this problem is an important consideration due to the wide upper and lower bounds in the search space and depends on the computing facilities used, in particular in industrial design environments. Therefore it is also worthy to use both parallel computing and multi-fidelity approaches. In this case it is desired to use a multi-objective parallel EAs (MOPEA) and select the HAPEAs approach which has all these capabilities but other EA approaches would perform well also like NSGAII or VEGA [5] for example.

A.3.4 Software Needed for Solving the Optimization Problem and Computer Used

The following software modules necessary to implement the test case are:

- The aerodynamic characteristics of the candidate aerofoils are evaluated using the MSES [6] software. This solver is based on a structured quadrilateral streamline mesh which is coupled to an integral boundary layer based on a multi-layer velocity profile representation. Details on MSES can be found in Drela [11].

Fig. A.17 Sequential procedure of the HAPEAs algorithm

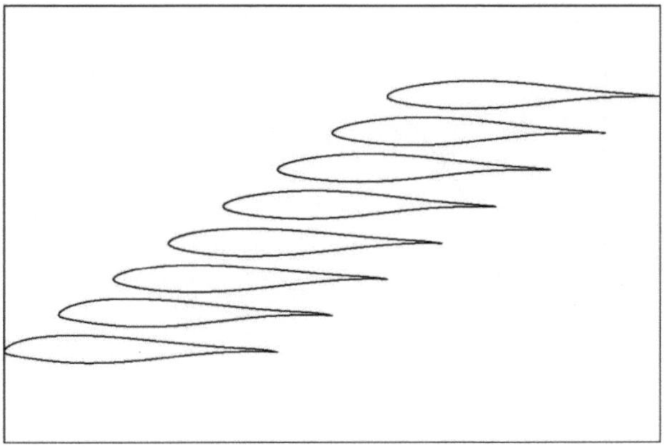

A.3.5 *A Step-by-Step Design Optimization Procedure*

With HAPMOEA The MSES solver is utilized and the following specific parameters are considered for the evolutionary optimizer using a hierarchical topology (HAPMOEA) as shown in Fig. A.17.

1st Layer	Population size of 10 and intermediate recombination with tolerance residual 1×10^{-6} of the MSES solver
2nd Layer	Population size of 15 and intermediate recombination with tolerance residual 1×10^{-5} of the MSES solver
3rd Layer	Population size of 15 and intermediate recombination with tolerance residual 1×10^{-4} of the MSES solver

A.3.6 *Analysis and synthesis of results obtained by HAPMOEA software*

This test case was allowed to run 2000 function evaluations on the top level and took approximately 8 h on a cluster of four 2.8 GHZ processors. Figure A.18 shows the set of airfoils in the Pareto front. From this front, three airfoils are selected: objective one optimal, objective two optimal and compromised airfoils from the middle of the front. These geometries are shown versus the NACA 4415 airfoil baseline in Fig. A.19.

A compromised airfoil is considered from the middle of the Pareto front for further evaluation. Figures A.20 and A.21 show the *Cp* distribution for the two flight conditions. Figures A.22 and A.23 show the comparative drag polars for Re = 1.480 × 106 that for Re = 1.085 × 10 6. But in this case a lower cm is obtained

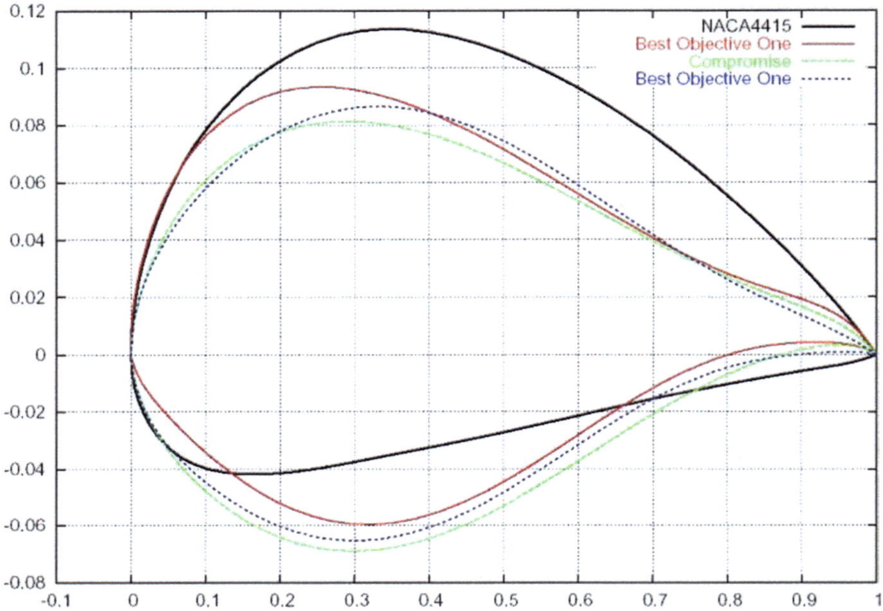

Fig. A.18 The set of Pareto Airfoils—Cm constrained

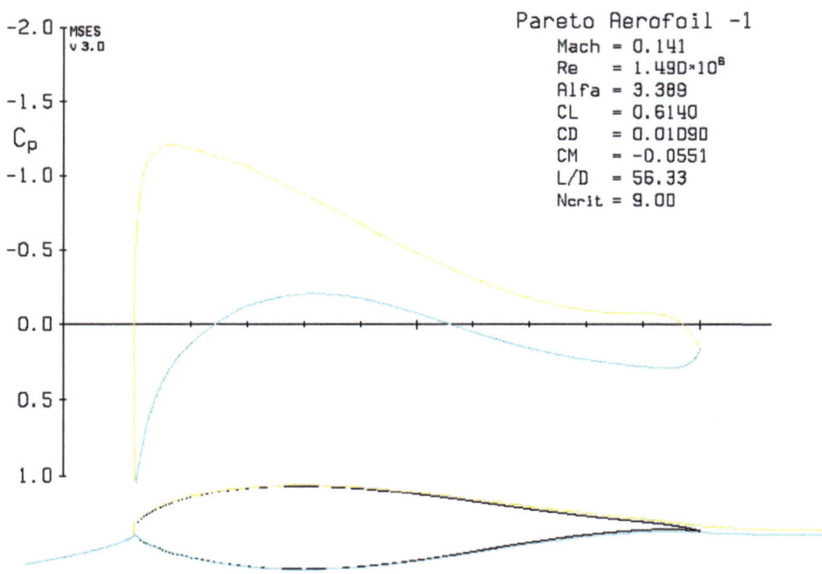

Fig. A.19 Comparison of selected geometries—Cm constrained

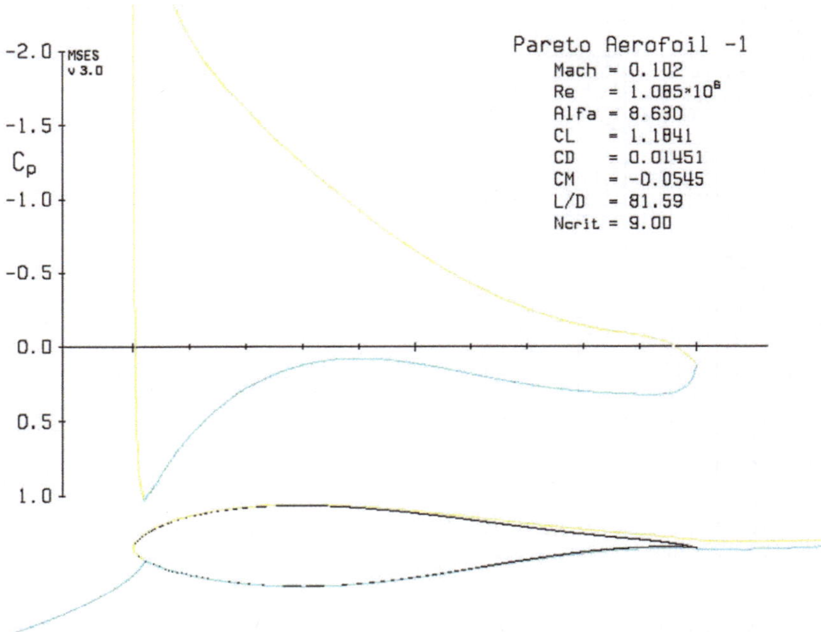

Fig. A.20 Pareto member 01 flight condition one—Cm constrained

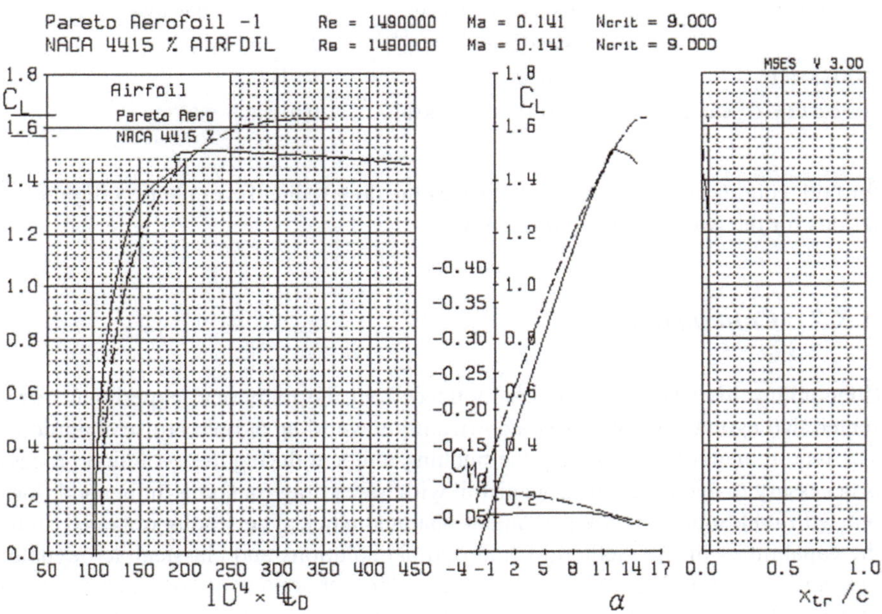

Fig. A.21 Pareto member 01 flight condition two—Cm constrained

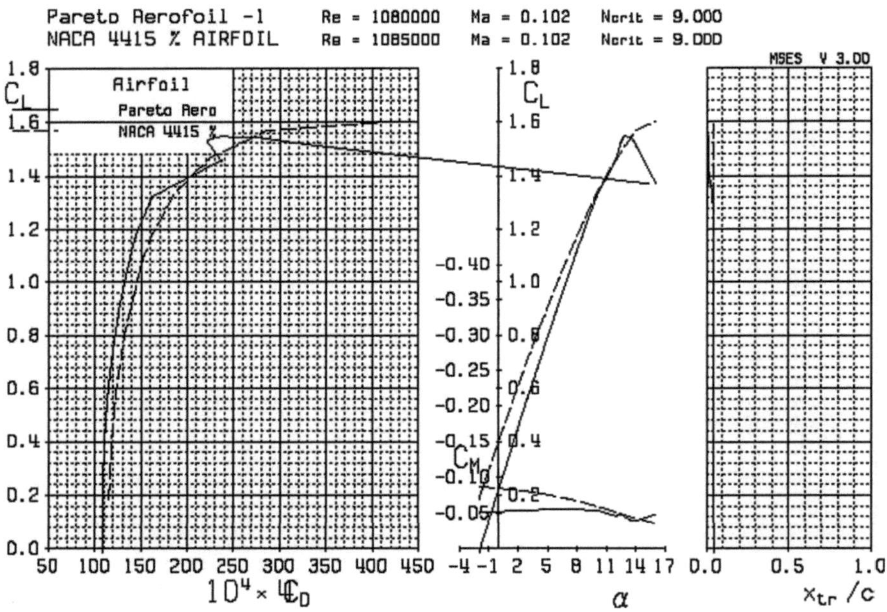

Fig. A.22 Comparative polars—compromised airfoil and NACA 4415 re = 1.490 × 10**6—Cm constrained

Table A.4 UAV drag reduction at two operating conditions

Description	Flight condition one	Flight condition two
	Slow cruise	*Fast* cruise
NACA 4415	0.01621	0.01173
Pareto 01—C_m constrained	0.01451 [−10.48 %]	0.01090 [−7.07 %]

for both conditions. Table A.4 summarizes the drag reduction at the two flight conditions for the two test cases considered.

A.3.7 Conclusion

Concluding this test case, it is can noticed from the numerical experiments presented that evolved airfoils offer significantly lower drag at both cruise conditions. The requirement of constraining the pitching moment during the evolution process is necessary to avoid obtaining an airfoil with lower drag for some flight conditions but with undesirable pitching moment characteristics. The results obtained show the capabilities of the method to find optimal solutions and classical aerodynamic shapes for flow drag reduction. However the importance of sound engineering judgement before, during and after the optimization cannot be under-emphasized:

a proper definition of constraints before performing the evolutionary optimization and final results need to be evaluated to obtain feasible designs.

References

1. GiD Manual CIMNE. http://gid.cimne.upc.edu/support/manuals. Accessed 05 Nov 2010
2. Mohammadi B (1994) Fluid dynamics computation with NSC2KE: a user-guide: release. Release 1.0. Rt-0164, INRIA
3. Flores R, Ortega E (2005) PUMI: an Explicit 3D unstructured finite element solver for the Euler equations, Technical Report. CIMNE, Spain. 2005
4. Woo A (1992) Gnuplot quick reference (1992) http://www.gnuplot.info/docs_4.0/gpcard.pdf. Accessed 2 Feb 2014
5. Schaffer JD (1985) Multiple objective optimization with vector evaluated genetic algorithms. Lawrence Erlbaum, Pittsburgh, pp. 931–1000
6. Drela M (1993) A user's guide to MSES V2.3, Feb. 1993

MIX
Papier aus verantwortungsvollen Quellen
Paper from responsible sources
FSC® C105338

If you have any concerns about our products,
you can contact us on
ProductSafety@springernature.com

In case Publisher is established outside the EU,
the EU authorized representative is:
Springer Nature Customer Service Center GmbH
Europaplatz 3, 69115 Heidelberg, Germany

Printed by Libri Plureos GmbH
in Hamburg, Germany